我的生涯与省思

My Life's Journey

Reflections of an Academic

我的生涯与省思

My Life's Journey

Reflections of an Academic

Wai-Fah Chen 陳惠發

University of Hawaii at Manoa, USA

World Scientific

NEW JERSEY · LONDON · SINGAPORE · BEIJING · SHANGHAI · HONG KONG · TAIPEI · CHENNAI

Published by

World Scientific Publishing Co. Pte. Ltd.

5 Toh Tuck Link, Singapore 596224

USA office: 27 Warren Street, Suite 401-402, Hackensack, NJ 07601

UK office: 57 Shelton Street, Covent Garden, London WC2H 9HE

British Library Cataloguing-in-Publication Data
A catalogue record for this book is available from the British Library.

MY LIFE'S JOURNEY
Reflections of an Academic

ISBN-13 978-981-270-811-3
ISBN-10 981-270-811-1

Printed in Singapore by World Scientific Printers (S) Pte Ltd

滚滚长江东逝水，

浪花淘尽英雄。

是非成败转头空。

青山依旧在，

几度夕阳红。

Preface
前言

After I stepped down as dean of engineering at the University of Hawaii, some of my colleagues and friends suggested that I should write a book on my career, engineering, and higher education based on my 40 years of experiences as a researcher, educator, and engineer. I have been thinking about that suggestion from time to time. Recent advances in computer technology have made it easy for one to write and produce a book with the help of "*Word*" software, digital photos, and Internet communications. When combined with the desire to reflect on my own career and life, and my struggle as an American immigrant; as well as to document some of my students' careers, their achievements and family lives led to the preparation of this book.

This book describes and evaluates three main events: *my career* and my path of establishing an engineering reputation; *my life* as an eye witness of the rise of China; and *my struggle* in achieving an American dream. I understand there is a limit to the extent to which a personal autobiography like this can be truly objective. The idea of writing this book is therefore to share with the reader some of my personal thoughts, experiences, and perspectives on these events through my own life journey and reflections over the last 40 years.

My engineering career basically involved three topics of interactions: *mechanics, materials,* and *computing.* I have witnessed the tremendous growth and interaction in these three topics over the years because of the rapid changes of computing power. It has had a profound effect on my research and our engineering practice in general. Changes over the 40 years of my career have been dramatic and exciting. More change is inevitable and more so in the years to come. So we must continuously adjust to change, and to learn. My life journey in the path of engineering science and structural engineering in rapidly changing times is described and evaluated in this book.

I was born in China during the Sino-Japan War（中日戰爭）; and grew up in Taiwan during the Civil War（國共內戰）. I was amongst the first group of Chinese-American scholars who visited China right after President Nixon's historical visit（中美建交）. I gave lecture tours in China right after its open-reform policy（改革開放）. I happened to be in Beijing during the students' demonstration in Tiananmen Square on June 4, 1989（六四運動）; and witnessed the rise of China over the last decade（和平崛起）. My life journey with a changing China has provided an eye witness on these historical events that will become part of modern Chinese history.

I came to America as a graduate student studying civil engineering. I have learned continuously throughout my life and appreciated the American values: *freedom, democracy,* and *market economy.* My American education has a profound effect on my aptitude for problem solving; on my attitude toward world affair, and on my scale of values for life and humanity. It has become part of the better me. My life's journey, I hope, will provide an incentive to new immigrants learning the English language and Western culture. For these are indispensable tools to become a global citizen for the understanding of the world of today and tomorrow.

I dedicate the book to my wife, Linlin Hsuan Chen（陳宣玲玲）, and to my sons, Eric（陳中傑）, Arnold（陳中毅）and Brian（陳中宇）; and to my students who have contributed much to my career achievements. I hope they will find the book interesting, easy to read, entertaining, and inspiring. To the younger generation and international students, my personal experiences and shared wisdom will, hopefully, provide them with aspirations to achieve their American dreams.

September, 2006   W. F. Chen
Honolulu, Hawaii   陳惠發

Contents

1

The Civil War
國共內戰

1.1 My Childhood 童年

My twin brother, Wai-Kai（惠開）and I were born in Nanking（南京），the capital of Nationalist China in 1936. It was the beginning of the Sino-Japanese War (1937-1945)（中日戰爭）. My father retreated with the government to Chongquing（重慶），with my eldest brother, Hollis（惠青），while my mother took us back to her parents' home in a remote village near the county Ching-Tien Hsien（青田縣）in Chekiang Province（浙江省）. Since China is a large mountainous country, it was a safe and realistic alternative for the women and children to return to their remote village in the mountain to escape the war. At the same time, it was patriotic for the men to go with the government to save China.

It is difficult nowadays for foreigner to truly appreciate the depth of patriotism and passion of Chinese people felt towards their country during this time. This passion was brought about by China's long history of humiliation by the western powers and Japan. It had begun more than 150 years ago; with the defeat of Ching Dynasty（清朝）by the British in the Opium War（鴉片戰爭）of 1841. When the Sino-Japanese War started in 1937, the Chinese were fighting back, and were not defeated. To save China was probably the most abiding idea for almost all Chinese at the time.

After the Sino-Japanese War in 1945, our family reunited in Nanking, and we settled down in a very nice house near the old palace. I was about ten years old and my youngest sister, Helena（惠美），was born at this new house. Wai-Kai and I attended a primary school for the children of the Chinese Air Force personnel near home; and my elder sister, Eileen（惠齡），entered the famous King-Ling Girls Junior High School（金陵

1

女中），which was considered to be an elite private school at the time. There was great excitement and expectation among many Chinese people that they would now be able to live in a peaceful life going forward; and to start rebuilding the country under the Generalissimo Chiang Kai-shek (蔣介石).

Wai-Kai (惠開) and W.F. attended the primary school in Nanking (南京) in 1948, seating from front row, third from left with a booklet, and standing third row, third from right. The class teacher Chen (陳岳如) standing at the left end.

This was the first time I could recall that we were living together with my father and siblings except my eldest brother, Hollis (惠青). Hollis was still living in a boarding school in Chongquing (重慶) at the time. It was my happy years at home; and we attended regular school for the first time in my memory. Eileen was staying in a dormitory during the weekdays; and came home every weekend. Unfortunately, these good things never last.

The Civil War between Nationalist and Communist started immediately after the Sino-Japanese War. Everyday, we read the headline news about the Civil War that was going on in Manchuria. The tragic events on starvation of people in the besieged cities became more and more frequent. In the school, teachers did not teach us much; but kept talking about the war and its casualties. We practiced more frequently in school on how to act when the insurgents attacked the city. Bad news continued. Again, my father was forced to plan another family retreat and family separation followed shortly.

The situation seemed to deteriorate very fast, and the People's Liberation Army (PLA)（人民解放軍）defeated the Nationalist in a major battle in Shu-Chow（徐州）（淮海戰役）; and they were ready to cross the Yangtze River（長江）to capture the capital, Nanking. In a hasty decision, my parents sent us back to the coast city, Wen-Chow（溫州）, to stay with our Aunt（姑媽）, my father's elder sister. My parents bought this house for my Aunt to live in; since she was single and never married. The city Wen-Chow was about 100 miles from our native county, Ching-Tien Hsien. Eileen was the only child staying with my parents in Nanking（南京）. My siblings – Hollis（惠青）, Wai-Kai（惠開）and I with my younger brother, Wai-Sun（惠森）, and younger sister, Helena（惠美）– boarded a passenger ship in Shanghai（上海）. We arrived at Wen-Chow in the summer of 1948.

In 1981, 33 years later, I had the first opportunity to return to China and visited this house again with my eldest son, Eric（中傑）. I showed Eric how we lived there from 1946 to 1948 when I was at his age. I was amazed to see at that time that nothing had really been changed since the Communist Revolution after Chairman Mao's Culture Revolution（文化大革命）, and followed with Teng's（鄧小平）open-and-reform policy（改革開放）. There were five families that lived in this house during my first visit in 1981.

The big house in Wen-Chow（溫州）was on the bank of a river. There was a large backyard with a huge orange orchard outside the house. We had a good time at the big house: swimming in the river, fishing on the pier, playing seek-and-hide games in the orange field, and using the hand-made Y-shaped bow with rubber bands to shoot birds. Sometimes our skills were practiced so well that we could even shoot down a bird standing on the ridge of a roof with confidence. Wai-Kai（惠開）and I entered the fifth grade of a private primary school. Hollis entered as a senior of a public junior high school. We had two teenage maids doing housework for us. One was bought by our parents and followed us from Nanjing（南京）to Wen-Chow, and the other was bought by our aunt to take care of our ailing grandfather, who had a major stroke and was totally paralyzed. They came from very poor peasant's families; because their parents could not afford to feed them.

In 1949, the PLA（解放軍）crossed the Yangtze River（長江）, captured Nanking, and entered Beijing and my parents retreated hastily to Taiwan with the Nationalist government（國民政府）. The city, Wen-Chow（溫州）, was still under the government control at the time. In the summer of 1949, my parents paid a fee in Taiwan; and hired an agent to come to Win-Chow to take us back to Taiwan. My aunt decided that the three elder brothers should stay in the big house to guard the property. Three of them – Aunt（陳英）, Wai-Sun（惠森）and Helena（惠美）– would go first with the agent to Taiwan. They left in one of the early mornings, boarded a small boat, and sailed directly to Taiwan. Since there was basically no effective government in Win-Chow at the time of confusion and transition, there was no security check at all, and everyone was on their own.

One morning in the fall of 1949, Wai-Kai（惠開）and I went to our school as we usually did every day. We heard loud gun fights and sounds of machine guns shooting near the river bank and on the hill nearby. We saw clusters of Nationalist soldiers in small groups passing us, some with wounds and blood. They looked like they were in a hasty retreat. We continued to cross the streets under gun fights and tried to reach our school. We suddenly saw soldiers with different uniforms. They were PLA soldiers. They told us to go home; there was no school today, they said. So, we crossed the battle line again and returned to our big house. We saw the retreating Nationalist soldiers looting and they took whatever they could grab during their retreat. The next day, the local Nationalist government announced that they had surrendered to the new government.

The PLA troops that just crossed the Yangtze River（長江）required a rest and re-supply in Wen-Chow（溫州）before their next battle. A

platoon of soldiers was assigned to stay in our big house.   Their
discipline was very impressive and their moral was super high. Every
morning they were doing their usual routine: singing songs, roll calls and
doctrine lectures. They dared not take anything from the house without
our permission. The two teenage maids were so impressed by the PLA
Army that they decided to join them. They said goodbye to us when the
platoon moved out a few weeks later. I saw the two maids dressed up in
PLA uniforms on the day of departure with high spirits. That was an
impressive scene. I still remember vividly some words from those songs
they kept singing and singing every morning and every evening:
"Do not take a needle and a thread from the people"
  (不取人民一針一線）；
"East is red, sun rises"
  (東方紅，太陽升）；
"China has a Mao Zedong"
  (中國出了一個毛澤東）；
"He is the savior of Chinese people"
  (他是人民的大救星）.

1.2 The Escape 逃出

In the winter of 1949, my parents again hired an agent in Taiwan; and
asked him to escort three of us back to Taiwan. Although Wen-Chow
(溫州) was under the communist government at the time, for all
practical purposes the government was not really established; and it was
more or less still a free port. There were open advertisements on the
newspapers for sharing the boat costs to sail to Taiwan. Hollis decided
to sell our rice grains in the storage bins and also some valuable
belongings; and he bought some gold bars for us to carry. So, in one of
the early mornings, we went with the agent to board a boat to sail
directly to Taiwan. It turned out this was just the beginning of an
unexpected and quite eventful journey to Taiwan. It seemed more like a
good movie script than real life.
    Our neighbor, a young lady, Yen Hong-Ying (嚴紅英), with her
infant son, decided to go with us to Taiwan. Her ultimate destination was

Amsterdam, The Netherlands, since her husband had a restaurant business there. They had just married in Win-Chow not too long ago. The fall of the Nationalist government was so fast that we were all caught off-guard.

When we boarded a Chinese junk in Wen-Chow（溫州）, before we reached the near-shore island, Ta-Chen（大陳）controlled by the Nationalist government at the time, we encountered a pirate ship. I remember vividly that we were hidden behind the dry bamboo leaves in the cargo bay. When they discovered this, they were furious, shouting, and yelling with guns pointing to us. After we reached Ta-Chen, we were wondering around the area for quite a while. Then one day, we landed in an island and stayed there for a week or so. During this period, they caught a barber, accusing him a communist spy and shot him in front of us.

Our boat finally docked at Ta-Chen（大陳）, which was under the Nationalist Army（國軍）control; and the strait was patrolled by the Nationalist Navy（海軍）. We were told at the time, there was a military exercise going on around the region; so no boats were allowed to move around freely. We waited in the boat for several days; while our agent went on shore and tried to find a way out. Fortunately, since our last name and our native town were the same as the four-star general, Chen-Chen（陳誠）, who was the commanding general on the site at the time, we were somehow treated as his relatives. Perhaps, the agent told the authority so. So, we were told we could aboard a large ship nearby and sailed to Taiwan with them. After the transfer to the big ship, our agent disappeared and from there on we were on our own.

After a few days, our ship, named J-Phone（吉豐輪）, left the port and continued the journey on an open sea. We were surprised to see that this ship was well equipped with large machine guns and small canons. The military personnel would not hesitate to shoot other civilian cargo boats and seize their cargos. We found out this ship belonged to the Nationalist intelligence office（情報局）; and they had to be self-sufficient since there was no re-supply forthcoming. They acted like a pirate ship on the open sea. The ship did not sail directly to Taiwan. They docked near another small island. We all went on shore and lived with a garrison of troops on the island. There were also fishermen on the island; so our regular foods were fishes, all kinds with and without shells. Since cooking oil or meat was not available on the island, the fish was simply cooked with boiling water as the main meals, no rice. There was not much taste at all for the meals.

One morning when we got up, we found we were the only few left on the island alone; all the garrison troops and our ship disappeared. After a day or so, they returned. We were told later that they received an intelligence that the PLA troops planned to land on the island that day; so they retreated to avoid the confrontation. We were shocked to learn this fact afterwards. They simply abandoned us. If they did not return to island, then we had to survive on our own.

Since there were no assigned spaces on the ship for us, we stayed on the walkway outside the officer's cabins. It was quite windy during the sailing. Each of us had only one blanket to cover ourselves during the night. One day, there was a storm; Hong-Ying (紅英) used all our blankets to help protect her infant baby from the storm. It was an unforgettable night for all of us. In 1984, when I took a sabbatical leave in Germany with my family, we made a special trip and drove to Amsterdam from Kassel, Germany to visit Hong-Ying (紅英). She was so happy to see us and treated us royally with best dishes at her famous restaurant in downtown district. It was a five-story building with an apartment on the fifth floor for her; and a kitchen on the first floor. The foods were lifted up and down through a lifter to serve the customers. Her husband passed away many years ago and her baby son, now, a grown up young man, married and owned several restaurants himself. We talked a lot about our adventure some 35 years ago. It was still fresh in our mind even after so many years.

Finally, our ship sailed directly to Keelung Harbor (基隆港), Taiwan. There were soldiers everywhere. They were just retreated from various parts of Mainland China. Our ship was docked near the harbor and we were allowed to leave the ship at night. My parents were totally unaware of our where about, and were pleasantly shocked when we appeared at their Taipei house. The Taipei house was a much smaller than that in Nanjing or Wen-Chow (溫州). Even in this house there lived two families. The first order of business for us was to take a bath and change to clean clothes. We finally reunited as a family in Taipei.

1.3 Experience in Taiwan 台灣經驗

Our formal education started in Taipei; when my twin brother, Wai-Kai and I entered as sophomores in the Junior High School of the Taiwan Normal University. We were assigned to Class 27. In fact, there was another pair of twins in the same class, John and Henry Mee (米明瑯, 米明琳). In 1999, after nearly 50 years of separation, we met again in Honolulu with John Mee (米明瑯) and his wife, Grace Mee (張瑋寶), when I took the dean's position at the University of Hawaii. The Mee's family lived in Hawaii Kai and we lived in downtown Honolulu, a half hour driving between our two homes.

Left: In front of Yen Hong-Ying's restaurant （富貴酒樓） in Amsterdam, The Netherlands, 1984. (Arnold (中毅), Linlin (玲玲), Hong-Ying, Eric (中傑), Shui-Tan (水丹), Brian (中宇), and Restaurant Manger).
Right: Inside Yen Hong-Ying (嚴紅英) Son's Restaurant. (Arnold, Linlin, Hong-Ying, her Son 荷凱/granddaughter/grandson, W.F., and Brian).

In Taiwan, most of the public schools were unisex schools. During the transition, however, our school had a few girls in our class. Most of our classmates came from the families retreated from Mainland China.

The local Taiwanese students were mostly assigned to other unisex schools. The traditional culture at the time was to keep us apart for as long as possible – thus avoiding the nightmare of sexual contact – and trained us to fulfill our traditional gender roles. It was a socially conservative approach to education and discouraged any interaction between boys and girls till the graduate school. As a result of this educational system, we were often dumb ourselves down or retreated into the woodwork when we were around girls. Keeping boys and girls apart obviously did not give us the best shot at reaching our full social and intellectual potential.

In 1955, after I graduated from the Senior High School of the Taiwan Normal University, I passed the national examination and entered the National Cheng-Kung University (NCKU 國立成功大學) in Tainan, Taiwan. At the time, Taiwan was at a crossroads; land reform and economical development were the focus of the government, a top priority for the regime survival. With an infusion of American aids, educational reform followed; and higher education became very much Americanized. Purdue University in West Lafayette, Indiana was selected as the counterpart by the US State Department for NCKU. Purdue served as our role model for curriculum reform, teacher training, and re-education. Many of the existing faculty members were sent to Purdue for advanced degrees. Several Purdue faculty members came to NCKU to teach a variety of courses and also guided the top administrators for the university system reform.

Those days, we used for the first time the original English textbooks as our official textbooks. Since the University had only a few original books from overseas; they were very expensive, the University set aside a room for us to read these "original English books" in the library. Later, these original textbooks were reprinted in Taiwan with an affordable price. Several of my classes were also taught directly by the Purdue professors. There were good interactions between us. It was a struggle, of course, for all of us to learn to read, write, and understand English quickly since we were not prepared at all for such a drastic change.

This turned out to be a blessing for many of us who came to the US later for advanced degrees. As a result of this relationship, I had a deep impression about Purdue and knew about its great engineering school in the US since my undergraduate years. Who could predict that, 20 years later, I would become a faculty member at Purdue, promoted steadily to the Head of Structural Engineering in 1980, and became the first George E. Goodwin Distinguished Professor of Civil Engineering in the School of Civil Engineering ever in 1992?

Receiving NCKU (成大) Distinguished Alumnus Award (傑出校友) in 1988 with Civil Engineering faculty and top administrators (from left: Chairman Tang (譚建國), President Ma (馬哲儒), W.F., former President Nee (倪超), and Professor Shi (史惠順)).

1.4 Remembering Father 紀念父親 – 陳又超

My grandfather was the principal of a primary school in my native town, Ching-Tien Hsien (青田縣), and my aunt was a teacher. As far as I know there are many more of our relatives and close friends who were also teachers. The teacher was a well respected profession in rural China at the time since most of the population was peasants. It was a big deal for their kids to attend school. The purpose of schooling in the old China was to pass government examinations to become a civil service official. My elder brother, Hollis (惠青), was a professor at Ohio University in

Athens, Ohio; and my twin brother, Wai-Kai（惠開）, was a professor at the University of Illinois at Chicago campus. My younger brother, William（惠森）, was once a professor at Morehead University, Ohio. So it was just natural for me to follow their footsteps and I aspired to be a teacher as well.

So, it had to be a dramatic event for my father not to follow the footsteps of the family tradition; and to instead join the newly established Chinese Air Forces Academy as a fighter pilot. In those days, just like Senator Kerry said recently during the 2006 US mid-year election: "*If you study hard, do your homework, and make efforts to be smart, you will be rewarded handsomely and realize your dream. If not, you will be stuck in Iraq*". This was precisely the same attitude in China about joining the military at the time.

What was the motivation for my father to join the military? For my father who was raised in an intellectual family, the most abiding ideal the younger generation had at the time was to save China, particularly during the gathering storm of the Sino-Japanese War which occurred later in 1937. The depth of patriotism and compassion to save China was brought about by China's long history of humiliation by the western powers and Japan in particular. It is difficult nowadays for us to understand; but it was much more important than any other ideal at the time.

In a recent search, I collected some historical records on my father's war records during the Civil War and Sino-Japanese War in which his name was either clearly identified or may be implied to be a part of the action team at the time during his active service period. As my tribute to my father, some selected highlights were reprinted in the following in its original Chinese format.

陳又超簡介 – A Brief Bio of Yu-Chao Chen
秀才陳克書長子, 兄弟三人, 姊妹二人均糸知識份子, 望重一時. 幼從父讀經書, 十五歲畢業于青田縣立敬業小學(今人民小學), 即考取省立十一師範學校. 畢業後去廣州考入黃埔軍校第六期. 1932年轉考筧橋航空學校（後改中央空軍軍官學校）飛行科第一期. 畢業後歷任飛行員, 隊長 , 科長, 浙江衢州空軍總站少將總站長.

1944 在重慶任航空委員會航政處少將處長等職.

中華民國空軍抗戰史-A Glance of Air Battles, 1934 to 1942

別以為我們中華民國空軍在抗日時期完全捱打，以下便是一些我軍在抗日戰爭時期的英雄事蹟。

The book "*One Hundred Chinese Generals from Ching-Tien Hsien* (青田縣)" was published by the Committee on Historical Records in (浙江省).
Photo: four-star general, Chen-Chen (陳誠), middle. Father Yu-Chao Chen (陳又超), right.

時間：民國二十三年　　地點：江西廣昌
　　　是役空軍參戰人員：轟炸第二大隊隊長王勳（叔銘）、副隊長王伯嶽、教官王衛民、分隊長王星垣、孫仲華、李賜楨。隊員張森樵、范伯超、朱天寶、**陳又超**、趙家義、彭允南、羅中揚分別駕機輪炸大羅山，摧毀其防禦工事。

八一四筧橋空戰

　　時間：民國二十六年八月十四日下午四時
　　地點：杭州筧橋

八一五杭州空戰
　　時間：民國二十六年八月十五日上午七時
　　地點：杭州

1942年浙江軍民援救美軍飛行員紀實 - Rescue of the U.S. pilots led by Lt Col. James H. Doolittle for the historical first U.S. air raid of Japan in 1942. The 16 B-25 bombers took off from the deck of the aircraft carrier Hornet. Plans originally called for them to land at airfields in China during the day. But the bad weather changed the plans. The pilots were forced to ditch their planes or bail out over or along the Chinese coast at night.
Father Yu-Chao Chen (陳又超) was the Director General of the Airport in (衢州，浙江省), a coast province, at the time. It was too late for him to receive the order from the highest authority in Chongquing (重慶), Capital of China during the WWII, to open the airport facility for the returning U.S. B-25 bombers to land in China.

2005-6-16 14:33:42 作者： 來源： 不詳
　１９４２年４月１８日上午，美國１６架B－２５轟炸機在杜特利爾中校率領下，從距日本６５０海裏的"大黃蜂"號航空母艦上起飛，轟炸東京、大阪、神戶等地。這是日本發動侵略戰爭以來，本土首次遭敵國機群轟炸，損失巨大，朝野震驚。B－２５機群返航時，由于和衢州機場地面聯系不上，在油料耗盡之後，機組人員被迫棄機跳傘，損失慘重。這一曆史事件，影片《東京上空３０秒》和《珍珠港》中都有反映。

加油未遂
對這次轟炸東京歸來的美國飛機未能按計劃在衢州機場降落加油，曾有一種說法：由于天黑雨大，衢州機場當局誤以爲日機侵犯而關閉機場。

　　爲了弄清事實真相，上世紀九十年代初，我訪問過當年任中國農民銀行衢縣辦事處主任、１９８９年從台灣回衢州定居的戴允銘先生。戴先生對我說，那天（１９４２年４月１８日）確是雨天。晚上七八點鍾，衢州有警報聲，也有飛機聲，但這飛機聲不同于日本蚊式飛機的聲音（日機一般晚上不出動）。次日，才知道是美國飛機。這些飛機轟炸東京航歸，計劃在衢州機場加油。"大黃蜂"號航空母艦將此事通知華盛頓最高指揮部，華盛頓再通知重慶中國最高當局。重慶將這一指令通知衢州機場，已是午夜１２時。盡管衢州機場立即開放，但未見美機降落－－原來，當晚８時至１０時，返航美機飛到浙江上空，已油盡而墜地，有的落入海中了。

戴先生說，當年衢州機場的負責人是陳誠的侄兒**陳又超**先生。戴允銘與**陳又超**時有往來。事發後，**陳又超**曾同戴先生講起過，當時空軍官兵也聽出不是日本飛機，但沒有重慶最高指揮部命令，即使知道是美國飛機，也是不敢擅自開放機場的；擅自開放機場，主官是要殺頭的；等來了命令，卻爲時已晚。

當時，１６架Ｂ－２５轟炸機中，１架（８號）迫降于蘇聯遠東地區，５位飛行員被蘇聯當局羈押。另外１５架墜毀在浙江、安徽、福建境內（浙江１０架、安徽２架、福建３架），機上７５位美國飛行員中，除５人喪生、８人被日軍抓獲外，其余６２人由當地軍民救助而脫險，其中４３人由駐軍、行政機構或遊擊隊護送到衢州空軍第十三總站集中。他們稍事休整，即由衢州機場登機直飛重慶。出發前，他們還請衢州照相館的攝影師到駐地拍了合影。

５０年後的紀念活動 The 50-Year Anniversary Activities in 1990: A Re-Visit of the Site by Some Rescued Pilots

杜特利爾（four-star General Doolittle）于１９８５年晉升爲四星上將。１９９０年，杜特利爾的朋友、原美國西北航空公司副總裁穆恩組織一支５人考察團來浙江、安徽等地尋訪當年參加救護美國飛行員的中國老人。

在"杜特利爾行動"５０周年紀念活動上，美國總統喬治·布什（President George Bush）對這段曆史作出了高度評價："在突襲以後，那些善良的中國人不顧自己的安危，爲我們的飛行員提供掩護，並爲他們療傷。在具有特殊意義的時刻，我們也向他們表示崇高的敬意，感謝他們作出的人道主義努力，是他們的幫助才使我們的飛行員能夠安全返回。杜特利爾行動雖然已經過去了半個世紀了，但這些英雄們一直受到美國人民的敬仰和尊重。我們永遠不會忘記他們作出的偉大功勳，也永遠不會忘記爲自由和正義事業作出貢獻的中國人". (We will never forget the humanitarian efforts of the great Chinese people to risk their life to help our pilots to return home safely.) By President George Bush.

2

The Wisdom of Class 33
附中智慧
What Said and Did
言行表現

On June 18, 2005, I attended my 50th high school class reunion in Los Angles for a six-day celebration. It was the eighth class reunion since our graduation from the High School of the National Taiwan Normal University (國立台灣師範大學附屬中學) (師大附中) in Taipei, Taiwan in 1955. The friendship of our Class #33 had lasted more than a half century with many memorable events. During the last 50 years, we grew up from 15 to 50 at our first Chicago reunion (1987); to 60 at our forty-year celebration in Seattle (1995); and to 70 at our last year reunion. It was a time tunnel in reverse; it brought back a lot of memories and accumulated wisdom on life, marriage, health, finance, friendship, and family issues. This chapter is intended to capture some of our life experiences and wisdom in a nutshell. Wisdom comes from experience, and experience …, well, that comes from poor judgment. If there is any secret of success in life, it lies in the ability to absorb these experiences from you and from others.

As General Macarthur said in his famous farewell speech to the U.S. Congress, *"an old soldier never dies, he just fades away"*. I think it is appropriate to quote here a counterpart of a Chinese poem from the famous *"Three Kingdoms Story"* (三國演義) to reflect the colorful life of our classmates at this stage of our life:
"滾滾長江東逝水, 浪花淘盡英雄。是非成敗轉頭空。青山依舊在, 幾度夕陽紅。"

My old classmates, just like everything else in life, as time goes by, will fade away too. But the beautiful memory will last forever in our hearts. To put it in perspective, I quote another Chinese poem for the occasion: "一壺濁酒喜相逢, 古今多少事, 都付笑談中。"

15

2.1 The Legacy 傳承風格

For my classmates, I can not help but cite the title of a famous Chinese movie "*Crouching Tiger and Hiding Dragon*" (臥虎藏龍). They are indeed a group of highly intelligent individuals with diversity, talents and visions. They brought honor to our class, served as role models for our next generation, and preserved the true tradition of our Chinese culture. Their life stories were characterized by their desire to learn, to serve, to contribute, and to create, and to hard work.

Our class leader, Yung-Chen Lu (魯永振), affectionately called *Monkey*, was a case in point. He was the key figure in our class to organize almost every reunion with endless patience and tolerance. He was amongst the first group of Chinese scholars returning to China to serve the mother land in the 1970's; but was disappointed by the reality of the situation that had developed in China at the time. He was super-active in community services in greater Columbus, Ohio during his teaching career at the Ohio State University.

He organized a medical service team for the poor and Asian immigrates in particular. He was the founding member of the Asian-American Community Service Council; and suggested the first Asian Festival in Columbus in 1994. He served as a member of the Board of the Greater Columbus Art Council in order to build a bridge among a variety of cultural and ethnic groups. I sometime wonder where he got all these energies to be so super-active to the community services. Columbus Asian community was indeed lucky to have a citizen like "*Monkey*".

Peter and Grace Wang (王成釗, 羅惠芳) a perfect couple, was another example of our classmate to whom I had the privilege to associate with. They made a good fortunate in a real estate investment in California in the 1970's. Despite their wealth, they lived simply; but had a great vision and ambition for building a bridge between US and China. They had two major projects in mind: Poverty alleviation in China, and Language program to overcome culture barriers between US and China. (Source modified from http://www.wangfoundation.net/fp_news.htm). They spent a full time effort in the last few years to develop and implement realistic programs to achieve these lofty goals.

Concerning the cause of poverty relief, Peter said, *"First, we must have faith. Second, the cause must be fully funded. Third, it takes time."* He was quite prepared in all three aspects. To him, poverty alleviation in China was a life long devotion. *"Poverty alleviation should not be used for self-interest or public image building. It requires concrete and consistent efforts. This is not a job that can be completed in a couple of years. It would take eight or even ten years to formulate a workable pilot methodology which, if implemented properly, would enrich the lives of millions, or even hundreds of millions of people."*

In 2006, Peter decided to found a platform by persuading 53 Christian schools to launch summer study overseas programs in China for their students. Christian schools were dedicated to instill in their students a habit of lifetime service rather than an occasional commitment to the society. Peter expected these students would make even more of a difference, compared to their predecessor American students who studied in China simply to land a good job in a multi-national company. The American students would work in rural areas and small towns, giving them the chance to understand China from a comprehensive perspective.

Peter never missed a chance to encourage American philanthropists to join his Tsinghua University's poverty relief efforts. Peter's blueprint for poverty alleviation in China had attracted interest from officials from the State Department and even the World Bank. Ten prestigious universities in California, a consortium of 1,500 members from top Mexican University *Tecnologico de Monterrey*, and many other higher education sectors had entered into discussions with Peter intending to get involved in his Tsinghua University's Poverty Alleviation programs.

In October 2006, the Nobel committee selected Muhammad Yunus of Bangladeshi as the winner of this year's Nobel Peace Prize. Yunus, like Peter, was an American-trained entrepreneur. He developed a micro-credit lending business model to pull tens of millions of people out of poverty by helping them help themselves, especially for women. As noted by the Nobel committee, *"Charity is not an answer to poverty; it only helps poverty to continue"*. Peter believed in education and self motivation. Education and self interest can unleash energy and creativity of each human being. That is the answer to his poverty alleviation program. We wish him success and to be recognized in the future as the winner of the Peace Prize for our Class 33.

Peter and Grace Wang (王成釗，羅惠芳) in 1987.

2.2 The Cardinal Rules for Senior Life 四老哲學

In July 2 to 9, 1989, our second class reunion was held on a Caribbean cruise ship named Jubilee. A total of 17 families attended the event along with our class master teacher, Yang Lau Shi (楊淑玉老師) among various topics discussed; the group wholeheartedly endorsed the four cardinal rules for good senior life covering four critical aspects of golden age: health, spouse, finance, and friends:

- o 老命 – 以動養身; 以靜養心
- o 老伴 – 愛其所同; 敬其所異
- o 老本 – 取之有道; 用之有度
- o 老友 – 以誠相見; 以禮相待

A direct English translation:

- o Senior Health – active for physical; meditation for mental
- o Senior Spouse – love the alike; respect the difference
- o Senior Finance – earn rightfully; spend appropriately
- o Senior Friends – meet with sincerity; treat with courtesy

Some more thoughts on life, especially the senior life:

Life should be fun, and what fun is life if you are controlled by an obsession? Obsessive exercise, food, diet, etc. Exercise is important; but there are other things just as important. More really isn't better, but getting people to understand that message may be a challenge. Avoid extremes; zhong yong zhi dao（中庸之道）is the way to go.

In life, the most joyful experience is "giving"; the most loss is the loss of "self respect". The deadliest weapon is the "tongue"; the most valuable asset is "faith". The most worthless emotion is "self-pity"; the most beautiful attire is the "smile". The most destructive habit is "worry"; and the natural resource is "youth".

We understand that a lot of people, as they age, tend to become dependent. They get in the way. That creates resentment on the part of young people. If the older people, to the extent they can, would continue to be active and independent; they would be respected. That is important in our senior life.

2.3 On Marriage 婚姻

The US Census Bureau in its 2005 American Community Survey, reported a sea of change in every facet of American life including for example, for the first time, unmarried households reign in the U.S. There were more unmarried men and women living together alone than the traditional households with married couples. It showed that the concerted efforts made in the past years by the government to shore up traditional marriage and families through tax breaks, special legislation and church-sponsored campaigns bore little fruit. It is difficult for us to maintain the traditional values of our community, of our family in particular; while watching over divorce rates in the US reaching 50 percent over the last 35 years. It is unthinkable to us but all certain to happen; that cohabitation and temporary relationships between people are likely to dominate America's social landscape for years to come. How will this affect our next generation? How we parents have to learn to adjust to this new reality in the coming years.

During my Class #33 (附中高 33 班) times, it was a boy school. In some adjacent classes, there were only a few girls for some reasons at the time. Basically, boys and girls were separate and were not allowed to date each other. Very few of us knew any of the girls. *"Girls watching"* was one of our favorite pastimes. We would watch those few girls on campus as they went from class to class. Their appearances were analyzed but none of us dared to approach. That was the environment we were in about 50 years ago.

Most of us were married right after our graduation with a traditional church wedding: a simple ceremony with a clergy man in a university setting or in a local church, simple reception without an elaborate banquet, and a small group of attendees of relatives and schoolmates. Most wife-to-be's were introduced through a family relation or classmate connections. Their family background was well known before the dating game began. In Class 33, almost all of us were happily married; and only one or two couples, as far as I knew, had some marriage problems.

W.F. (惠發), Linlin (玲玲) and Susan at the University Chapel in June 1966.

I think the reasons for the success; and the logics behind this traditional dating and marriage is that the marriage represents the joining of two families. The parents consider the long term challenges of a marriage, based on their own experiences, and will ask who will be a good partner. The children trust their parents. They know their children better than anyone, and are concerned about their long term happiness. Since the parents and other members of the extended family participate in the selection process, they have an investment in the success of the marriage. When the inevitable difficulties arise, the couple has a concerned support group whose members will offer various kinds of help.

On the other hand, in the US, people put a lot of efforts on selecting a marriage partner through dating. But it is difficult to predict the future and the compatibility of the couple, especially the relationship between two families. In China, there is more emphasis on the development of a long term relationship after the marriage and the supporting group. Without a strong supporting group, the individuals have less willingness to put up with an unhappy marriage; especially when the couple spends more time these days for their companies. The current woman's earning potential has also had a major impact on the traditional marriage in unexpected ways.

Overall, we provided the following marriage rules for building a long lasting relationship:

The following are negatives to be avoided:

- o   criticism
- o   contempt
- o   stonewall

The following are positives to be encouraged:

- o   listening
- o   humorous
- o   intimate

My second son, Arnold (中 毅), married Lin Ng (黃 慧 琳), a Malaysian Chinese, on September 29th, 2001 at the Four Seasons Resort Hualalai, the Big Island of Hawaii, with a more Western style self-direct wedding ceremony: Ceremony held at the beautiful beach with a sunset scene, read each other their self-prepared vows, delivered intimate speeches by close friends, and hired a wedding coordinator to manage the details.   As parents, we were truly bystanders; and served as the witnesses for the whole formality without a real active involvement.   It was a relaxing time for us to be there; and to enjoy the companionship and friendship.   It reminded me of my attendance to a convocation ceremony for the College of Engineering.   We just followed the script; everything was well planned and well prepared; and moved forward smoothly, right on schedule.   It reflected the modern efficiency in conducting business meetings in the Internet age.

My youngest son, Brian (中宇), married Christine de Asis, a Philippine American, on April 2, 2005 in Austin, Texas at the Star Hill Ranch with a specially decorated wedding gazebo.   The wedding ceremony was basically Western style; but added one scene of Chinese ceremony for the newly wed to bow to their parents to receive their blessings.   I delivered a speech at the banquet; while Linlin delivered a speech at the rehearsal dinner.   The organization and the logistics was quite challenging since the wedding ceremony and reception were held at the Ranch, the rehearsal dinner was held in a different part of the city; and the hotel was in downtown Austin.

Lin (慧琳) and Arnold (中毅) at a sunset scene of their wedding ceremony in the Big Island of Hawaii in 2001.

I prepared a speech with a Power Point Presentation. I thought it was appropriate; but initially, there were some objections on my proposed seminar-like format at a wedding banquet environment. It turned out to be a quite memorable occasion. The following was a summary of my slide talk: one in Chinese and the other in English. My five-minute speech was reprinted in Appendix 2.1; while Eric's speech was very nicely done and quite emotional with tears and full attention from the audience. His ten-minute script was reprinted in Appendix 2.2. For Brian's wedding ceremony, we were fully involved and actively participated. Even for my twin brother, Wai-Kai (惠開), he was asked to serve as the master ceremony for the Chinese portion of the formality.

Blessing from the Parents for the newly wed
April 2, 2005
Brian and Christine
- o Today, you have just married as the perfect couple.
- o From now on, you will be together for a lifetime.
- o Enjoy each other's company and happiness.
- o Destiny will make this marriage last forever.
- o Your future will be lit by a lucky star.
- o Hold hands and work together through life.

o   Be thankful, for it is the will of God.

o   There is nothing more to ask from God.

Love from Dad & Mom

給 新 人 的 祝福

Brian (中宇) and Christine

來自父母對你們的愛

April 2, 2005

今日成佳偶

日後長廝守

夫唱婦隨樂

良緣是永久

福星照前路

途上手牽手

不負上蒼意

此生復何求

In January 7, 2006, my nephew, Herman Hsuan (小龍), married Junghae Suh, a Korea American, in San Diego, CA. They met each other through an Internet dating service and communicated extensively through emails. Eric (中傑) was Herman's adviser and trusted big brother. It was a love at the first sight with a full blessing from me as Herman's close uncle. I was initially quite cautious about this very untraditional dating game; but the more I learned about their preferences and backgrounds, the more comfortable I felt in supporting their fast track relationship building. It appeared to be a god-send, perfect couple, as far as I could tell.

As dean of engineering, I regularly received announcements of faculty appointments in engineering schools around the nation; I was pleasantly surprised to read in May the appointment of Junghae Suh as a new assistant professor of bioengineering at Rice University in Houston, Texas. It instantly removed any of my concerns about the background of Junghae; since her brief biography was so impressive that I sure would offer her the same faculty position at the University of Hawaii. What a coincidence! It was indeed a small world.

Brian (中宇) and Christine bowed to W.F. (惠發) and Linlin (玲玲) during the part of Chinese ceremony in Austin, Texas, 2005.

Dr. de Asis gave her daughter, Christine, away at the outdoor ceremony.

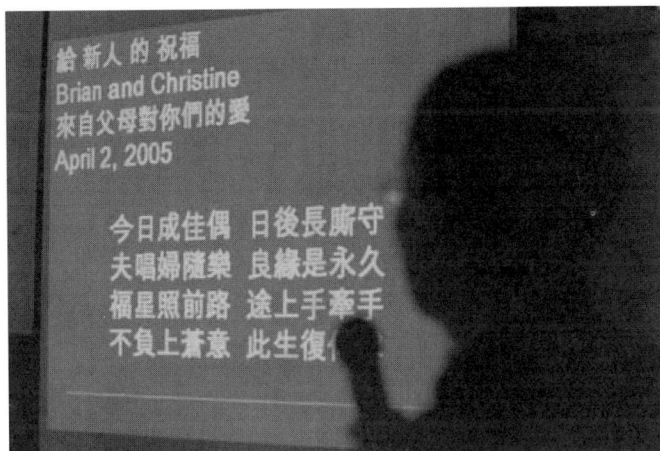

W.F. (惠發) made a Power Point presentation at Brian (中宇) and Christine's wedding reception in Austin, Texas in 2005.

More surprises were coming! Herman and Junghae asked me to serve as their wedding master and I conducted the entire wedding ceremony on their behalf. Their wedding was held at the Museum of Photographic Arts (MOPA) in San Diego, CA. It was a beautiful setting and I was responsible for the Western portion of their wedding ceremony at the Museum. This would be followed later with a Chinese ceremony; then a separate Korean ceremony. In addition, an important local Cantonese custom with a special tea ceremony would also be held in the hotel after the Korea ceremony. Since each of these ceremonies was dressed in its local traditional dress; they were all very colorful. The relatives of the two families met at the ceremonies for the first time. It was a very joyful occasion. Everyone was very excited; and the events were so fresh; all had a wonderful time.

Another new idea followed. The honeymoon would be a continuation of the celebration on a cruise ship in San Diego. All wedding party participants were invited to join the newlyweds. What a creative idea! It was an eye opening event for everyone; and a good learning experience for our older generation. I quote here from a Spanish dramatist Pedro Calderon de la Barca: "*When love is not madness, it is not love*".

Looking at these marriages from my Class 33 time to the present Internet time, I really feel that the changes are quite dramatic and are now gathering acceptance in our community. These changes are taking

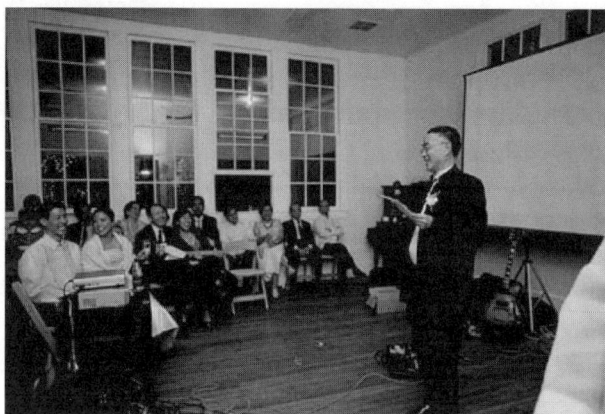

W.F. (惠發) shared a light moment with Brian (中宇) and Christine at their wedding presentation in Austin, Texas, 2005.

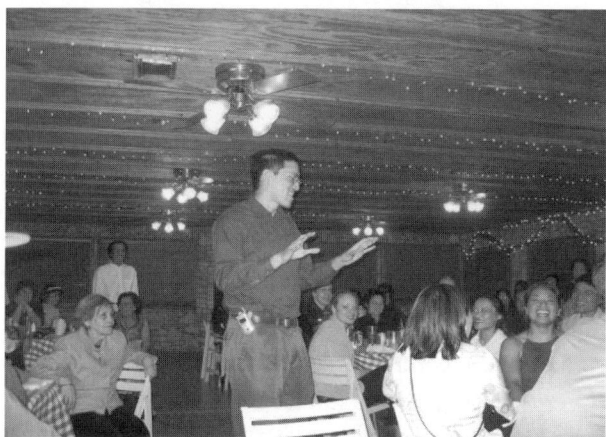

Arnold (中毅) gave his remarks at Christine and Brian's (中宇) wedding rehearsal dinner party in 2005.

place in a more persistent and deepening manner in wider areas of social and economic life than a couple of years ago. These changes are going to affect some of the social and cultural aspects of our Chinese family tradition in the years to come.

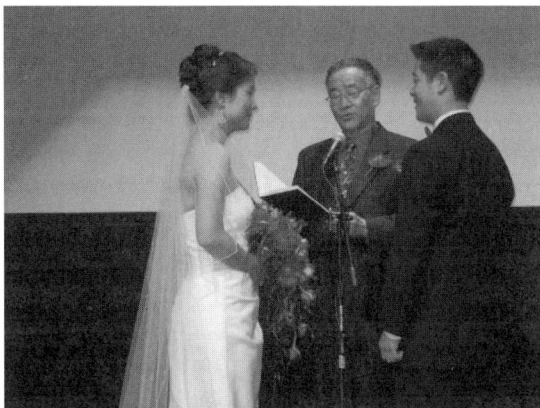

W.F. (惠發) conducted the wedding ceremony for Herman (小龍) and Junghae at the Museum of Photographic Arts in San Diego, California in 2006.

Herman (小龍) and Junghae bowed to their four cousins (Arnold (中毅), Johnny, Brian (中宇), and Eric (中傑)) at their Korean ceremony.

2.4 On Family 家庭

When we were in West Lafayette, Eric (中傑) and Arnold (中毅) were in Urbana, Illinois, and Brian (中宇) was in Austin, Texas. During this time, each of them was frequently getting that urge to come home again and again to visit during these years. For them, West Lafayette was the home where they grew up and where most of their friends were. Over the years, as they graduated from schools; and found jobs and relocated around the country, they still returned to West Lafayette more frequently than they ventured elsewhere. For us as parents, for a long time, we considered this as a matter of course. The three boys were close to us since their childhood as we traveled frequently in our big van from Disneyworld in Florida to the Disneyland in California, especially around holidays.

Herman (小龍) and Junghae bowed to each other at their Chinese ceremony in San Diego, CA in 2006.

Junghae served tea to W.F (惠發) and Linlin (玲玲) at their Cantonese ceremony.

When we moved to Honolulu and bought a house in Fremont, California, this tradition continued but the family home now moved to, Avalon - the big house, as we called it. The planning of a big house in Avalon was intended for a frequent big family reunion in the future with grandchildren; and enough rooms for our extended family. Since my job is in Honolulu; we plan to move to Avalon home when I retire from the University of Hawaii. The reason we chose Fremont was based on three points: Arnold's job was in Fremont, Eric loved Los Angles; he intended to transfer to Los Angles when he completed his training in Raytheon, and Linlin's extended family was in Bay Area.

There was another strong argument from Linlin (玲玲, Lily) too: we were living in Bethlehem, Pennsylvania and West Lafayette, Indiana far from her family in the Bay Area because of *my* job. Linlin felt that a fair price for her sacrifice was to move close to her family when I retired. Since I prevailed on where we lived for the last 40 years, she felt she should have a more say in where we spent on our retirement years. So, the Avalon home, a big house, seemed to be a logical conclusion. For many years, during vacation season, we flew directly to San Francisco from Honolulu, Brian flew to San Jose from Austin, and Eric flew to California from Boston, and Arnold lived in Fremont with friends. This became a routine for us before Arnold married in 2001.

Of course, too much of a good thing could not be lasting. After Arnold (中毅) married, the routine vacations gradually lost their luster. To us, we were all coming to a new location, the new home, fresh and exciting. But to Arnold's family, these were not vacations in the traditional sense. They wanted to take off for other places and have new adventures for their precious holidays. As was often the case with couples, the lure of family visits to renew family bonds had to be balanced with the pull of the trailing spouses. So, the planning of family reunion was anything but a routine matter. Now, family vacation planning was filled with so much tension.

When Brian (中宇) married in 2005, we now have three separate families, and a long trip to Hawaii from mainland for all families at the travel peak season was out of the question. This would consume the greater chunk of their travel budgets and vacation days. Now, there was travel conflict – within each family and between us - for good reasons. Linlin felt a need to routinely reconnect with her family, visit her parents in Fremont, and join the annual family Christmas dinner with her brothers, cousins, nieces, and nephews. The spouses felt obligated to go to their own homes or desired to be somewhere more exciting rather than seeing the same routine, sleeping in the same bedroom, and eating at the same restaurants. With more grandchildren coming, there was an urge for each young family to start their own family traditions; and venture off someplace fresh with their own siblings and close friends.

Most of my classmates faced a similar problem: How to work out a vacation plan with the spouse and how to share their children's vacation time with their other families every year? There had to be a way to combine vacations with the obligatory family visits and new excitements. We also noted one common problem with family vacations. Our kids were all grow up adults by now; and most of them had responsibilities for major decision making in their companies. Yet, when they returned home, they wanted to enjoy their own kids as their parents – not as another kid in the household under their own parents.

The way this works now is by default: We continue operating as we have done for years – the children come to Honolulu or Avalon when the urge strikes one of them; and we quickly adjust our routine to receive them, or join their vacation plan. Sometimes, Linlin goes home when there is a need or urge with or without me. I will do the same if she does not object. I don't argue because I know it means so much to her, or to the children; and the timing must be okay for them with their life style.

As my classmate John Mee (米明琳) kept reminding us, "I was ranked number two in importance with my son before he married. After his marriage, I became number three. After the birth of my grandson, my ranking was reduced to number four or even five depending on where my in-laws stood". This is the reality of life; and the early we recognize this fact, the better the future relationship with our children will be.

2.5 On Aging Parents 奉養

One of the subjects facing our classmates was on the caring for an aging parent. We are not young anymore. Since we are at age 70, an aging parent requiring caring is in his or her 90's. The Chinese tradition of caring for ailing parents is for the parents to live with one of the children under the same roof. Even for a healthy parent living together is important since he or she may fall around the house and sometimes forget to take their prescriptions. For an ailing parent, she may be on a dozen or more prescriptions. It is inevitable for a more serious ailing parent to lose their independence and require assisted living facilities. Sooner or later, some families reach their breaking point and realize they cannot do it by themselves. For a big family with several siblings, the current practice in the Chinese community is to share the responsibility by rotating the caregiver through each family for a fixed period of time.

W.F. (惠發), Linlin (玲玲), Eric (中傑), Arnold (中毅), and Brain (中宇) at their West Lafayette's big house celebrating Eric's high school graduation in 1986.

We are not from the baby boomer's generation; but they are coming not far behind us. They will come with a huge demand for caring for their aging parents soon. It can be a major societal problem requiring different levels of services. For parents living alone, they require bill-paying, doctor-appointments, medication delivery, and planning for future nursing home. According to available statistics, uncompensated family members provide two-thirds of the caregiving in the US at the present time. More families are expected to turn to geriatric care managers, home care agencies, home healthcare aids and other professionals in the future.

For us, the Chinese American, it is a major cultural hurdle to overcome before we feel comfortable admitting our ailing parents to a caregiving facility. Deep down, we know that the aging parents will be better cared for by the professionals; but the perception of "abandonment" by their children is something that cannot be avoided in its appearance.   Sibling rotating caregiving can go on for a while but sooner or later, it will reach a point and we realize the fact that we cannot do it ourselves.

W.F.'s (惠發) mother, S.T. Chen (陳夏水丹), on the Waikiki Beach, Honolulu in April 2005.

For most of my classmates as far as I can tell, they continue operating as they have done for years in one of the two working models: rotating caregiver with each sibling equally; or staying in one location with a major estate planning incentive or compensation for the caring family. We are still talking and looking for a workable solution; that can combine the Chinese culture of homecare with the Western culture of professional care for our aging parents.

2.6 On Economics 101 老本

The economy has changed over the past centuries; and accompanying this change has been a major power shift. It began with aristocrats who owned land as royal. They had dominating power. Then, the industrial revolution in the 19th century created the industrialists; who owned factories. They become the new power group at the time. Now, we have a new economy; based on knowledge and information. The power group has shifted again to technologists; who own Intellectual Property (IP). In the information technology age, we must learn some basic rules of Economics 101 in order to manage our finance or make informed decisions. These simple but useful rules are described briefly in what follows.

Mother Shui-Tan (水丹) with grandsons, David (中平) front row, Arnold (中毅), Jerome (中堯), Brian (中宇), Hollis Jr. (濤濤), Joe, and Eric (中傑) in Orlando, Florida in 2000.

Estimate your real estate Net Asset Value (NAV):
NAV = (rental income – maintenance costs) × (1 + growth rate) / (cap rate)
Example:
Real estate annual growth rate = 10%
Cap rate = 4.4% 10-year Treasury rate
For the Hawaiki Tower Condo unit:
Monthly rental = $3000
Condo fee = $430
Use the formula above, the value of this condominium is estimated:
NAV = (3000 ×12 – 430 × 12) × (1 + 0.10) / 0.044 = $771,000

The actual costs and profit of homeownership - an example
Assume the following data for owning a home:
$200,000 home purchase
$40,000 down payment
Six years later sold at $282,000 with 5.9% annual increase based on the Federal National Mortgage Co
Nominal gain $82,000

Costs:

6% mortgage 30 year fixed rate

Equity: $13,763

Interest: $55,305

Tax deduction at 25% rate

Total after tax costs: $41,479

Net profit:

82,000 + 13,763 − 41,797 = $54,284

Other costs:

5% broker commission = $14,100

Closing cost: ?

Insurance: ?

Property tax: ?

Maintenance: ?

Inflation: ?

Maximum profit:

$54,284 − $14,100 = $40,184

Summary:

About half of the nominal profit at most

Double initial investment in six years or roughly 13% annually

Benefit: Free housing

Fed model on stock and Treasury bond

Example:

10 year Treasury: 4.7 %

S & P: $P/E = 20$ (price/earning)

$E/P = 1/20 = 5\%$ (earning per share)

Assume 7 % earning increase annually

The earning per share will reach 10% at the 10th year

So, the stock's average yield over the ten year period is

$(5\% + 10\%) / 2 = 7.5\%.$

Compare with the Treasury yield: 4.7% annually.

The difference $7.5\% − 4.7\% = 2.8\%$ is the premium to compensate for the market risk, since Treasury bond is considered risk free.

Fed, Interest, Bonds, and Mortgage

Fed jacks up interest rates to fend off inflation when economy grows fast.

Steep Yield Curve – Example:

Federal rate is at 1%

10 year Treasury is at 4%

The spread is at 3%

The average spread is 0.84% historically

The inflation now is between 1 to 2%

The market predicts higher inflation and asking for 2% more.

Yield goes up 1%:

10 year Treasury price drops 8%

Mortgage Rate – Example:

The spread between 30 year fixed and one year ARM is 2.3%

Compared with historical average 1.8%

ARM gives more saving

Most mortgage last an average 7 to 9 years

15 year mortgage is 3/8% lower than 30 year fixed on the average

A bad economy or soaring mortgage will send housing price down – cannot sell the house

Savers:

3 to 5 year notes offer a good mix of risk and return

If rates go up slow, bonds and money market funds will do the same

Borrowers:

ARM offer most savings than usual

Soaring mortgage rates send housing prices down

15-year mortgage is cheap right now

Some Profitable TIPS on Bonds

Everyone should have at least some fixed incomes guaranteed to beat inflation. Among the best inflation hedges are the federal government's Treasury Inflation-Protected Securities known as TIPS, which are currently paying 2.4 percentage points above the consumer price index (CPI). The interest is exempt from state and local taxes. The alternative is to buy the government's inflation-protected savings bonds known as I-bonds. I-bonds now on sale earn 1.4 percentage points more than the CPI. Its interest is not only exempt from state and local taxes, but also deferred for federal tax until redemption.

Both TIPS and I-bonds can be bought directly from Federal Reserve Bank using the Treasurydirect.gov online free of commissions. The key to lock in top rates and reduce interest risk is to build a "ladder" of bonds with varying maturities and rates of return. When the first matures, you simply buy another at the far end of your investment horizon. By repeating the process each time the next comes due, you create rungs on the ladder that average out your overall return. Since federal tax for TIPS is due each year on interest and gains, they are best held in tax deferred accounts such as Individual Retirement Account (IRA).

Financial Statement - Margins

Income Statement
Net sales (revenues) - say 4.2 billion
COGS - cost of goods sold (cost of goods or services) - say 2.5 billion
Gross profit = 4.2 – 2.5 = 1.7
Gross Margin = costs/sales = 1.7/4.2 = 0.4 = 40 %
Remaining costs – operating costs (salaries, utilities, advertising) - say = 623 M.
Operating Margin = 0.623/4.2 =15%
   (Profitability of the company)
Other Expenses
Tax & interest = 742 M
Net income = 1.7 – 0.623 – 0.742 = 0.335
Net Profit Margin = 0.335/4.2 = 8%
(How much of $1 of sales a company keeps as a profit)
This is known as the bottom line in the income statement.

2.7 On Health 老命

We have two prominent medical doctors, George Ting (丁兆治) and Fan Kan (范抗), as our class health consultants. Through questions and answers via Internet, we have benefited a great deal from their advices from time to time. Some of the highlights of these discussions and information were edited in what follows:

As Dr. Fan Kan (范抗) stated in his email to classmates dated on September 4, 2003: "*You are what you eat*".

Although genetic plays a major role in heart diseases, such as hypertension (high blood pressure); but there are some things we can do to lessen the risks of heart attacks. The most devastating heart problem is acute occlusion (blockage) of the coronary arteries that supply oxygen to the heart muscles. Over the years, the lumen of the arteries will gradually become narrowed due to deposition of cholesterol and LDL (low density lipoprotein) related components; which subsequently become partially calcified. The trick is to prevent such build-ups through reasonable and intelligent selection of what you eat and if that is not enough, preventive medication such as now widely available cholesterol-lowering drugs can help.

The saturated fat content in our diet is the major source of heart unfriendly component. The dietary fat can be generally classified as saturated fat (like animal tissue fat), polyunsaturated fat (vegetable oils) and monounsaturated fat. American diet and some Chinese dishes are full of saturated fat. The worse of the fat actually is the so-called Trans-fat. Small amounts of Trans-fat exist naturally in meat but the majority, if not all, of dietary trans-fat are man-made. Polyunsaturated fat is generally heart-friendly and provide HDL (the good cholesterol or high density lipoprotein). In our body, we make our own cholesterol, exclusively by liver.

Trans-fat is made by bubbling hydrogen gas through vegetable oil (polyunsaturated fat). Food industries use Trans-fat to stabilize the oil, for deep-frying and prolong the shelve life of packaged food and solidify the oil, as for making margarine and baked goods. They use the term "*shortening*" to describe the content.

For brief purpose without detailing the nitty gritty here, Trans-fat is known to clog the arteries even more so than saturated fat. FDA probably soon will force the food industries to label their products with data about the Trans-fat content. You should look for it on the containers.

In the meantime, don't buy foods that have labels show high saturated fat content (but low saturated fat content does not necessarily mean low Trans-fat contents, it is not labeled yet). The rule of thumb: When you see something like "*shortening*" or "*hydrogenated*" on the label, don't buy it, and don't eat it.

Good advice from Fan Kang: Eat vegetables, Salmon (natural, not farm raise type) and more fish, and fruits. Some lean meats are good for you too.

Ten super foods you should eat:
Moderation, Variety, All colors
1. Broccoli - Breast, colon, stomach cancers
2. Spinach - Age-elated blindness but not for people with iron over load condition
3. Tomatoes - Prostate cancer
4. Blueberries - Heart, brain function
5. Salmon - Cholesterol, aging, Omega-3
6. Nuts - LDL, HDL, E, high calories
7. Oats - Fiber, cholesterol
8. Garlic - Heart, cholesterol, antibacterial
9. Green tea - Stomach, esophageal
10. Red wine - HDL, arteries

Eight foods you should never eat:
1. Entenmann's Rich Frosted Donut
2. Nissin Cup Noodles with Shrimp
3. Frito-Lay's Wow Potato Chips
4. Oscar Mayer Lunchables
5. French Fries
6. Campbell soups (regular or high salt type)
7. Bugles
8. Contadina Alfredo Sauce

Low carbs vs. low fat diet
Low - carbohydrates life style avoids sugars/starches; rich in proteins and fats. Carbohydrates are the sole source of vitamins, minerals, antioxidants and protein.

Blood Pressure Guidelines:
            (Systolic to Diastolic)
Normal   (120-80)
Pre-Hypertension (120-139 to 80-89)
Change health habits:
   o   Lose weight (if heavy)
   o   Reduce salt intake
   o   Eat more fruits/vegetables
   o   Get more exercise
   o   Reduce alcohol
Stage 1 Hypertension   (140-159 to 90-99)
Blood pressure medication
Stage 2 Hypertension   (>160 to >100)
Take 2 blood pressure medications.

Appendix 2.1
W.F. Speech for Brian's Wedding
Saturday, April 2, 2005
惠發演溝

It seems to be just yesterday. It is still very fresh in my memory; when Brian was just born. I was deciding whether to accept an offer from Purdue University.

We postponed one year and moved to Purdue the following year. We built a big house in Lafayette. At that big house, I can still hear the voice from Brian "*I don't want Walter Cronkite to say good night from the CBS evening news*", because that was the time he has to go to bed.

I can still see Brian in my mind setting up a lemon table at our drive way to sell his juice to the students going to the football game.

I can still feel his excitement when he first bought his moped driving around the block. He was so excited to get his first driver's license and then was so sad to see him to get his first ticket the same day.

For so many years at Purdue, he was our only child at home while his two brothers went to schools elsewhere. We enjoyed so much his company. He was our "*little Brian*" as his Mom affectionately called him.

I loved to tease him. How come you are not reading book. Brian! You know "*why Johnny can't read*?"

I still remember vividly how surprised I was when a civil engineering staff member asked me to bring a document back to Brian.

The document certified that Brian was ranked number 2 in his graduating class in the whole College. I was really impressed because I did not expect that.

He participated in all kinds of activities. On several occasions, I bumped into him in our faculty meetings and was pleasantly surprised to find out that he was there to receive an award.

I am a structural engineer and Brian is a structural engineer. Inevitably, in university or in business, people somehow will connect us together.

Brian told me in several occasions, some of his professors will tease him saying: Hi, Brian, why don't you ask your Dad to answer this question?

I still remember our conversation during his junior year. "Since you have almost completed all your credits, why not try to finish up in three years?" "Why rush?" he replied "I want to enjoy life".

Sure enough, he had a real good time during his senior year. He enjoyed the fraternity live, "*The Triangles*", and participated in all kinds of activities.

During his growing up years, we sometimes argued about "*why plays the margin*". I keep telling him that at my age "*I can not leave anything to chances*".

So, I want to be at the gate before the boarding time, but not Brian. He sometimes showed up at the last minutes. I felt nervous but he believed in schedule and showed up at the last but right time.

In one of the trips, we were making an around the world trip. Believe me, that was an experience.

The trip started from Purdue airport and connected at Chicago. So, it is critical that we do not miss the Chicago connection.

We arrived at the Purdue airport. It was crowded. I lined up first and Brian joined me later. A few ladies were behind us.

When I entered the gate, Brian was behind me and suddenly he let the ladies go first. Then, Brian was blocked by the agent outside the gate. The flight was overbooked and no volunteers were forthcoming. So, the last few had to stay.

I was standing between the airplane gates and boarding gate. I refused to board without Brian. I refused to return to the boarding gate because of our critical connection in Chicago.

Finally, they let Brian in and sit in the steward seat to solve the impasse. That was an experience; it still looks like yesterday's event. That is our "*little Brian*" all right.

I never knew that my words had so much influence on little Brian. He never shows it. When he applied for graduate schools, he asked me. I told him that UT Austin is a rising star in civil engineering, but not Stanford. That was my big mistake.

We were very excited when he received a $45,000 fellowship from Stanford. But to our dismay and big surprise, he decided to go to Austin.

He declined Stanford and asked me to mail that letter for him. I held the letter for more than a week before I had the heart and courage to mail it.

He loves Austin. He used to own his condo here, received his degrees here, met Christine here, made friends here, and now chose Austin to tie the knot.

Austin must be a lucky city for Brian and Christine.

For so many years, Brian has always been a "*little Brian*" in our mind. He is a "*Crown Jewel*" in our hearts.

It is very hard for us to "let it go". But time has come and we have to let go. So today, we will pass on our "*Little Brian*", our "*Crown Jewel*" to you, Christine.

I hope that you will always take a good care of our little Brian. And, Brian, I know you will take a good care of your little Christine too.

To celebrate this happy occasion, Linlin and I wish to give the some blessing words to our Little Brian and to his Little Christine.

The blessing words were written in Chinese and can best be read and expressed in Chinese. I will show these Chinese words in the following slides.

To understand the meaning of our blessing words, we also put their English translation side by side with our Chinese words.

If you can read Chinese, it sounds like a poem (see page 22-23). The meaning of this poem reflects our value on the marriage. And I hope you two will also share our value on your marriage too.

Appendix 2.2
Eric's Speech for Brian's Wedding
中傑的演講

Winter

From the earliest recorded times, it was Eric and Arnold
One day in Pennsylvania mom went to visit the hospital, and you came out crying.
After that, it was Eric, Arnold and Brian
We'd go outside and build forts in the snow
We'd have snowball fights you and I, with Arnold, Mat and Chester Ho
We donned moon boots and snowsuits, and the excitement would build
With Arnold driving the sleigh, we tried not to get killed
Mom covered us with mittens, scarves and a hat
To keep the boys warm she sure did it right
Hooray we would cheer, and we'd sleigh with delight
At the end of the driveway the sled would spin and roll
Except for that one time, Brian steered us straight into that pole
(Did not read)
Sometimes dad would come out and shovel the snow
Eric and Arnold would grab the red and silver shovels, and away we'd go
We'd plow down the driveway till we'd hit ice and dad would yell halt

Then Dad and Brian would come by with a big bag of salt
Back inside we run towards the showers while tearing off our suits
With Arnold and Brian yelling "I've got the hallway bathroom, I've got the master bedroom bathroom" with me in pursuit

Summer

During the summer the boys would read and swim and play
But 5pm always marked the end of the day
At 515pm dad would approach the driveway, with his familiar yellow briefcase in his hand
We could hear his footsteps walking up the wooden steps, it was all rather grand
The door would crack open and dad's voice would echo with a boom, "Boys I'm home..." and Brian would be off with a zoom
"Dadee Dadee..." Brian would scream as ran down our long, narrow laundry room
Dad lifted him up and carried him back to his room
"Brian how was your day..?"
Brian would then proceed to describe his day
The start of summer vacation meant no more schools
The boys ran around the house and mom said, "Dad, we'd better have some rules"
In the morning the boys studied English and in the afternoon rode bikes,
They hadn't had so much fun since dad had taught them to fly kites
One summer the boys found out that dad was writing a book
He showed them where their names were, they jumped up and down and the house nearly shook
During those early summer days, dad often worked in his study
Brian really missed him, and just wanted to be his buddy
Sometimes, when dad was writing Brian would secretly crawl into the study on the floor and hide in front of his desk
Dad pretended not to notice, but couldn't help but crack a smile
Sometimes Brian jumped out to scare him, and sometimes he waited a while
One summer night dad taught the family how to play Russian poker
We played in the family room on the plush green carpet

We ate potato chips with mom's famous dip, sour cream with French
onion
We played with cheap plastic chips, red, blue and white
We loved to eat and play and laugh, late into the night
One summer the boys went to their first summer camp, Camp Tecumseh
Brian called home every night
Summer camp, it was such a fright
After 2 days, around the corner we saw a familiar red van
Inside was Brian, drinking his favorite pink fruit punch
Ready to head home, for an air conditioned lunch
Some afternoons Eric and Arnold would go to the gym
Mom took Brian to the pool, where he learned how to swim
Weekends in the summertime meant it was time for a family trip
The boys jumped up and down, and piled into the red van
We drove and drove and drove and then the fun really began
Turkey run state park – hiking, climbing and our first tandem biking
Michigan City - sand dunes, sand castles and burying Brian in the sand

Autumn

(Did not read)
In the autumn, on Saturday afternoons, Purdue football was near
The stadium so close to our house, we could hear the fans cheer
The fans were walking and clapping you could hear the band
At the end of the driveway boys set up their lemonade stand
One year the Chen family visited Kassel, Germany
Dad, Arnold and Brian arrived first.
Dad took the boys to the local German restaurant for their first German
meal.
He was so proud.
The three Chen men decided to order chicken soup
They were not entirely successful.
15 minutes later, after the waiter had brought those 3 diet Fanta sodas,
They noticed a red dot on Brian's head
I don't feel so well said Brian
Maybe it's the non-homogenized milk you drank said Arnold
Brian's head looked rather red

Red as a fox, oh no, he had chicken pox
Later, when Brian was resting in bed
Dad gave Brian a special bell
Whenever he rang it, Arnold would appear – to get him water, towel, anything to make him feel well

Spring

On Tuesdays in the spring, I remember mom used to love to hum and dance and sing
On Tuesday nights she'd go to Purdue, to teach and cook and sing
On these nights, Brian she could not bring
When Brian awoke from his naps and found out mom was gone
Brian's natural coping mechanism for this, was to of course… start crying
The boys tried to cheer Brian up when he would cry
But it didn't matter if Eric would pretend to swim or if Arnold knew how to fly
One thing that always helped Brian feel better though, was learning how to write all the letters of the alphabet
He used these letters to spell his name, even if some of the letters were written backwards and he used these letters to write his name, all over the walls of the house.
Favorite memories
Brian, in red, furry pajama suit with a Winnie the Pooh patch.
The boys loved to play in the basement, where Brian learned how to catch things

Bedtime

The boys climbed into their suits with white crinkly feet
They were soft and plush, but they sure could overheat
Bedtime for Brian meant a warm glass of milk
A hug and a goodnight from mom
He was off down the hallway; you could hear the patter of feet
His bedtime ritual all but complete
At night Brian loved to talk
Sometimes we would even find him in the hallway where he liked to

sleepwalk
When it was tuck in time for Brian he was always a little sad
Until of course he got his bedtime story from dad
The days the boys were sick and stayed home from school
Mom would bring them jello and applesauce and their favorite Mac & cheese for fuel
The boys sat in the family room in their PJs and pillows
And watched TV under blankets and pillows
After lunch we played games with long forgotten names,
Like marbles & bubbles
Lincoln logs, matchbox cars & trouble

Life Poem

From the earliest recorded times, it was Eric and Arnold
One day in Pennsylvania mom went to visit the hospital, and you came out crying
After that, it was Eric, Arnold and Brian
1-2 you learned to buckle your shoe
White hard leather shoes with white shoe laces
Brown leather sandals
Squeaky white high-tops
3-4 close the door
We lived in Mohican court near Burtsfield, in a house named Delleur
5-6 pick up sticks
You learned to walk and waddle
After all these years, you still loved your bottle
7-8 looking great
Your first big winter, you're ready
With a panda bear named Teddy
9-10 the boys becoming men
We lived in a German town, named Landwehrhagen.
11-12
Balance was always such a fright
We spent the summer in Hayward,
And Ja Ja taught you how to ride a bike
13-14

Brian and Imran sitting in a tree
Playing basketball in the backyard 1-2-3
15-16
Our 2nd summer in Kassel, it was all like a dream
A dog named Bruno, a house named Suzman and Brian mesmerized by a
remote control yellow submarine
17-18
Brian in high school, decides to learn German you know
Then's he's off to Germany, to visit his best buddy named Flo
21
Brian's off to Texas, suddenly he's all grown up
He meets a friend named Hong
Right from the beginning they always got along
25
We all have our favorites
For Eric it was a mouse named Mickey
For Arnold, it was ducks named Donald and Daffy
One day, Brian arrived at Avalon with his favorite - a dog named...
Kaffee
29
One night swing dancing, Brian meets a queen
Hello, nice to meet you, my name is Christine
Tennis, dancing, climbing they go
Italy, NY and a Broadway show
One summer in china they climb a great wall
Brian on his knees, suddenly Christine looks really tall
Bring presents her with a ring
A symbol of unity from above
Now they'll be partners in doubles, dancing and love

Reflections

One weekend, after mom and dad decided to move off to Hawaii
The boys were home packing
I find my hands full of memories
Red plastic floaters Brian used to wear when he learned how to swim
Boxes in the basement wareroom that used to form the boys childhood

fort

A faded yellow poster of a giant ruler

Hanging on the wall next the family room

Filled with pencil marks where dad marked the heights of the boys at different ages

All souvenirs of childhood and brotherly adventures

Older sibling

Camaraderie, parenthood

Grow older, things change, never be the same

At the end of the journey of eyore, pooh and piglet,

Eyore and piglet hugged pooh

We'll miss you pooh bear they said

We'll miss our special moments

No more pillows above your door

No more tickle fights laughing so hard, till we'd fall onto the floor

30 yrs

Care; look out, even if it didn't seem like it at the time

Quote spoke to me

"Don't be sad that it's over,

Smile because it happened"

2 simple rules

For Brian:

No such thing as being right when you're married

Good things small packages

For Christine:

When Brian is under duress/stress, coping mechanism start crying. Bottle

When traveling with Brian, feed fiber

Drink, Steal, Swear and Lie:

Drink – from the fountain of knowledge. Learn something new every day and make yourself better

Steal – a few minutes each week, and do something nice for some less fortunate than you. In a way that they'll never find out it was you who had helped them.

Swear – that you'll make today, the best day of your life… for it may be your last

Lie – there at night, and thank god you live in a country where you're free to pursue your dreams, and that you've found the person lying beside you who you're going to spend the rest of your life with.

Eric (中傑) gave his speech at Brian (中宇) and Christine's wedding reception in Austin, Texas in 2005.

Appendix 2.3
Suh and Hsuan Wedding Ceremony
小龍婚禮祝辭

Museum of Photographic Arts (MOPA) San Diego, CA
2:00 pm, Saturday, January 7, 2006

[Uncle]: Please be seated. Good afternoon. My name is Wai-Fah Chen and I am Uncle of Herman Hsuan. It is indeed a pleasure and honor for me, on behalf of the families of Suh and Hsuan, to welcome all of the family members and friends to this ceremony celebrating the marriage of Junghae Suh and Herman Hsuan.

As family members and friends, you should be justifiably proud of this beautiful couple here to unite. They have just completed one of the most challenging high tech dating programs in the Y-generation in our family tree. They have demonstrated not only a high level of intelligence in selecting their life-long partners; but also the desirable personal characteristics of determination, perseverance, and belief in hard work in pursuing their careers and balancing their personal and family life.

To Jung, congratulations, on nearly completing your post doctoral work. To Herman, congratulations, on nearly completing your MBA program. The long study sessions, many homework problems, and reports will soon be over in a year or so.

To Jung, you will start your teaching career soon. To Herman, you will start your business career soon. From this day on, hold hands and work together, for a common future. Be thankful, it is the destiny of your life.

As we join together Herman and Jung in this marriage, let us decide to share our knowledge of ideas of love, loyalty, trust, fidelity, and forgiveness as they start this journey together.

[Uncle]: Do you have the rings?

(Herman gets rings from Eric)

[Uncle]: Behold the symbol of marriage. The perfect circle of love, the unbroken union of these souls united here today. May you both remain faithful to this symbol of true love. Herman, please place the ring on Jung's finger and repeat after me.

(Herman puts ring on Jung's finger)

[Prompted by Uncle, repeated by Herman]: I, Herman, take you, Jung, (*pause*) to be my wife, my constant friend (*pause*), my faithful partner (*pause*) and my love from this day forward (*pause*). I promise to love you unconditionally (*pause*), to support you in your goals (*pause*), to honor and respect you (*pause*), to laugh with you and cry with you (*pause*), and to cherish you (*pause*) for as long as we both shall live.

[Uncle]: Jung, please place the ring on Herman's finger and repeat after me.

(Jung puts ring on Herman's finger)

[Prompted by Uncle, repeated by Jung]: I, Jung, take you, Herman, (*pause*) to be my husband, my constant friend (*pause*), my faithful partner (*pause*) and my love from this day forward (*pause*). I promise to love you unconditionally (*pause*), to support you in your goals (*pause*), to honor and respect you (*pause*), to laugh with you and cry with you (*pause*), and to cherish you (*pause*) for as long as we both shall live.

[Uncle]: These rings mark the beginning of a long journey together. Wear them proudly, for they are symbols which speak of the love that you have for each other

[Uncle]: Herman and Jung are now going to light their Unity Candle, a symbol of their marriage. The candles from which they light it are lit by their parents to represent their lives to this moment.

The lights, representing the faith, wisdom, and love they have received from their parents, are distinct, each burning alone. Herman and Junghae will light the center candle to symbolize the union of their lives. As this one light burns undivided, so shall their love be one.

(Candlelight ceremony)

May these candles burn brightly as symbols of your commitment to each other, and as a tribute to your parents' lasting and loving marriages. From now on, your thoughts shall be for each other rather than for your individual selves. Your joys and sorrows shall be shared alike. May the radiance of this one light be a testimony of your unity?

Now you will feel no rain, for each of you will be shelter for the other. Now you will feel no cold, for each of you will be warmth to the other. Now there will be no loneliness, for each of you will be companion to the other. Now you are two persons, but there is only one life before you. May beauty surround you both in the journey ahead and through all the years, May happiness be your companion and your days together be good and long upon the earth.

I now pronounce you husband and wife. You may kiss.

(Kiss)

Ladies and gentlemen, it is a pleasure and privilege for me to introduce to you Mr. and Mrs. Herman Hsuan.

Class 33 at the hotel lawn - 50th Reunion in Los Angeles with wives, in June 2005.
(2005 年師大附中高 33 班同學畢業 50 週年聚會紀念)

台北師範大學附屬中學的校訓:
人道、健康、科學、民主、愛國

我們台北師大附中的老師:
楊淑玉, 蕭端靜, 張薇雲, 林民和, 唐玉鳳, 邱維成, 江芷,
程敬扶, 王清源, 金承藝, 熊公哲, 吳貴壽, 向玉梅, 黃徵。

一九五四年九月，高三十三班座位表

楊淑玉老師

1. 喻應祖	2. 魯永振	3. 程儀賢	4. 張梁圻	5. 賴其明	6. 方欽鐘	7. 容樹藩
8. 張曉春	9. 林鋤雲	10. 張春义	11. 毛鐘靈	12. 江作舟	13. 張逢猷	14. 丁兆治
15. 余　元	16. 陳松齡	17. 李建勛	18. 米明榔	19. 李熊飛	20. 陳建寧	21. 孫鳳岡
22. 夏志雲	23. 石　型	24. 米明琳	25. 吳大銘	26. 姚克定	27. 楊學奇	28. 韓純武
29. 范　抗	30. 唐　亢	31. 林豈凡	32. 許渝生	33. 鄒多星	34. 韓祖武	35. 劉前覺
36. 曹永齊	37. 彭楚欽	38. 游復熙	39. 彭蔭萱	40. 陳惠發	41. 孫長貴	42. 高遠普
43. 錢致平	44. 王成釗	45. 劉平鄰	46. 于建中	47. 王明遠	48. 周洪輝	49. 牛振鏞

我们高卅三班的嫂夫人們 (按同学座位次序)：
張雪梅, 郭家靜, 王良如, 鍾辰珠, 陳碧蓮,
吳瑞麟, 甘蘭敘, 劉光華, 周海萍, 夏開莉,
陳寶瑜, 陳淑美, 譚靜如, 張瑋寶, 何宜賓,
婁璧輝, 劉西琴, 谷若芳, Miriam, 李賽熙,
孫莉莉, 趙蜀寧, 馬渝光, 鄭海蘭, 鄭嘉玞,
谷樹薰, 吳質芳, 焦寶進, 鄭月蓮, 李念慈,
季光容, 楊瑪琍, 宣玲玲, 沈淬華, 李錦花,
羅惠芳, 馬麗容, 鄭麗波, 曹秀苓, 潘庶慶.

Class 33 banquet photo – 50th Reunion in Los Angeles with Teacher Yang Lau Chi
(楊老師), in June 2005.

台北師大附中的校歌:
附中、附中、我們的搖籃。
漫天烽火, 創建在台灣。
玉山給我們靈秀雄奇, 東海使我們擴大開展。
我們來自四方, 融匯了各地的優點。
我們親愛精誠, 師生結成了一片。
砥礪學行, 鍛鍊體魄, 我們是新中國的中堅。
看我們附中培育的英才, 肩負起時代的重擔。
附中青年, 決不怕艱難, 復興中華, 相期在明天。
把附中精神, 照耀在祖國的錦繡河山。

3

The Voyage to America
漂洋過海

3.1 The Start 準備

In the spring of 1961, I was discharged from the Taiwan Navy Marine Corps（海軍陸戰隊）after serving a two-year term as an ROTC officer （預備軍官）in the military. I received a scholarship for one year of study at Stanford University in Palo Alto, California. I also received a research assistantship from Lehigh University in Bethlehem, Pennsylvania. Stanford's scholarship was not enough to cover my living costs; while Lehigh's assistantship was adequate for my graduate study for the first year. If I could make enough money during the summer months in San Francisco, I might be able to stay on the West Coast and attend Stanford. This was the plan and my wishing thinking.

I passed the required national examination conducted by the Ministry of Education（教育部）for all students who wanted to study abroad for graduate degrees. The next major hurdle was to secure an American visa, which required a certificate of bank deposit of $2,400 in US dollars. This was a major problem for most families in Taiwan at the time. I recall the average salary for a middle class family in Taiwan was around NT$2000 per month or about $50 ($1 = NT$40)（新台幣）. $2,400 US dollars was an astronomical figure for most of us. We all dreamed of hitting a lottery jackpot of NT$200,000 known as the Patriotic Lottery （愛國獎券）at the time. The lottery was issued and run by the central government; and the total amount of the first prize or jackpot（頭獎） was worth about $5,000 US dollars at the time. To most of us, this was considered to be a huge amount at the time.

Fortunately, my elder sister, Eileen（惠齡）, who was married at the time in New York, helped and loaned me the required amount through her connections with a distance relative in Long Island, New York. It took almost six month time for me to complete the formal process of going abroad:

1. Form I-20 graduate admission certificate issued by the University (入學許可);
2. Study abroad certificate issued by the Ministry of Education for granting permission to study abroad (留學証書);
3. Health certificate issued by a public hospital for a complete physical check up (體格檢查);
4. Honorary discharge certificate issued by the Ministry of Defense (退伍証書);
5. Passport issued by the Ministry of Foreign Affairs (出國護照); and finally
6. US visa with the $2,400 certificate of deposit (留美簽証).

In Taiwan we described this long and endless process with a Chinese Proverb: "Pass five blockades; overcome six hurdles" (過五關, 斬六將).

3.2 Voyage to San Francisco 漂洋過海到金山

Since the airfare was very expensive, I could only afford the fare for a cargo ship. The ship was allowed to carry 12 passengers according to the Maritime laws. Our ship, New Kaohsiung of Taiwan (新高雄輪), was an old cargo ship. Since the ship was docked in a Southern city, Kaohsiung (高雄), I took an early train from Taipei (台北) and arrived at Kaohsiung around the evening time to board the ship. I had two large luggages: one suitcase and one bamboo case containing books and some personal belongings. My parents said goodbye to me at the front door; and I went to the main train station in Taipei all by myself. I felt quite lonely at the time; and somewhat scared too. It was indeed an adventure for me, to a new world and an unknown place, something like the feeling described in a song "*There is a river; but this river is of no ret*urn". (風蕭蕭兮易水寒, 壯士一去兮不復還).

Fortunately, I found a few traveling companions on the ship. George Lo (羅金陵) was one of my college roommates. He was also a civil engineer and a graduate of the National Cheng-Kung University (成功大學). Ta-Liang Teng (鄧大量) was a graduate of Class #35 of the High School (附中高 35 班) of the National Normal University (師範大學) in Taipei, Taiwan. I was from Class #33 (高 33 班) graduate of the same high school in Taipei. We both served in the Taiwan Navy Marine Corps (海軍陸戰隊) during our ROTC service (預備軍官). Their parents and relatives were at the port to see them off for the long journey. Ta-Liang had a research assistantship from the California Institute of Technology in Pasadena, California. His major was in geology. George had an admission from a smaller college somewhere but not in California. He had no financial assistance from the university. He planed to transfer to the University of California at Berkeley when he arrived at San Francisco. We both needed summer jobs to cover the gaps for living costs.

On the cargo ship "New Kaohsiung" (新高雄), touring Yokohama (橫檳), and steering at the control room in 1961.

All the cabins for passengers were near the captain's cabin. The cabins had two-tier beds, a wash basin with hot and cold water and a mirror. All passengers dined with the captain at the same time each day. Since the room was on the upper deck, the location resulted in more severe motion that induced seasickness, and the screws caused perpetual noise. To make matters worse, the air was always bad. The ventilation was inadequate for most cargo ships that carried some passengers. The thing that impressed me the most was the TV set in the dinning room. This was the first time I saw a TV set. It was not on since there were no TV stations in Taiwan at the time.

The food on the ship was plentiful and good, since we ate with the ship captain together. Although there were some passengers that felt seasick during the voyage, I did not miss a single meal, and could not recall that I had any serious seasickness, nor George (金陵) or Ta-Liang (大量). The ship left the dock in Kaohsiung (高雄) in the morning. The first stop was Yokohama (橫檳), Japan to pick up more cargos. The journey to Japan took about two days. We spent first day exploring the ship; but not much else to do afterwards. I brought with me the famous Chinese novel *"The Dream of Red Chamber"* (紅樓夢) to read; but soon discovered that Tai-Liang was studying his calculus book. It reminded me that I better prepare myself for the upcoming challenges in a very competitive American university for my graduate degrees. At the end of the day, I spent most of my time on my textbooks and much less on reading the novel.

When the ship docked in Yokohama, most of us went on shore for some sight seeing. This was the first time in my life to be in a foreign country. Yokohama was about one hour drive to Tokyo (東京) and we did not dare to go that far. We could not afford to spend any money for the famous Tokyo's strip tease show we had heard about so many times with vivid descriptions from friends and from reading articles. The stop at Yokohama was short and our next destination was Vancouver (溫哥華), Canada. From here on it was a long journey. It took several weeks for us to reach the North America.

During the long journey on the ship, walking the decks and talking with fellow passengers became the principal activities to pass times and to avoid the boredom. My command of English was poor; but I had no opportunity to improve my English because we all spoke Mandarins (國語). However, the interaction and pastimes among the passengers were good friend-building. From our conversations and information-sharing, it helped me understand more about America. Once we arrived at San Francisco, each of us would go on towards our destinations. Some would continue their journeys to universities in the East Coast, where most of the best American universities like the Ivy League schools were located. Some were headed to universities in the Midwest, the heartland of America.

One of the passengers I forgot his name had a family friend in San Francisco whose name was Y.C. Yang（楊裕球）. Yang was a classmate of the famous structural engineer, T.Y. Lin（林桐炎）. They were both graduates from the famous Shanghai Jiaotong University（上海交大）in Shanghai, China in the 1940's. Yang was the chief engineer of the T.Y. Lin Associates in San Francisco at the time. We were told by this passenger friend that we could try our luck and go with him to Mr. Yang's home to stay a few days when we arrived at San Francisco. George and I took the invitation quickly and deeply appreciated the generous offer, since both of us knew no one in San Francisco; and did not really want to spend money, nor were we able to afford to stay in a hotel before we could find a summer job there. Tai-Liang was a lucky guy; because his sister was working as a librarian at Stanford University in Palo Alto, CA. He was fully supported by an assistantship with additional supports from his family resources. He could just focus on his study and prepare his research and be ready for the upcoming enrollment in the fall.

The ship docked in Vancouver for a few days to unload some cargos, and we went on ashore for sightseeing during this period. This was my first time to visit Canada. The things that impressed me the most was the beautiful houses on the hill, which looked just like the Christmas cards which we were so familiar with for a long time. It was really a beautiful setting compared with the constructions and landscape in Taiwan at the time. You felt the real difference between a rich country and a developing country like Taiwan at the time. When we were in Taiwan, we were bombarded with the popular slogans. Something like *"Overcome the difficulty"*（克難）, *"people can achieve the impossible"*（人定勝天）which tried to give us hope for the future. But here in Vancouver, I saw for myself a dream society really existed in the real world of human life. It was like a dream. Our next stop was in Seattle, Washington.

We docked in Seattle for a short time to unload more cargos. I was impressed to see the sailors using a simple cage with dead fishes to catch a lot of crabs. There seemed to be plenty of crabs in the harbor water and it was very easy to catch them in large quantities. The sailors threw most of their catches back to sea just for fun. This really impressed me and I witnessed first hand the abundance and richness of American life. I could not imagine this if this had happened in Taiwan, there would have been hundreds of people coming to the port to make a living through the crab trade. During our short stay in Seattle, my sister-in-law-to-be, Donna Liu (劉蕙君), came to the ship for a short visit with me. She was staying with her uncle in Seattle at the time before continuing her journey to Syracuse, New York, to join my eldest brother, Hollis Chen (陳惠青), for his graduate study and her librarian job at Syracuse University. In Seattle, we also began the formality of entering America though the customs and immigration office. I was on the American soil for the first time. It was a lifelong dream of coming to US and it was finally realized. It was a very exciting day.

The last days on the ship were spending on sailing to San Francisco. We got up early to watch our ship entering the port. It was a beautiful, clear morning. The world famous Golden Gate Bridge (金山大橋) came into sight first, then the skyscrapers of San Francisco city. The whole voyage took more than a month. Too long I wrote to my parents in my first letter from America. The first thing to do in San Francisco was to send back the $2,400 certificate of deposit to my sister, Eileen (惠齡). It was a big relief for me and also for my parents too. Now, everything was on my own. There was no more safety net in helping me if there were any unexpected financial needs that would be tough luck.

Mr. Y.C. Yang (楊裕球) came to the ship to pick up his friend's son, our new acquaintance on the ship. Mr. Yang was a very generous person, and he happily accepted us and took us to his home. We were very grateful. My first impression of San Francisco was its traffic, car after car, as well as the fascinating cable cars mixed with the traffic and up and down hilly scenery. It was unforgettable and it was my first day in America, the new world for me. We stayed at Yang's home for about a week. During the week George (羅金陵) and I found a temporary job of cleaning windows of an apartment building for an old lady. We were very eager to work and sometimes did some acrobatic postures in order to clean some outside but hard to reach windows on the high rise. The lady was very satisfied with our work, and asked us to be careful next time, and not to take too much risk in cleaning the outside windows.

The reason we were asked to move out of Yang's home at that particular week was that Professor Ti Huang (黃棣) and his family from New Mexico were coming for a visit. They were good friends. It turned out that five years later, I joined Lehigh University in 1966 as a faculty member in the department of civil engineering. Professor Huang was also recruited the following year by Lehigh as a senior faculty in the same department. We became colleagues for exactly nine years before I moved on to Purdue in 1976. In fact, we became close neighbors and lived in the same newly developed area in Bethlehem, Pennsylvania. What a small world!

Y.C. Yang (楊裕球) (front), Linlin (玲玲) (seat, middle), Ben Yen (顏本正) (left of Linlin), W.F. (惠發) (camera), Koh (standing) at El-Salvador conference in 1978.

In 1978, I attended an Earthquake Engineering Conference in El Salvador with my wife, Linlin and we met a group of delegation and old friends from Taiwan. We had a dinner party together in the Sheraton Hotel. To my surprise, Y.C. Yang was also there and we sat together and chatted about my short stay at his house. I thanked him again and again for his help at the most critical time in my life. Later, I learned from the Taiwan Ambassador to El Salvador that Yang (楊裕球) was a community leader in the Bay Area; and had helped many overseas Chinese; and was truly generous. As a structural engineer myself, I knew very well that Yang was an excellent engineer with innovation and creativity and had produced many landmark structures around the world as part of the T.Y. Lin (林同炎) consulting firm's signature products.

3.3 Summer Jobs in New York 紐約的花花世界

After we moved out of Yang's home, I decided to continue my journey to New York for a summer job for two reasons: My sister Eileen (惠齡) lived in Manhattan, and Lehigh University's civil engineering had a much better reputation than I was aware of. For example, Stanford at the time was described only as the West Coast Harvard, while the University of California at Berkeley was just another state university, nothing really special. George Lo (羅金陵) visited Berkeley and got his graduate admission quickly; and I was told by many including Yang that Lehigh was a top school in steel construction in the nation. I did not know that; and all we knew in Taiwan was only a few well-known names like Harvard, Cornell, Columbia, MIT and Yale. Because of this new information, I decided to go to New York to earn some money; and then enrolled at Lehigh without worry on my financial supports.

George decided to stay in San Francisco and found a temporary job at T.Y. Lin Associates and working under Yang (楊裕球). I teamed up with a few fellow passengers and took a Greyhound bus to New York. I was told it would take a few days to cross the country. The seats in the bus were spacious and could be reclined. One could sleep quite comfortably. Every few hours, there were bus stops; and we could get out for a stretch and use bathrooms and eat ice creams and snacks. While we were at each bus stop, I loved the ice cream and always ordered one. Since I only knew one brand and its English name: chocolate ice cream, so this was my order all the way from San Francisco to New York. The food was good compared with what we had at the university dormitory or the food in the Marine Corps. I remember well the serving of almost unlimited quantities of milk that was consumed by us almost as fast as the servers could bring it to the table. The servers were amazed to see us adding a lot of sugar to each glass and finishing it quickly. I was impressed with the efficiency of the bus driver who handled almost everything for the passengers.

When I finally arrived at the Manhattan bus station, my brother-in-law, Ian Chiang（蔣惺怡）, picked me up at the station and drove us back to his uptown apartment. I was surprised to learn that this was the same day my sister, Eileen（惠齡）and her baby boy, Joe Chiang, came home from the hospital. It was a two bedroom apartment; but there was no parking garage and everyone had to park their cars along the sidewalk. For the convenience of street cleaning, the cars had to park on the left side for, say, even days and right side for the odd days during the night. I always wondered what if you had to be away for a week or so. Who would take care of your left and right parking rules in the night?

The next morning, Eileen (惠齡) gave me a Manhattan map and some directions to Chinatown's employment agencies to look for a summer job. I was told the subway would be the excellent way to get around and costs were very reasonable. The fare was paid only on entering the subway domain through a turnstile. All lines were interconnected so that one did not leave the system until reaching the final destination. Once I procured a map, riding the subways presented no problems. My first job was as a kitchen helper. I was hired right away because I told the agency that I had experience.

I reported to the restaurant to work. The chef asked me to cook the rice, and I poured too much rice in a big pot. When it boiled, it spilled all over the floor. I was fired after only one of day work. Subsequent jobs accumulated enough experience for me. I learned quickly how to break eggs in a group of six, to peel off the skin of onions without tearing the eyes, to wash dishes a dozen at a time, and to clean the shells of jumbo shrimps with high speed, amongst other things. The key in American kitchen work was efficiency and speed, not saving the materials. The hours were long and the working environment was unpleasant: hot and smelly. Your meal time was irregular and off the normal schedule in order to fit with the restaurant business hours. You did not see the sunlight for the whole day; and worked all day in the basement or backrooms.

Front: Father Yu-Chao (又超), Helena (惠美), and Mother Shui-Tan (水丹).
Standing: Wai-Sun (惠森), W.F. (惠發), Hollis (惠青), and Wai-Kai (惠開).

The kitchen work provided me with a strong incentive and motivation to earn money not with labor but with brainpower. This is American society, the center of capitalism, where you can make money easily with money; but very hard with labor. Education is the key to success in American. There is a Chinese saying *"Laborious work does not necessary make good money; and making good money does not really need laborious work"* (吃力不賺錢, 賺錢不吃力). You have to be smart and use your talents to make money. Timing is critical. Knowledge is power enabling you smart. Education or self-learning provides you the knowledge to be competitive.

3.4 Settled Down at Lehigh 安身在里海大學

Lehigh University is located in Bethlehem, Pennsylvania. Bethlehem is a steel town where the Bethlehem Steel Corporation was formed in 1904 and later became American's second largest steel company. Lehigh's Fritz Lab, named after the original university trustee and Bethlehem Iron Works founder (which later became Bethlehem Steel), received a major addition in 1955 and was hailed as the world's most impressive structural testing facility. L.S. Beedle was director of the Lab when I enrolled as a graduate student in the department of civil engineering. Beedle was also acting Chairman of the department at the time. We all worked under him.

The stipend for a half-time research assistantship was about $2000 per academic year at the time. For all graduate assistants, our tuition was waived. Most of the research and tests were carried out in the Fritz Lab, which was a five-story steel framed building housing a five million pound universal testing machine, the largest one in the world at the time. While the structural research facilities took up the largest space, several other research and instruction laboratories were located throughout the building. Offices, classrooms and various smaller laboratories were located on the second and upper floors.

Two-thirds of civil engineering faculty members at Lehigh were in structural engineering; and steel research activities through the sponsorship of the American Institute of Steel Construction (AISC) for steel industry were the main source of revenues. A revolutionary method for steel design known as "*plastic design*" for steel structures was the main theme of research at the time. Lehigh University was well known for its engineering program in general, structural engineering in particular. The nickname for Lehigh's football team was "*The Engineers*". It seems that my life journey and fate are always associated with engineering-focused schools. The first one was the National Cheng-Kung University (NCKU) (成功大學) in Tainan, Taiwan, where I received my BSCE degree in 1959. NCKU's original name was "*Tainan Institute of Engineering*" (台南工學院). The nickname of Purdue University, in West Lafayette, Indiana, where I spent 23 year as a faculty member, was called "*The Boilermakers*". What a coincidence.

I stayed in a dormitory during the first year. The food was very good and plentiful. I was impressed with the unlimited supply of milk and ice cream and three regular meals. The only problem with dormitory living was that I had to move out to another dormitory during the holiday period, so the university could lock up most of its facilities except one for all foreign students. My monthly pay check was about $200. It was enough for my simple living and I began to send $50 cash as a start via registered mail to home to support my parents. This practice has lasted for the last 45 years. I remember that I spent about $750 to buy an engagement diamond ring for Linlin (玲玲) in 1966; it was almost equal to my four month research assistant salary.

John Badoux and my former doctoral student, Xila Liu (劉希拉), on Tsinghua University (清華大學) campus, China, 2006.

In the second year, I moved out and rented a house near the campus with three friends. Since everyone's schedule was different, we did not share cooking and grocery shopping. I spent most of my time in my office. My officemate was an American-Korean student, Hai-Sang Lew. My immediate research supervisor was John Hanson, a doctoral student working on his Ph.D. thesis and I helped him test concrete specimens. His thesis was on shear strength of reinforced concrete beams. The second year, a new student from Switzerland, John Badoux, joined our research group and I became his immediate supervisor. I received my MS in structural engineering from Lehigh in 1963 and decided to transfer to Brown University in Providence, Rhode Island to focus on theoretical work of solid mechanics. Since 1963, each of us was on our own way and we had not seen or communicated to each other for nearly forty years. We were reconnected somehow again through our professional society activities or student-related connections. We were all reaching our retirement age. Although our time together was relatively short, our friendship and relationship has been continuing for nearly half century.

3.5 Looking Back 回顧

Looking back to the three classmates at Lehigh during my two years as a graduate student in 1961-63, John Hanson first worked as a researcher at the Portland Cement Association, then joined Wiss, Janney, Elstner Associates (WJE) as a practicing engineer, and finally returned to the academic world as a distinguished professor of North Carolina State University. He was elected to the National Academy of Engineering in 1992. I was elected to the same academy three years later in 1995.

John Badoux returned to the Swiss Federal Institute of Technology, Lausanne, as a faculty; and rose steadily to become the President of the Institute (1992-2000). Hai-Sang Lew joined the National Bureau of Standards (NBS) in 1968 as a structural research engineer. He successively became Chief of the Structures Division. He was a member of the National Academy of Engineering of Korea. This is in a nutshell a brief description of the life journey of three of my former classmates, met nearly 45 years ago at Lehigh University in the steel town called Bethlehem. But the Bethlehem Steel Corporation has since become long gone and bankrupted as the world becomes more and more flat and global. Lehigh is still going strong and continuing to produce good leaders in the field of structural engineering.

3.6 Joining the Elite Group 列入學術名流

It was a great honor to be included in a recent book, *Giants of Engineering Science* (Troubador Publishing Ltd, UK, 2003), which is a "biographical monograph examining the life and works of ten of the world's leading engineering scientists." Author O. Anwar Bég of United Kingdom, examined the contributions of engineers worldwide in diverse areas of engineering science and selected those that he felt have made the greatest contributions to their fields.

Engineering science is a branch of applied physics involving a mathematical formulation of basic differential equations to engineering problems and their solutions. It covers a wide variety of subjects such as biomechanics, structural and geotechnical mechanics, fire dynamics, heat transfer, hydrodynamics and computational mechanics, among others. Only one person in each area is featured in the book. Other notables include the late Chancellor C. L. Tien (田長霖) of the University of California at Berkeley, Society of Engineering Science founding father A. C. Eringen from Princeton University, and the father of bioengineering, Y. C. Fung (馮元楨), from the University of California at San Diego.

In selecting individuals discussed in his book, Author Beg stated that "all had prolific career in their various disciplines. Many have written in excess of two hundred research papers and revolutionized their academic fields. Only one research (along with A. C. Eringen) however has had a truly prolific in both research publications and research books – W. F. Chen. He is arguably the most active geotechnical/structural engineering scientist in the world today".

According to Bég, "*Dr. Chen's enormous input into the area of advanced plasticity theory and design has won him worldwide recognition. His mammoth monograph on Non-linear Analysis in Soil Mechanics co-authored with E. Mizuno (1990) is the definitive treatise for modern geotechnical engineering science. It typifies the excellence and clarity of communication and thinking that Professor Chen has brought to the constantly evolving disciplines of geotechnical and structures, both foundations of the civil engineering profession*".

Author Beg concluded this chapter with the following statement "*Dr. Chen has been a principal figure worldwide in the development of structural and geotechnical engineering sciences from the mid-1960s to the modern era of advanced scientific computing and smart structural systems. His influence on engineering science is undisputed and unquestionably he ranks as one of the most prolific, famous and respected civil engineering scientists worldwide – a true giant of engineering science*".

4

My Lifelong Companion – Linlin
鶼鰈情深

On June 11, 1966, Linlin (玲玲, Lily) and I were married in Providence, Rhode Island. This was the time we both graduated and received our degrees. Linlin received her B.S. degree in mathematics from University of Illinois at Urbana, Illinois; and I received my Ph.D. degree from Brown University in Providence, Rhode Island. We had known each other since 1961.

4.1 The Romance 戀愛

In the summer of 1962, I had just finished my first-year graduate study at Lehigh University in Bethlehem, Pennsylvania, and attended my twin brother Wai-Kai's (惠開) wedding in Tarrytown, New York. Wai-Kai, his wife Shirley (筱玲), Shirley's brother David (陳筱聰), and his wife Gwen (鄧元玉) had a double wedding together. After the wedding, David and Wai-Kai asked me to take Linlin back to her friends' residence in downtown Manhattan, New York. I was told that Linlin's family and Shirley's family knew each other well. Their mothers were old mahjong friends. Linlin also knew Shirley's two brothers well: Stanley Chen (陳筱雄) and David Chen (陳筱聰) since their childhood.

Our first encounter was in Syracuse when we both attended the wedding of my elder brother Hollis and his wife Donna in 1961. I had a deep impression of her liveliness and openness, typical of girls raised in Western culture. The first time we were together was when we took a train from Tarrytown to Manhattan. At the time, Linlin was just ready to go to college, and I was just completing my first-year of graduate study in America. To me, her English was just perfect like an American, while I was still in the process of learning conversational English. I considered her as an American teenage girl but with deep traditional Chinese culture. I recalled that the ticket cost was about three dollars, and that she insisted to pay her own fare. This was our second encounter.

In high schools in Taiwan, we had separate boys' and girls' schools, but for some reason at my high school we did have a few girls on campus. Very few of us knew any of the girls. Even in College, very few of us had girlfriends. But this did not diminish our interest in them. "*Girl watching*" and "*who's who girls*" were the favorite topic and good pastime on campus. We watched the girls between classes, went to the campus cafeteria, especially in the evening when the boys and girls took walks on the spacious campus and said goodbye to each other at the front doors of girl's dormitories. Their dresses and appearances were analyzed and new information was shared. We talked a lot about girls, especially attractive girls but none of us dared to approach them.

In the summer of 1964, after I had registered at Brown University and was settled down in my daily routine. Shirley told me that Linlin was living in Boston and staying with her distant Aunt Joyce Chen. Joyce owned a famous Chinese restaurant in Cambridge near MIT. Linlin had a summer job there. Shirley (筱玲) gave me her address. I felt that she was giving me someone special and a precious opportunity for me to follow up. I found that the Joyce Chen Restaurant in Cambridge was about one hour driving distance from Providence. So I shared this information with my two close friends C.T. Liu (劉錦川) and King-Ning Tu (杜金陵) and asked them to accompany me for the trip, since I had done similar things for them too. At the time, we all lived in a cluster of commune-like houses nearby with four to five students forming a cluster. So one day in June, I called Linlin (玲玲) at Joyce Chen's home and she agreed to see me and my friends Liu (劉) and Tu (杜) on the coming Saturday. After more discussions, we decided it was better to have only one of them to accompany me. That was Tu (杜).

However, we did not realize that Saturday was a big day for Joyce Chen's family clan to attend a graduation ceremony for one of their family friends. So, we all went to the graduation ceremony at Boston University stadium without really knowing who was graduating and at what location at the time. It did not really matter of course. Joyce Chen surprised us by treating us to a very nice lunch afterward at her home with barbecue steak prepared by her husband. I was encouraged and invited to visit them again. That was my first dating.

I drove to Cambridge regularly every Friday evening. I usually waited and sometimes worked as a host in the Joyce Chen restaurant till Linlin completed her duties either as a cashier or as a waitress. Afterward, we drove to the nearby Fresh Pond Park to take a walk. It was a quiet and peaceful setting in the evening. This scene has remained with me all these years. We also drove to the nearby beach and watched the sunset. In these days, many times when we walked on the Waikiki beach in Hawaii and watched the sunset, it always reminded me this scene in our Boston beach walk. The first movie we watched together in Cambridge was a James Bond's movie *"From Russian with Love"* in a theater near Harvard University.

With such a romantic beginning, what could one expect but a successful love story and quick marriage? But it took much longer; and what followed was a sequence of steps I learned from Linlin about how a romance would proceed in a typical American college setting at the time. That is, it should start with *"dating"*, advance to *"steady"*, achieve *"pinned"* status, then reach *"engaged"* and finally achieve the goal *"married"*. In each step, there were clear understandings and certain commitments from each side about the relationship. It was an eye opening process for me. And she kept telling me that I was always one or sometimes two steps ahead of her. The process may take years to develop. This was quite different from my expectation of dating and marriage. I was expecting the dating and engagement process to be short and quick; but the formal marriage could be postponed to the time of graduation.

As a graduate student, I could not afford the time to go through this long process. Being a foreign graduate student, there was tremendous pressure that domestic students would not generally face. With a graduate research assistantship, I had to generate results quickly for my adviser's project. As a doctoral candidate, I had to pass my qualifying examination and foreign language requirements in the first year; and my preliminary examination showing good progress in my thesis work in the second year. My thesis adviser was expecting us to graduate in three years because the competition for financial support was quite fierce. At Brown, for example, a typical first year graduate student would take three graduate courses with a financial support of about $3000. If your grades were dropped to two A's and one B, the next year's support would be reduced to $2000. There was a tremendous pressure for us to keep our assistantship to survive in a very competitive environment.

W.F. (惠發) and Linlin (玲玲) at the University of Illinois Assembly Hall 1965.

After the summer of 1964, Linlin returned to Urbana to continue her study in mathematics. Through the grape vine, I learned that she was popular and active on campus; had a number of male acquaintances; and her reputation was that she was *"hard to get"*. Through the Chinese student community, I learned that she was really a *"good"* girl, not really *"Americanized"*. *"Good"* in Chinese means kind, family-focused, and conservative. Despite her early year's education in American, she still enjoyed reading the Chinese classic novels like *"The Dream of Red Chamber"* (紅樓夢) which is similar to *"Gone with the Wind"* (飄) in American literature. I was very impressed with her hard work ethic in American culture and yet still quite conservativeness in her relationships with boys. To me, she was a perfect mixture of Eastern and Western cultures. This provided a lot of interests for me; and I decided to approach her more aggressively. From then on, our relationship advanced steadily without a hitch. At the time, she was a junior and had some tuition and room and board scholarship.

In the winter of 1965, Linlin moved to Providence and enrolled in Rhode Island College to complete her senior year credits for graduation. I completed most of my doctoral requirements and was working on finishing my dissertation, preparing for job-hunting, and anticipating graduation in 1966. Linlin (玲玲) lived in a dormitory-like housing near the Brown campus for girls only. She became a part of Brown's student community and knew most of the Chinese graduate women students at the time. More than 40 years have passed from the day we were married until today; we had never had a fight or even raised our voices. Forty years of marriage is a blessing for me and is one of the great successes of my life.

In the summer of 2006, the year of our forty-year anniversary, we attended the Academia Sinica meeting in Taipei; we met for the first time since our first date, my old classmate and dating companion, K.N. Tu (杜金陵) and his wife (喬芹). They remembered vividly the events and seeing Tu served as a "*light bulb*" in my first dating. Tu was elected to the Sinica in 2004; but was unable to attend because of the illness of his wife at the time. C.T. Liu (劉錦川), my best man at my wedding, was elected to the US National Academy of Engineering last year; and C.H. Liu (劉兆漢) who was already married during the time of our graduate study, and whose apartment was frequently used as our gathering place for the "*Gang*", was elected to the Sinica the same year as I was in 1998. Linlin knew all of them and their wives because the wives were all part of the "*housemates*" or "*Brown gang*" at the time.

Left: Engaged in August 1965 at Brown. Right: Graduated from Brown in June, 1966. Linlin's (玲玲) Father, H.C. Hsuan (宣錫鈞) (Left).

W.F. (惠發), 喬芹, Linlin (玲玲), and K.N. Tu (杜金陵) at Academia Sinica Meeting (中央研究院院士會議) July 2006, Taipei, Taiwan.

4.2 The Wedding 婚禮

Our marriage ceremony was held right after my graduation ceremony and was entirely arranged by ourselves with the help of our friends. At that time, my parents were in Taiwan and they could not come to my wedding. Linlin's (玲玲) parents were also in Taiwan and her father came to Brown to attend our wedding. My siblings in the United States at the time included my elder brother Hollis (惠青) and his wife Donna (劉慧君) from Syracuse, New York; my elder sister Eileen (惠齡) and her husband Ian Chiang (蔣悍怡) from Caldwell, New Jersey; and my twin brother Wai-Kai (惠開) and his wife Shirley (筱玲) from Urbana, Illinois. All of Linlin's (玲玲) siblings were still in Taiwan. So they invited some close relatives for a dinner banquet in Taipei afterwards.

The wedding ceremony was a standard church wedding held in the University Chapel as most Chinese students did at the time. The reception followed in the student activity center. C.T. Liu (劉錦川) was my best man and Susan Chang from New York was Linlin's maid of honor. The whole process was completed in less than half a day. Since

Providence was a seaport, we bought several dozen of fresh crabs to entertain the guests, friends, and relatives. Most people came by car with very few coming by air. I bought a "*big*" diamond ring with all my savings from a New York jewelry store for Linlin. It was a big ring all right; but not in high quality by today's standard. Since we did not know much about the diamond, it did not really matter much at the time.

Linlin, her father and I went to the honeymoon together after my graduation ceremony. I drove my old Buick car for the long trip. The sightseeing included the familiar spots in New York City, Washington D.C. and Niagara Falls, among others. We did not even make any hotel reservations and just played it by ear at the time. There was very little advanced planning, partly because of the rush in graduation, moving, job-hunting, and wedding, and partly because of our lack of experience in traveling. It was a good and long-waited reunion for Linlin with her father since she left Taiwan at the age of 16. This was the first time they saw each other again after nearly a decade of separation. It sounds incredible to take the father-in-law together for the honeymoon in today's culture; but it seemed to be the right thing to do at the time.

This was the first time I met my father-in-law. It was at the Providence airport. I still remember vividly, his reaction to seeing Linlin at his first sight was first to stare at her for a long while, and then he said to her with a big smile "*You need to add some weight*". His first reaction to me was to my old Buick "*Does the government let you drive this car?*" My father-in-law, H.C. Hsuan （宣錫鈞）, was a successful business man, first in Shanghai, then later in Taiwan. His expertise was in the textile and garment industry. Later, he built a chemical company in Taiwan producing soda ash. He was thinking and possibly negotiating with Pepsi headquarters in New York to open the Pepsi Cola in Taiwan when he came to our wedding. He foresaw the potential of a big market for brand name soft drinks when Taiwan's economy would take off in the future. I was impressed with his vision and his forward looking thinking.

4.3 The Hsuan's Family 宣家

I feel that in order to understand Linlin's personality and emotion, it is necessary to begin with her family. Her mother's home town was in Hangzhou, a paradise in China, a place where many Chinese beauties were born as frequently told by many Chinese fairy tales. Her father's hometown was about 100 miles from Hangzhou（杭州）. He was initially

Wedding party on Brown campus in June 1966. From left: Wong (汪), Wu (吳), Susan, Linlin (玲玲), W.F. (惠發), Liu (劉景川), Kong (孔).

Wedding held at Brown University Chapel on June 11, 1966.

reluctant to go through an arranged marriage; but when he was ready; the girl suddenly became sick and passed away. Mr. Lu (盧), the father of Linlin's mother, liked Linlin's father, arranged their marriage quickly afterwards. That was a common practice in China at the time. Parents decided your marriage.

Auntie Chao (趙姑媽), my father-in-law's younger sister, ran away from her hometown to join my father-in-law in Shanghai, in order to avoid an arranged marriage. She later married Uncle Chao (趙姑父) who was a business partner of my father-in-law. Linlin knew Auntie Chao since her birth and was practically raised by her. They were very close just like mother and daughter. The Hsuan's family and Chao's family were very close, and Uncle Chao and my father-in-law were lifelong business partners since their adulthood. They all had a happy family life and marriage. With a family background such as this, it is not necessary to explain why they are all good wives and mothers. All of them are personable and loving people.

The Hsuan's family has one daughter and four sons: Jason (建生), Eddie (建華), Bakey (建國) and Jim (建敏). Linlin is the eldest and the only daughter. My father-in-law was a perfectionist and paid attention down to every detail. He had endless energy and drives to continue improving his existing business. He was always looking for new opportunities to invest. He constantly encouraged his sons to seize opportunities to establish new businesses. He was happy to see Jason build a major TV monitor manufacturing company in China. He was also happy to see the establishment of a major garment factory in China by Falton Chao (趙天星), the son of Uncle Chao, with some of their former business partners. He was a lifelong learner and had a high respect to scholars. He was very proud of my election to the Academia Sinica and purchased a big announcement in the Chinese newspapers. He even posted the congratulation letter of my election by at the time Taiwan's Prime Minister (行政院長蕭萬長) in his living room.

4.4 Linlin's Business Adventure 創業

Linlin (玲玲) has been consistently supportive of my aspirations, encouraging me in my studies and career. When I started my teaching career at Lehigh University, she found a teaching job at the nearby community college. She was enjoying the teaching of mathematics despite the fact that some of the students were poorly prepared and not ready for her class. The Northampton Community College was a short drive from our Bethlehem's home. She also enrolled as a graduate student in the mathematics department at Lehigh University for her M.S. degree. It was a busy time for her.

Falton (趙天星), Jim (建敏), Bakey (建國), Eddie (建華), Jason (建生) and Linlin (玲玲).

Front row: Wen-Mei (盧文美) and H.C. Hsuan (宣錫鈞) and Auntie Chao (姑媽).
Second row: Brian (中宇), Arnold (中毅), Linlin (玲玲), W.F. (惠發) and Eric (中傑).

Our three boys were all born in Bethlehem: Eric in 1968, Arnold (中毅) in 1970, and Brian (中宇) in 1975. Linlin was very busy raising the boys: schooling, music, and Sunday Chinese class, among others. She also started helping her family immigrate to U.S. The first arrival was Eddie (建華) and his fiancée, Verna (溫柔燕) in 1968. They enrolled in the graduate school at Lehigh University for their M.S degrees. Then, Jason (建生) came and enrolled in Boston University first and later transferred to Brooklyn Polytechnic University, New York. Then, Jim came and enrolled in a Texas university and later came to Bethlehem and lived with us. Finally, Bakey (建國) and Helen (海倫) came. They started a small business painting color pictures on chinaware. Then, the Eddie and Bakey (建國) families decided to move westward to Indonesia and San Francisco respectively for better business opportunities. During this period, Linlin (玲玲) provided loving encouragement for me, a big sister role for her siblings and a loving mother for the three boys, and a supportive daughter-in-law for my parents. She was playing a super woman at that time for our family.

From Left: Eric (中傑), Arnold (中毅) and Brian (中宇).

She decided to open an ice cream franchise store, Carvel, with the arrival of her youngest brother Jim (建敏). Since Jim decided not to continue his graduate study in U.S. This was her way to help Jim. It was a new experience for all of us. They went to Carvel headquarters in New York to receive their basic training in production and management. It was exciting and fun for all of us. We ate a lot of ice cream at the time. My sister, Eileen (惠齡), lived in New Jersey, a one hour drive from Bethlehem, and came frequently to visit us since my parents were living with us at the time. Linlin's mother also came over to join the excitement. Everyone had a good time but everyone gained weight; because we ate ice cream not in cups but in bowls. Linlin (玲玲), like her father, was a perfectionist and was very serious when it came to running the store. Her attitude and seriousness still has a deep impression on me. Although she looked like her beautiful mother, her character was 100 percent of her father's image.

Because of her long time association with Joyce Chen in Cambridge, she knew the restaurant business inside and out and knew every trick and pitfall of running a restaurant, a Chinese restaurant, in particular. She profoundly dislikes the business, long hours, particularly the pungent restaurant smell, and vicious behavior of the kitchen staff and chefs. It is not a business she wants to touch. She is a gourmet cook and can cook tasty and beautiful Chinese dishes. She has deep interest in cooking and in fact taught a cooking class at Purdue, which was very well received by her students. She was quite talented in many things and sometimes her friends wondered why she chose mathematics as her major.

When we moved to Purdue in 1976, we designed and built a "*big*" house at "*1021 Vine Street*" near the campus. We acted ourselves as the prime contractor and hired a builder to build it. The builder ordered the materials and the stores billed us directly. We then paid the bills directly to the stores. As a result, we paid a large sum of state sale taxes on various orders. Linlin (玲玲) handled all these billings. In that year we were audited by the Internal Revenue Services for claiming high state sales tax deductions. Linlin showed the auditor a booklet of these receipts and it took the lady nearly an hour to add them up. She was amazed and looked at us unbelievably and said *"You are right"*.

In West Lafayette, Indiana, where Purdue University is located, we began some real estate investments. It was fashionable at the time. At the end of the day, we owned an apartment complex in downtown Lafayette and three rental houses in West Lafayette near our house. Linlin（玲玲）handled all the rentals, maintenances, and collections. There was a lot of tedious work. It is not for the faint of heart to manage such a business. Although we had good tax deductions, it came with a lot of headaches. We also invested with our friends and relatives in Arizona, Georgia, Hawaii, and California. They were good hands-on lessons for us to learn Real Estate 101 in the real world of wealth building. To this end, I will say there is no free lunch. It is not easy to make money in real estate; most of the time when you won, it is your good luck.

4.5 The Chen's Family 陳家

Linlin has been a caring mother to our three boys. When Eric（中傑）was born, she took him back to Taipei for a family visit for the first time ever since she came to U.S more than a decade ago. It was a big event for the Hsuan's and Chen's family. Eric was the first grandson in the Hsuan family and was treated like a prince in Taiwan. When he returned to Bethlehem after the visit, he forgot the changes, thought I was the maid Ah Wah（阿花）, and ordered me around naturally and loudly. I think this visit gave him a better feeling of Chinese culture and appreciated more the richness of family relationships among the three boys. I still remember singing along with him the popular Taiwanese song "綠島小夜曲" every morning and every evening as Eric affectionately called it *"The Green"*. This scene has been remained with me all these years.

Eric（中傑）, like his Grandpa Hsuan（宣）, was born to be a good business man. He listens to your conversation carefully, focuses on the issue and bigger picture, and responds to you with a possible solution. He writes well and communicates clearly. He, like Linlin（玲玲）, took almost everything seriously. He accompanied me to Nanjing（南京）, China at the age of 9; shortly after China had just adopted the open door policy to the world. He even took one semester off from his work and enrolled in a classical Chinese class from Harvard University in Cambridge to improve his knowledge in Chinese literature. He is the only grandson in the Chen family who write goodwill letters to Grandma Chen in Chinese on a regular basis. He has improved his Chinese speaking ability and learned to appreciate the richness of friendships and relationships among the Chen and Hsuan siblings. These opportunities together with his conscious efforts gave him much better knowledge of Chinese culture.

Left: Front from left-Eric (中傑), Melissa (美珊), and Jerome (中堯). Second: Yu-Chao (又超) and Shui-Tan (水丹). Right: Father Yu-Chao and Mother Shui-Tan.

Eric（中傑）was born on June 15, 1968. He graduated from West Lafayette High School in 1986 and received his B.S., M.S., and Ph.D. degrees in Electrical Engineering from the University of Illinois at Urbana, Illinois in 1990, 1992 and 1996 respectively. After working nearly five years at MIT's Lincoln Lab as an engineer, he liked more the business aspects of his work, and entered the University of Chicago Business School and received his MBA degree in 2003. His first post MBA job was with the aerospace giant Raytheon in Lexington, Massachusetts in its Merges and Acquisitions unit. He later rotated to California and loved Los Angeles living and moved to another aerospace giant Northrop Grumman in Redondo Beach in 2005. He has now settled down happily and serves as the Director of Finance in Capital Market Investor Relation dealing with Wall Street analysts. I sometime teased him that he got paid to read Wall Street Journal in his office. This is what I want to do. What a treat!

From left: Arnold (中毅), Eric (中傑) and Brian (中宇).

Arnold（中毅）was a very smart and intelligent boy and learned things quickly. He listens to you attentively and gets the key points quickly. It was always a pleasure to teach him the materials either from his school work or from his Chinese class. He is bored quickly if not challenged either intellectually or physically. He joined the high school swimming team and quickly received the most improved award from his coach. He had *"perfect ears"* according to his music teacher. He played violin well. We were seldom surprised when we heard or received compliments from his teachers and coaches and our friends about his achievements.

He was a very active boy and full of energy. We had to watch him constantly to make sure that he would not get lost in shopping malls or during any stopover in our car travel. In January 1975, we were in Sheraton Waikiki Hotel attending a Tall Building Conference in Honolulu. Arnold was about 4 years old at the time. When I was at the conference, Linlin called me in a panic and told me that Arnold disappeared on the hotel beach. All the life guards and hotel personnel were searching for him. It was a very tense moment for all of us, especially Linlin（玲玲）; she was crying. It turned out that Arnold already returned to our 11th floor hotel room all by himself; when he realized he was lost on the beach. He described to us how he had to jump to push the elevator key because he was too short to reach it, and why he had to sit outside the hotel room to wait for us because he needed a key. We were impressed for a four year old boy who could remember the hotel room number, figured out a way to find the room, and wait for us outside the room; while we were desperately searching for him on the beach and in the hotel. He has generally a quiet personality; but can be very talkative if the conditions are right. He does not express much of his thoughts or opinions unless asked.

Arnold（中毅）was born on October 21, 1970. He graduated from West Lafayette High School in 1988 and received his B.S. degree from Purdue University's Electrical Engineering Department in 1991. He finished his Bachelor's degree in just three years with ease. He transferred to University of Illinois and received his M.S. and Ph. D. degrees in 1993 and 1996 respectively. His first job was with Rockwell Science Center in Thousand Oaks near Los Angles as a research scientist. During the Internet bubble time, he joined the start up company Genoa Corporation in Fremont, CA in 1998. He was a member of the founding core technical team at Genoa, the inventor and manufacturer of the Linear Optical Amplifier, the world's first single-chip linear optical amplifier. The technology was superior to what was available on the market at the time; but the timing was bad and the company eventually merged with another company in 2002.

He joined another start up company Infinera in Sunnyvale in 2003. This company has developed a low-cost way to convert light to electrons and back to light for use in voice and data networks. This conversion can be achieved by two chips it developed that are designed to replace more than 50 optical parts currently in use by most telecommunications carriers. The two chips cost less than half of the price of the current optical parts. Arnold（中毅）was responsible for the production of these chips for the company. His optical work was started at Purdue when he was an undergraduate student; and continued in his graduate research and thesis work. He is one of a few rare engineers capable of doing both experimental as well as theoretical work in the real world of engineering. He loves what he is doing; that is his talent; that is also his passion too. Great match!

He was the first son in our family to marry in the beautiful State of Hawaii in 2004. His wife, Lin Ng（黃慧琳）, a Malaysian Chinese, was a graduate of UCLA Business School and is currently working in the Business School of Stanford University. Lin has talent in languages and can speak three languages fluently and is quite knowledgeable and well experienced in financial planning and business management. Our granddaughter, Chloe（秀敏）, was born on December 10, 2006, the day we arrived at San Francisco airport to attend my 70th birthday for a greater family celebration reunion in San Antonio, Texas by the end of the year.

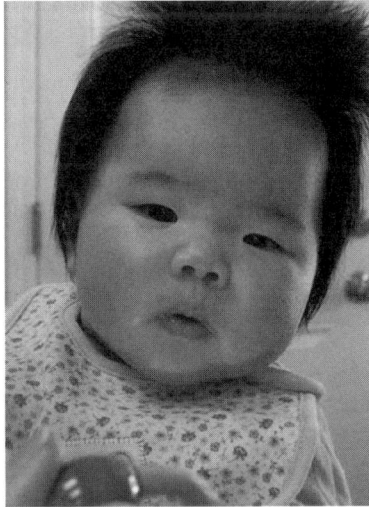

Chloe Chen (陳秀敏) was born in Sunnyvale, California, on December 10, 2006.

Brian （中宇） was born on October 2, 1975, a five year gap after Arnold's birth. He was the most personable boy in the family; and was always surrounded by girl cousins and baby sitters by girls during his early up years. He was the most well rounded son in our family and has a very lovable personality. It was always a pleasure to travel with him and enjoy his companionship. His interests were quite diverse. He was talented in many things including cooking, dancing, artwork, and training dogs, among others. Once, Linlin and I had a rare opportunity to attend a Structural Stability Research Council meeting in Baltimore, Maryland a few years ago. We were really impressed by his technical presentation on his research work on highway bridge girders. It was the best presentation I had attended in recent years: clear, concise, and to the points. Even Dr. Beedle - Director of Fritz Laboratory, the big boss at Lehigh University when I was a graduate student there, and the guru of public speech - came to me after the presentation and congratulated us for his talk. The SSRC Executive Committee voted his paper as the best student paper; and the audience generally agreed that his presentation was the best presentation. He is indeed very talented and meant to be a teacher, and a future academician. He has a lot to offer and will be a great structural engineer in the years to come. We are lucky to have a son like him.

From left: Brian (中宇), Eric (中傑), and Arnold (中毅) at Ja Ja Hsuan's wedding, Fremont, California in 2005.

The year Brian （中宇） was born; I accepted an offer from Purdue University. So, Brian was really a Purdue boy, starting from Purdue's Nursery to Kingston Elementary School (1980-87), to West Lafayette Junior and Senior High Schools (1987-93), and to the receipt of his B.S. degree from Purdue's School of Civil Engineering in 1997. He had the second highest grade average in the Schools of Engineering at Purdue that year and received several honors. Like father like son, he chose structural engineering as his profession and received his M.S. and Ph. D. degrees from the University of Texas at Austin in 1999 and 2002 respectively. He was the first family member to receive admission with a full graduate fellowship from Stanford University; but he chose to go to UT Austin. A decision even today, we are still debating.

He taught for one year at Bucknell University after graduation; and decided to go into engineering practice; and joined the famous failure investigation company Wiss, Janney, Elstner Associates (WJE) in Dallas,

Texas in 2003. WJE is known for its pioneering and high profile investigations such as the early TWA crash investigation in Long Island, the Katrina investigation in New Orleans, and the most recent "*Big Dig*" investigation in Boston. This type of work fits Brian's personality and expertise well: personable, teamwork, and hands on skills on modeling (theory), simulation (computing), and verification (experimentation). He has a lot of potentials for growth in the real world of structural engineering.

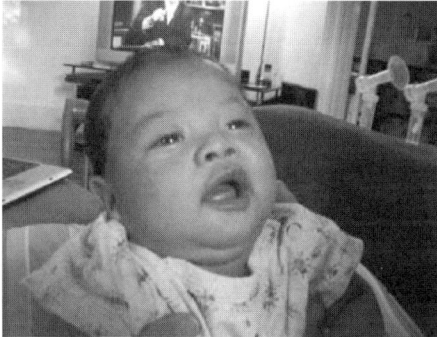

Emerson Alexander Chen (陳安祥) was born in San Antonio, Texas, on March 10, 2007.

Brian was a good dancer, swing dancing in particular. He married his dance partner and UT Austin schoolmate, Christine de Asis in Austin in 2004. They moved to San Antonio in 2006 after Christine completed medical school in Dallas, Texas. Christine has broad interests in almost everything and is quite knowledgeable on many subjects ranging from cooking, decoration, jewelry, foods, and health related issues. She is doing her medical residency in a San Antonio hospital; while Brian is continuing his structural engineering practice with WJE as a remote employee of their Austin office. Our grandson, Emerson Alexander (安祥), was born on March 10, 2007.

With a long history of Linlin's （玲玲） family background in merchant and business, I have been privileged to be loved and consistently supported by her to pursue my aspirations and passion of academic life and career. She is the one who provided the loving encouragement for me, which I solely missed as a child, which allowed me to advance in my career. She has been a peerless mother for the three boys; a loving daughter and daughter-in-law for her and my parents; and a caring big sister for her four brothers and their wives and children. She is indeed a true central figure in our Chen（陳）and her Hsuan（宣）family.

The greater Chen family reunion celebrating Mother Shui-Tan (水丹)'s 90th birthday banquet in Orlando, Florida in 2000.
First row: Shirley (筱玲), Melissa (美珊), Vicky, Miranda, Brian (中宇), and Desiree.
Second row: Linlin (玲玲), Glenda (笑梅), Shui-Tan (水丹), Eileen (惠齡), Mine-Yearn Shine (夏曼雲), Wai-Kai (惠開).
Third row: Richard, Betsy, Ian (惺怡), Arnold (中毅), and W.F (惠發).
Fourth row: Eric (中傑), Hollis Sr. (惠青), Hollis, Jr. (濤濤), Joe, and Jerome (中堯)
(2000 母親陳夏水丹 90 生誕慶祝宴會).

The Hsuan (宣) and Chao (趙) family reunion celebrating Father H.C. Hsuan (宣錫鈞)
50th wedding anniversary in the summer of 1989.
(50 週年宣錫鈞盧文美金婚紀念)

Linlin (玲玲) attended her nephew Fred Hsuan's wedding in Napa Valley, CA in 2000.

Top: Brian (中宇), Arnold (中毅), and Eric (中傑) were each giving a speech in Chinese to congratulate our 25th wedding anniversary banquet in West Lafayette 1991. Seating: S.S. Shu (徐賢修), Hollis Sr. (惠青), Helena (惠美), Linlin (玲玲) and W.F. (惠發) (clockwise).
Bottom: Linlin (玲玲) gave a speech as Helena (惠美) and W.F. (惠發) looked on.
(1991 年惠發與玲玲 25 週年結婚慶祝酒會紀念)

5

Academic Career at Lehigh
教學相長

During my 40 year academic career in America, ten years were spent at Lehigh University as a faculty member (1966-76). In addition, I spent two years at Lehigh's Fritz Engineering Laboratory as a graduate research assistant and received my M.S. degree in 1963. Well, now looking back, I can provide a better perspective on my own university and the education I received as an outsider, than I could as a full time faculty and insider. My profession has always made me a constant Lehigh watcher, and I feel that Lehigh's way of teaching and research during Lynn Beedle's period have been a success and have produced advanced and experienced students who later became leaders in education, industry, and the armed services and government. Let me share some of my own experiences and observations with my own students and their success stories during my Lehigh time.

5.1 The Leader - Lynn S. Beedle 領導者

When I was appointed as an assistant professor at Lehigh in 1966, Lynn was director of the Fritz Engineering Laboratory and directed research on the plastic design of steel structures. His research laid the groundwork for designing structures on the basis of their load carrying-capacity rather than their allowable stress. His students have gone on to become leaders in limit-state design, load and resistance factor design, and auto-stress design. Most of the experimental and theoretical work was performed at Lehigh's Fritz Lab and formed the basis of a series of new AISC specifications for structural steel building design from 1950's to the present time. In particular, Beedle was a force behind the 1959 AISC *Plastic Design in Steel* and the 1986 AISC *LRFD* Specifications, which

91

changed the whole design of steel structures.

Many of Lehigh's faculty members at the time were recruited by him. They came to Lehigh because of him and worked under him. He was a frequent international traveler and was a friend to everyone he came in contact with. The group of faculty at Fritz Lab were quite international including for example Ted Galambos from Hungary, Alex Ostapenko from Russia, L.W. Lu (呂立武) and Ben Yen (顏本正) from Taiwan, Lambert Tall from East Europe, and John Fisher and George Driscoll from US, among others. A good number of his former students were elected to the National Academy of Engineering in later years. Most of them are now leaders of steel research in Europe, Australian, Japan, as well as North America.

Beedle congratulated W.F. (惠發) on his election to the National Academy of Engineering in 1998. From left: Fox, Beedle, W.F. and Linlin (玲玲).

I was initially recruited by Professor L.W. Lu to work with him on the research project *"Columns under Bi-axial Loadings"*. Later, I was asked to work on the *"Beam-to-Column Connections"* under Beedle. He was truly dedicated to Fritz Lab and put his whole effort into it. He pushed his students and junior colleagues to do the best they could. He had a passion for promoting Lehigh's reputation and maintained contacts and relationships with his former colleagues and students over many years.

He remembered my birthday and sent me a congratulation card on a regular basis. He helped almost everyone improve their presentations and showed you how to prepare better slides (see Sec. 9.5, Chapter 9).

Beedle devoted a tremendous amount of his time serving on several professional societies. He headed the *Structural Stability Research Council* (SSRC) for almost 25 years. The council is credited with influencing most of the stability research and steel design around the world. He also founded the multidisciplinary, international *Council on Tall Buildings and Urban Habitat* (CTBUH) in 1969. In this Council, he brought together many different groups of professionals besides structural engineers to look various aspects of tall buildings from architectural, environmental, social political viewpoints. He invited us and encouraged us to get involved in these organizations and contributed our expertise. He was a firm believer that a true fulfillment of engineering research and education is *"a place in practice"*. This conviction has been deeply rooted for most graduates from Fritz Lab at Lehigh.

As a student, I used his book *"Plastic Design of Steel Frames"* to learn the plastic design methods. It was an easy to understand and concise book tailored squarely for practitioners. As an assistant professor, I worked along side him and contributed to the writing of the ASCE's Manual 41 (*Plastic Design in Steel*). The Manual became a standard reference book for teaching plastic design methods in the country. As a colleague, I contributed chapters to his first edited book *"Structural Stability: A World View"* and later the book *"Tall Buildings and Urban Environment Series"* published by McGraw-Hill. These books were truly international in contents and in authorships. He is indeed a giant in the world of structural steel design and a great mentor and my lifelong friend. He passed away on October 30, 2003 of pancreatic cancer. He was 85.

5.2 On Structural Connections 結點

In the development of plastic analysis methods for steel framework, the basic assumption is that the beam-to-column connections have sufficient rigidity to maintain its original geometric angle between intersection members until the member's maximum or plastic strength is reached. Afterwards, the ductility or rotation capacity of the connections must be maintained to provide the moment redistribution among beams and columns in the framework. Beam-to-column connections play a key role in the application of plastic methods.

Full-scale experiments are the only way to obtain the real moment-rotation characteristics of beam-to-column connections. Steel is usually assumed to be a homogenous, isotropic and ductile material for most structural analysis and design. However, at the connecting joints, none of these is true. In addition, the in-situ conditions such as residue stresses due to fabrication and installation, the heat affected zone, and notch effects and back-up bars during erection can all affect the analytical results.

Lehigh's Fritz Lab was one of the first schools of engineering in the country to establish facilities for research and student instruction. It had the world largest five-million pound universal testing machine at the time. Fritz Lab and its personnel provided the tradition and the facilities enabling one to become thoroughly immersed in studies of the broad subject of structural engineering. My assignment to the investigation of fully welded, welded and bolted steel beam-to-column connections over the years created a special niche for developing my own expertise.

The faculty members and staff with whom I was associated in early research were John Fisher, George Driscoll, R.G. Slutter, and Lynn Beedle. All of them were active in steel research and testing; and most of their works including the connection research were sponsored by the American Institute of Steel Construction (AISC), US Steel Corporation, Bethlehem Steel Corporation, among others. We attended the annual meetings of the Institute; published the results of our work in the AISC journal, the Welding Research Council Journal (WRC), and the ASCE Structural Engineering Journal. We all served on the Institute committees and contributed to the development of AISC specifications. The graduate students worked under my guidance were D.J. Fielding (1971), J.S. Huang (1973), J.E. Regec (1973), D.E. Newlin (1973), I.J. Oppenheim (1974), K.F. Standig (1975) and G.P. Rentschler (1979).

Most of the steel research at Lehigh was sponsored by the AISC. To organize and oversee such activities, the Institute established an engineering subcommittee chaired by John A. Gilligan, manager of plate sales for the U.S. Steel Corporation and I.M. Viest from Bethlehem Steel Corporation was a member of the committee, among others. John was a knowledgeable and hands-on chairman who reviewed the progress, provided guidance and initiated new study on some new projects recommended by the committee. Our first joint paper on the state-of-the-art on connections was presented at the International Conference on Planning and Design of Tall Buildings in 1972.

My first paper on moment connections was with J.S. Huang and L.S. Beedle *"Behavior and Design of Steel Beam-To-Column Moment Connections"* (1973). My first paper on the panel zone deformation was with D.J. Fielding *"Steel Frame Analysis and Connection Shear Deformation"* (1973). My last tests at Fritz Lab before moving to Purdue University were with K.F. Standig and G.P. Rentschler *"Tests of Bolted Beam-To-Column Flange Moment Connections"* (1976).

In 1976, when I moved to West Lafayette, Indiana where Purdue University was located, Glenn Rentschler helped drive my other car to West Lafayette and then we both flew to Madison, Wisconsin to attend an ASCE Structural Specialty Conference where we presented our joint paper. At the time my youngest son, Brian (中宇), was about one year old. That was the last time we saw each other. In 2006, Brian's company, Wiss, Janney, Elstner Associates, assigned him to work on a special project in New Orleans related to a failure investigation caused by the Katrina hurricane. The team leader turned out to be Glenn. They found out our relationship in a causal conversation during their driving to the site. What a surprise to meet this way with my son after 30 years of separation. Glenn joined WJE in 1983 and is now serving as unit manager and principal. What a coincidence! What a small world!

In 1986, the AISC specification committee approved the new Load and Resistance Factor Design Specifications (LRFD) for steel building design. This new LRFD Specifications designated for the first time two types of construction in its provisions: Type FR (Fully Restrained) and Type PR (Partially Restrained). Type FR corresponds to the traditional rigid connections as used in plastic design. If engineers want to use the

PR construction as permitted by the new specifications, the effect of connection flexibility must be taken into account in the analysis and design procedures.

Since the new design specifications for Type PR construction only provided broad principles in a quantitative manner, we had to develop design guidelines for engineering practice. Since full scale experiments were the only way to obtain the real moment-rotation characteristics of semi-rigid or PR connections, I started a systematic and ambitious research program at Purdue afterwards. The end results were the publication of two comprehensive reports by N. Kishi, a visiting scholar from Japan, entitled *"Steel Connection Data Bank Program"* CE-STR-86-18, and *"Data Base of Steel Beam-to-Column Connections"* CE-STR-86-26, Structural Engineering Report, School of Civil Engineering, Purdue University, 1986. The experimental connection test data collected in these two reports could be controlled by the Steel Connection Data Bank (SCDB) program from which three prediction equations were developed and recommended for general use. Practical implementation of the use of PR connections in structural system by designers was made easy with these and other related information (see Sec. 8.3.6, Chapter 8).

5.3 On Beam-Columns 梁柱

In structural steel design, one of the critical safety issues is the stability of the structural system when the system becomes more and more slender. In an attempt to answer some of the more important questions concerning the design of steel columns under a three dimensional loading condition, a program of systematic research known as *"Columns under Bi-axial Loadings"* was begun in 1966 at Fritz Lab. I joined the program at its inception and was responsible for its investigation. The project director was L.W. Lu (呂立武). The doctoral students working on this topic under my guidance were S. Santathadaporn (1971) from Thailand, and T. Atsuta (1972) from Japan.

My first paper with Santathadaporn on the subject was a state-of-the-art *"Review of Column Behavior under Biaxial Loading"* (1968). The first solution was achieved in *"Analysis of Bi-Axially Loaded Steel H-Columns"* (1973). The first design interaction equations were proposed with T. Atsuta in *"Interaction Curves for Steel Sections under Axial Load and Biaxial Bending"* (1974). The *"Design Criteria for H-Columns under Biaxial Loading"* proposed with Tebedge in (1974) was later adopted by the AISC specifications.

Beedle congratulated W.F. (惠發) for his AISC Lifetime Achievement Award in Baltimore, 2003. From left: Brian (中宇), W.F. (惠發) and Beedle.

In 1975 and 1976, a two-volume treatise on the *"Theory of Beam-Columns"* coauthored with Atsuta was published as our formal documentation of the extensive work completed over the decade. *The first volume* "In-Plane Behavior and Design" presented the basic theoretical principles, methods of analysis in obtaining the solutions of beam-columns, and developments of theories of bi-axially loaded beam-columns, and to show how these theories could be used in the solution of practical design problems. After presenting the basic theory we proceeded to solutions of particular problems. Both refined and simplified design procedures, along with their limitations, were presented.

This second volume *"Space Behavior and Design"* discussed systematically the complete theory of space beam-columns. It presented principles and methods of analysis for beam-columns in space which should be the basis for structural design and showed how these theories were applied for the solution of practical design problems. The two-volume provided the basis for a new section in the specification for the design of steel framed building columns in the 1986 AISC LRFD Specifications. The work was considered a landmark in the theory and design of beam-columns and has been well received by the structural engineering community (see Sec. 8.5.1, Chapter 8).

In the early 1970's, we had an energy crisis and long lines at the gas stations; offshore structural engineering became a hot topic of research at the time. The American Petroleum Institute (API) was in the process to update its specifications on the design of offshore structures in particular. I was responsible to carry out the large-scale tests on fabricated steel tubular members as used in offshore structural engineering; and to develop its design criteria. The doctoral student worked under my guidance was D.A. Ross (1978). Our first large scale *"Tests of Fabricated Tubular Columns"* paper was completed in 1977. The design criteria *"The Axial Strength and Behavior of Cylindrical Columns"* were submitted to API and recommended for general use (1977).

Left: Signing books for the National Taiwan University students. Right: W.F. (惠發) gave lecture at NTU, Dean Y.B. Yang (楊永斌), moderator.

5.4 On Limit Analysis and Soil Plasticity 極限分析與土壤塑性力學

In the early 1960's, the computer was in its infancy while the theory of plasticity was in its golden time. We had a beautiful theory but few practical solutions. We need simple methods for practical solutions.

Nothing can be more practical than a simple theory. As a result the simple limit analysis methods were developed and widely applied to steel structures. This is known as the plastic analysis methods for steel design in the terminology of structural engineering. When I returned to Lehigh to teach in 1966, I started to apply these limit analysis techniques to soil mechanics and developed simple methods for obtaining practical solutions.

The first paper with Drucker *"On the Use of Simple Discontinuous Fields to Bound Limit Loads"* was presented at the Engineering Plasticity Conference at Cambridge University, London in 1968. The proposed method was familiar to structural engineers and used their intuitions in the construction of safe solutions as they usually did in their design process. Graduate students worked under my supervision for their M.S. degrees at the time were C.S. Scawthorn (1968), M.W. Giger (1969), and J. Rosenfarb (1973).

Limit theory of perfect plasticity was so powerful at the time and provided a consistent scientific basis, from which simple and above all, clear models could be derived to determine the statical strength of many soil mechanics problems including such practical solutions as bearing capacity, earth pressures, and stability of slopes, among others. The success of these techniques motivated me to write my first comprehensive book in my life on *"Limit Analysis and Soil Plasticity"* in 1975. The book created a lot of excitement in the geotechnical engineering community; and was considered a *"milestone"* and *"classical"* text in soil mechanics (see Sec. 8.4.2, Chapter 8).

Charles Scawthorn was my first graduate student when I began my teaching career at Lehigh University. For me the graduate students furnished the highlights of my career and expanded the breadth of my technical interests. Their activities were characterized by their desire to learn, to create, and to work hard. It is a source of continuous inspiration and excitement. During my ten year tenure at Lehigh, nine obtained doctoral degrees under my guidance and eighteen others either obtained their M.S. degrees or reported to me for their special topics at the beginning of their graduate studies. The nine doctoral students were S. Santathadaporn (71), D.J. Fielding (71), T. Atsuta (72), J.S. Huang (73), A.C.T. Chen (73), H.L. Davidson (74), N. Snitbhan (75), D.A. Ross (78), and G.P. Rentschler (79).   All nine authored papers in professional journals, reporting the results of their studies (see the list in Reference). They listed me as a coauthor when appropriate.   Charles departed after his M.S. degree and received a doctoral degree from a major Japanese national university. It was quite rare at the time for an American student to have the courage to attend and compete in a foreign country like Japan. He is a great success story.

Charles Scawthorn is now Professor of Lifeline Engineering in the Department of Urban Management, Kyoto University, Japan, and a Principal in the firm of Scawthorn Porter Associates (SPA). A structural engineer and recognized authority in the analysis and mitigation of natural and technological risk, he heads the Earthquake Disaster Prevention Laboratory at Kyoto University. When he graduated from Lehigh in 1968, I lost contact with him. One day when I read a Wall Street Journal article on a disaster investigation team headed by him after the 1994 Northridge earthquake in Los Angeles; we established contact again and I congratulated him on his success and his achievements.

Scawthorn has been known for his specialty in risk analysis; natural hazards, post-earthquake fire spreading, and damage mitigation. We edited the first *Earthquake Engineering Handbook* in 2002. In addition, he contributed chapters to my *Structural Engineering Handbook* in 1997, and the *Bridge Engineering Handbook* in 2000 (CRC Press). He also contributed an article for the *McGraw-Hill Yearbook of Science and Technology* for which I served as a consulting editor for the civil engineering section.

In the meantime, in the 1970's, our computing power changed drastically with mainframe computing. The finite element methods were well developed and widely used in structural engineering. We were able to apply the theory of soil plasticity and the critical state soil mechanics to obtain large deformational behavior of soil medium with increasing confidence. It was the first time the advancement of soil plasticity is leading its counterpart of metal plasticity in developing more sophisticated theories along with their solution algorithms.

The doctoral students worked on these topics were H.L. Davidson (1974) and N. Snitbhan (1975). Davidson proposed *"Two Elastic-Plastic Soil Models for Numerical Analysis"* in 1976; *while Snithan provided an*

"Elastic-Plastic Large Deformation Analysis of Soil Slopes" in 1978. I lost all contacts with both of them after leaving the University but learned recently about their activities through Eleanor Nothelfer, the editor of the news letter of the Fritz Engineering Research Society (FERS). In 1994, I was honored to be elected to serve as an Honorary President of the Society.

5.5 On Concrete Plasticity 混凝土塑性力學

Concrete is a very old construction material in civil engineering but we were not even able to write down its stress-strain relation or the so-called constitutive relations under various combined stress and environmental conditions at the time. This relationship for characterizing the concrete material properties had to be developed before any finite element analysis could be carried out for its computer simulation. As a result, the existing design practice for reinforced concrete structures was a curious blend of elastic analysis to compute internal forces and moments in a structural system, and then plasticity theory was used to size up the members with empirical expressions for member strength based on full scale tests.

The early effort to develop plasticity models for concrete materials had been centered in the search for suitable failure surfaces. A failure criterion of Coulomb type with a tension cutoff had been used widely in engineering practice at the time. Based on the knowledge concerning the shape of failure surface, a variety of failure criteria had been proposed over the years. Most of these criteria were classified by the number of material constants appearing in the expressions as one-parameter through five-parameter models, all included the strong influence of the normal stress on the shear required in the plane of sliding.

The doctoral student worked on the development of a concrete plasticity model suitable for finite element applications was Andrie C.T. Chen (1973). He proposed a failure criterion and used the work-hardening theory of plasticity to drive for the first time the *"Constitutive Relations for Concrete"* in 1975. It was the beginning of the modern development of concrete plasticity. His constitutive model for concrete materials was included in my 1975 book on "*Limit Analysis and Soil Plasticity*" Elsevier; and subsequent developments by other students at Purdue University formed the basis of my 1982 book on "*Plasticity in Reinforced Concrete*" McGraw-Hill.

In the meantime, limit analysis methods were also applied to the design of reinforced concrete structures at the time. These included for example the ultimate strength design in the 1960's, the yield-line analysis for slabs in the 1970's, and the truss model for design of beams in the 1980's and for joints in the 1990's. The applicability of limit analysis to concrete materials was somewhat justified in my 1970 ASCE paper: "*Extensibility of Concrete and Theorems of Limit Analysis*". The M.S. students worked on this topic were M.W. Hyland (1970), J.L. Carson (1971), S. Covarrubias (1971), and B.E. Trumbauer (1972).

5.6 On Polymer Concrete 塑膠混凝土

In 1972, we conducted an interdisciplinary program on polymer impregnated concrete under the sponsorship of the National Cooperative Highway Research Program. The research team included John A. Manson, and John W. Vanderhoff, colleagues from the polymer laboratory of the Materials Research Center and from the Center for Surface and Coatings Research at Lehigh respectively. The research assistants were Einar Dahl-Jorgensen from Norway and Harsh Mehta from India. The purpose of this research was to develop technology to impregnate previously cured concrete with monomer, which was then polymerized within the pore system to give a composite comprising two interpenetrating network - polymer and cement.

Because Polymer Impregnated Concrete (PIC) effectively resists penetration by water and salt solutions, it is an ideal technology to prevent corrosion of the reinforcing rods in bridge decks. In addition, the PIC is several times stronger, stiffer and tougher than concrete, it is of potential interest for structural applications, e.g., in buildings. But the technological progress requires, however, the solution of certain fundamental problems. For example, successful impregnation of salt-contaminated bridge decks in the field to a depth of sufficient to immobilize the salt already present required basic knowledge of drying, impregnation, and polymerization, which did not exist in appreciable quantity before an extensive study, was undertaken at Lehigh.

The major research finding is that field impregnation of structurally sound, salt-contaminated concrete bridge decks with polymer to a depth of four inches (about 10 cm) is technically feasible. Concrete can be impregnated with monomer to any desired depth provided it is dried to that depth. The time required for drying depends on the heating rate and surface temperature attained. All impregnations used a 90:10 methylmethacrylate: Trimethylol propane trimethacrylate mixture with 0.5% AZO catalyst. The monomer mixture in concrete was polymerized in situ to high conversions by ponding hot water in the pressure impregnation chamber.

Using the foregoing monomer mixture gave a threefold increase in compressive strength and twofold increase in split tensile strength. The PIC concrete arrested the corrosion of the reinforcing steel, virtually eliminated freeze-thaw damage, and dramatically increased resistance to chemical attack. The mechanical properties of PIC can be varied systematically by the choice of the mixtures, e.g., from a ductile with slightly greater strength than that of the control concrete to a very strong, hard, brittle material. The PIC offers an economical solution to the deterioration of critical areas of concrete highways such as bridge decks; because of the longer service life and reduced maintenance costs engendered by impregnation.

5.7 On Lehigh Education 里海教育

I began my graduate study at Lehigh University in 1961 and obtained my M.S. degree in 1963. As a graduate research assistant, I was assigned to help John Hanson, a doctoral candidate worked on shear design of concrete, complete his extensive tests on concrete beams at Fritz lab. My officemate was H.S. Lew from Korea. In 1962, a new research assistant, John Badoux from Switzerland, joined the concrete research team as my junior colleague. I graduated in 1963 and lost all contacts with all of them. But through the FERS news letters, I learned over the years their careers and achievements.

Hanson received his doctoral degree in 1964. He contributed substantially to the structural engineering community through his long and very productive work career at the Portland Cement Association, Wiss, Janney, Elstner Associates (WJE), and North Carolina State University. He served as President of WJE for 13 years; and returned to academic as a Distinguished Professor of North Carolina State University afterward. He was elected to the National Academy of Engineering (NAE) in 1992.

W.F. (惠發) received his National Academy of Engineering membership in 1998.

Badoux received his doctoral degree in 1965 and returned to the Swiss Federal Institute of Technology, Lausanne, as a faculty; and rose steadily to become the President of the Institute (1992-2000). He has been President of the Swiss Academy of Engineering Sciences and the Swiss Society of Engineers and Architects. He met my former student, Xila Liu on Tsinghua campus during his China visit in 2006.

Lew joined the National Bureau of Standards (NBS) in 1968 as a structural research engineer. He successively became Chief of the Construction Safety Section (1978-1985), Chief of the Structural Evaluation Section (1985-1989) and Chief of the Structures Division (1989-1999). He is a member of the National Academy of Engineering of Korea.

I have maintained close contact with Ted Galambos, my mentor who taught us structural stability. He was recognized as the father of the AISC LRFD specifications; and was elected to the NAE in 1979. John Fisher who taught us bolted joints design, became the guru of fatigue and fracture of bridge structures, and was elected to the NAE in 1986. The University had indeed produced great leaders in education, industry and government as demonstrated in this simple illustration of three of my former classmates and two of my former teachers whom I had known personally since 1961.

Appendix 5.1
Doctoral Students - Lehigh University
里海博士生

1966 to 1975 (Total 9)

1. S. Santathadaporn (71)
2. D.J. Fielding (71)
3. T. Atsuta (72)
4. J.S. Huang (73)
5. A.C.T. Chen (73)
6. H.L. Davidson (74)
7. N. Snitbhan (75)
8. D.A. Ross (78)
9. G.P. Rentschler (79).

Appendix 5.2
Publications
論文集
1968 - 1980

Dr. Chen is the author or co-author of more than 595 articles in various refereed Technical Journals (347), Conference Proceedings and Symposium Volumes (248). The following is the list of journal articles during his years at Lehigh University (1966-75).

1968 - Journal Articles (3)

1. (With D. C. Drucker) On the Use of Simple Discontinuous Fields to Bound Limit Loads, Engineering Plasticity, edited by J. Heyman and F.A. Leckie, Cambridge University Press, London, March (1968) 129-145.
2. On the Rate of Dissipation of Energy in Soils, Soils and Foundations, The Japanese Society of Soil Mechanics and Foundations Engineering, Vol. VIII, No. 4, December (1968) 48-51.
3. (With S. Santathadaporn) Review of Column Behavior under Biaxial Loading, Journal of the Structural Division, ASCE, Vol. 94, No. ST12, December (1968) 6316-3021.

1969 - Journal Articles (5)

1. (With S. Santathadaporn) Curvature and the Solutions of Eccentrically Loaded Column, Journal of the Engineering Mechanics Division, ASCE, Vol. 95, No. EM1, February (1969) 21-40.
2. Methods of Computing Geometric Relations in Frames, Journal of the Structural Division, ASCE, Vol. 95, No. ST8, August (1969) 1789-1794.
3. (With M.W. Giger and H.Y. Fang) On the Limit Analysis of Stability of Slopes, Soils and Foundations Engineering, Vol. IX, No. 4, December (1969) 23-32.
4. Soil Mechanics and the Theorems of Limit Analysis, Journal of the Soil Mechanics Division, ASCE, Vol. 95, No. DM2, March (1969) 493-518.
5. (With D.C. Drucker) Bearing Capacity of Concrete Blocks or Rock, Journal of the Engineering Mechanics Division, ASCE, Vol. 95, No. EM4, August (1969) 955-978.

1970 - Journal Articles (8)

1. Plastic Indentation of Metal Blocks by a Flat Punch, Journal of the Engineering Mechanics Division, ASCE, Vol. 96, No. EM3, June (1970) 353-363.
2. Effects of Initial Curvature on Column Strength, Journal of the Structural Division, ASCE, Vol. 96, No. ST12, December (1970) 2685-2691.

3. Extensibility of Concrete and Theorems of Limit Analysis, Journal of the Engineering Mechanics Division, ASCE, Vol. 96, No. EM3, June (1970) 341-352.
4. (With C. Scawthorn) Limit Analysis and Limit Equilibrium Solutions in Soil Mechanics, Soils and Foundations, the Japanese Society of Soil Mechanics and Foundations Engineering, Vol. X, No. 3, September (1970) 13-49.
5. (With M.W. Hyland) Bearing Capacity of Concrete Blocks, Journal of the American Concrete Institute, Vol. 67, March (1970) 228-236.
6. General Solution of Inelastic Beam-Column Problem, Journal of the Engineering Mechanics Division, ASCE, Vol. 96, No. EM4, August (1970) 421-441.
7. Double-Punch Test for Tensile Strength of Concrete, Journal of the American Concrete Institute, 67, No. 12, December (1970) 993-995.
8. (With S. Santathadaporn) Interaction Curves for Sections under Biaxial Bending and Axial Force, Welding Research Council Bulletin, Bulletin No. 148, February (1970), 11 pp.

1971- Journal Articles (6)

1. (With J.L. Carson) Stress-Strain Properties of Random Wire Reinforced Concrete, Journal of the American Concrete Institute, Title No. 68-77, December (1971) 933-936.
2. (With S. Covarrubias) Bearing Capacity of Concrete Blocks, Journal of the Engineering Mechanics Division, ASCE, Vol. 97, No. EM5, October (1971) 1413-1431.
3. (With M.W. Giger) Limit Analysis of Stability of Slopes, Journal of the Soil Mechanics and Foundations Division, ASCE, Vol. 97, No. SM1, January (1971) 19-26.
4. Further Studies of an Inelastic Beam-Column Problem, Journal of the Structural Division, ASCE, Vol. 97, No. ST2, February (1971) 529-544.
5. (With H.Y. Fang) New Method for Determination of Tensile Strength of Soils, Highway and Research Board, No. 345, (1971) 62-68.
6. Approximate Solutions of Beam-Columns, Journal of the Structural Division, ASCE, Vol. 97, No. ST2, February (1971) 743-751.

1972 - Journal Articles (8)

1. (With T.D. Dismuke and H.Y. Fang) Tensile Strength of Rock by the Double-Punch Method, Rock Mechanics, Vol. 4/2, November (1972) 79-87.
2. (With B.E. Trumbauer) Double-Punch Test and Tensile Strength of Concrete, Journal of Materials, ASTM, Vol. 7, No. 2, June (1972) 148-154.
3. (With T. Atsuta) Interaction Equations for Bi-axially Loaded Sections, Journal of the Structural Division, ASCE, Vol. 98, No. ST5, May (1972) 1035-1052 (Closure, Vol. 99, No. ST12, 1973, pp. 2488-2493).
4. (With T. Atsuta) Simple Interaction Equations for Beam-Columns, Journal of the Structural Division, ASCE, Vol. 98, No. ST7, July (1972) 1413-1426 (Closure, Vol. 99, No. ST10, 1973, pp. 2210-2211).
5. (With S. Santathadaporn) Tangent Stiffness Method for Biaxial Bending, Journal of the Structural Division, ASCE, Vol. 98, No. ST1, January (1972) 153-163 (Closure,

Vol. 99, No. ST3, 1973, pp. 578-579.

6. (With T. Atsuta) Column Curvature Curve Method for Analysis of Beam-Columns, The Structural Engineer, The Journal of the Institution of Structural Engineers, Vol. 50, No. 6, June (1972) 233-240.

7. (With L. Tall and N. Tebedge) Experimental Studies on European Heavy Shapes, Proceedings of the International Colloquium on Column Strength, IABSE, Paris, France, November 23-24 (1972) 301-320.

8. (With N. Tebedge and L. Tall) On the Behavior of a Heavy Welded Steel Column, Proceedings of the International Colloquium on Column Strength, IABSE, Paris, France, November 23-24 (1972) 9-24.

1973 - Journal Articles (13)

1. (With J.S. Huang and L.S. Beedle) Behavior and Design of Steel Beam-To-Column Moment Connections, WRC Bulletin No. 188, October (1973) 1-23.

2. (With J.E. Regec and J.S. Huang) Test of a Fully-Welded Beam-To-Column Connection, WRC Bulletin No. 188, October (1973) 24-35.

3. (With S. Santathadaporn) Analysis of Bi-axially Loaded Steel H-Columns, Journal of the Structural Division, ASCE, Vol. 99, No. ST3, March (1973) 491-509.

4. (With D.E. Newlin) Column Web Strength in Beam-To-Column Connections, Journal of the Structural Division, ASCE, Vol. 99, No. ST9, September (1973) 1978-1984.

5. (With D.J. Fielding) Steel Frame Analysis and Connection Shear Deformation, Journal of the Structural Division, ASCE, Vol. 99, No. ST1, January (1973) 1-18.

6. (With T. Atsuta) Strength of Eccentrically Loaded Walls, International Journal of Solids and Structures, Vol. 9, (1973) 1283-1300.

7. Bearing Strength of Concrete Blocks, Journal of the Engineering Mechanics Division, ASCE, Vol. 99, No. EM6, Proc. Paper 10187, December (1973) 1314-1321.

8. (With C.H. Chen) Analysis of Concrete Filled Steel Tubular Beam-Columns, IABSE Publication 33-11, (1973) 37-52.

9. (With H.L. Davidson) Bearing Capacity Determination of Footings by Limit Analysis, Journal of the Soil Mechanics and Foundations Division, ASCE, Vol. 99, No. SM6, June (1973) 433-449.

10. (With J. Rosenfarb) Limit Analysis Solutions of Earth Pressure Problems, Soils and Foundations Japanese Society of Soil Mechanics and Foundations Engineering, Vol. 13, No. 4, December (1973) 45-60.

11. (With T. Atsuta) Ultimate Strength of Bi-axially Loaded Steel H-Columns, Journal of the Structural Division, ASCE, Vol. 99, No. ST3, March (1973) 469-489 (Closure, Vol. 100, No. ST10, 1974, pp. 2149).

12. (With T. Atsuta) Inelastic Response of Column Segments under Biaxial Loads, Journal of the Engineering Mechanics Division, ASCE, Vol. 99, No. EM4, August (1973) 685-701.

13. (With E. Dahl-Jorgensen) Stress-Strain Properties of Polymer Modified Concrete, Symposium on Polymers in Concrete, ACI Publication SP-40, (1973) 347-358.

1974 - Journal Articles (8)

1. (With E. Dahl-Jorgensen) Polymer-Impregnated Concrete as a Structural Material, Magazine of Concrete Research, Vol. 26, No. 86, March, London, (1974) 16-20.

2. (With G.L. Kulak, et al.) Survey of Research on Structural Connections, Journal of the Structural Division, ASCE, Vol. 100, No. ST12, December (1974) 2537-2540.

3. (With I.J. Oppenheim) Web Buckling Strength of Beam-To-Column Connections, Journal of the Structural Division, ASCE, Vol. 100, No. ST1, January (1974) 279-285.

4. (With T.A. Colgrove) Double-Punch Test for Tensile Strength of Concrete, Transportation Research Record 504 on Portland Cement Concrete, National Research Council, December (1974) 43-50.

5. (With T. Atsuta) Interaction Curves for Steel Sections Under Axial Load and Biaxial Bending, Engineering Institute of Canada, Transactions, EIC, Vol. 17, No. A-3, Mar./Apr. (1974) p. I-VIII.

6. (With N. Tebedge) Design Criteria for H-Columns under Biaxial Loading, Journal of the Structural Division, ASCE, Vol. 100, No. ST3, March (1974) 579-598 (Closure, Vol. 101, No. ST8, 1975, pp. 1705-1708).

7. (With J.L. Carson) Bearing Capacity of Fiber Reinforced Concrete, International Symposium on Fiber Reinforced Concrete, ACI Publication SP-44, (1974) 209-220.

8. (With E. Dahl-Jorgensen) Stress-Strain Properties of Polymer Modified Concrete, Symposium on Polymers in Concrete, ACI Publication SP-40, (1974) 347-358.

1975 - Journal Articles (9)

1. (With E. Dahl-Jorgensen et al.) Polymer-Impregnated Concrete: Laboratory Studies, Journal of Transportation Engineering, ASCE, Vol. 101, No. TE1, February (1975) 29-45.

2. (With H.C. Mehta et al.) Innovations in Impregnation Techniques for Highway Concrete, Transportation Research Record 542, "Polymer in Concrete," TRB, Washington, D.C., January (1975) 29-40.

3. (With H.C. Mehta et al.) Polymer-Impregnated Concrete: Field Studies, Journal of Transportation Engineering, ASCE, Vol. 101, No. TE1, February (1975) 1-27.

4. (With N. Snitbhan) On Slip Surface and Slope Stability Analysis, Soils and Foundations, The Japanese Society of Soil Mechanics and Foundations Engineering, Vol. 15, No. 3, September (1975) 41-49.

5. (With A.C.T. Chen) Constitutive Relations for Concrete, Journal of the Engineering Mechanics Division, ASCE, Vol. 101, No. EM4, Proc. Paper 11529, August (1975) 465-481.

6. (With A.C.T. Chen) Constitutive Equations and Punch-Indentation of Concrete, Journal of the Engineering Mechanics Division, ASCE, Vol. 101, No. EM6, Proc. Paper 11809, December (1975) 889-906.

7. (With N. Snitbhan and H.Y. Fang) Stability of Slopes in Anisotropic, Non-Homogeneous Soils, Canadian Geotechnical Journal, Vol. XII, No. 1, February (1975) 146-152.

8. (With M.T. Shoraka) Tangent Stiffness Method for Biaxial Bending of Reinforced Concrete Columns, IABSE Publication, Vol. 35-1, (1975) 23-44.

9.   (With J.A. Manson et al.) Polymer-Impregnated Concrete for Highway and Structural Applications, Proceedings of the First International Congress on Polymer Concretes, The Concrete Society, London, U.K., May 5-7, The Construction Press, (1975) 403-408.

1976 - Journal Articles (15)

1.   (With Mehta, et al.) High-Temperature Drying of Thick Concrete Slabs, Journal of Transportation Engineering, ASCE, Vol. 102, No. TE2, May (1976) 185-200.

2.   (With H.C. Mehta and R.G. Slutter) Sulfur- and Polymer-Impregnated Brick and Block Prisms, Journal of Testing and Evaluation, ASTM, Vol. 4, No. 4, July (1976) 283-292.

3.   (With J. Parfitt) Tests of Welded Steel Beam-to-Column Moment Connections, Journal of the Structural Division, ASCE, Vol. 102, No. ST1, (1976) 186-202.

4.   (With H.L. Davidson) Two Elastic-Plastic Soil Models for Numerical Analysis, Soils and Foundations, Japanese Society of Soil Mechanics and Foundations Engineering, Vol. 16, No. 2, (1976) 43-59.

5.   (With A.C.T. Chen) Nonlinear Analysis of Concrete Splitting Tests, Computers and Structures, Vol. 6, (1976) 451-457.

6.   (With D.A. Ross) Design Criteria for Steel I Columns Under Axial Load and Biaxial Bending, Canadian Journal of Civil Engineering, Vol. 3, No. 3, September (1976) 466-473.

7.   (With K.F. Standig and G.P. Rentschler) Tests of Bolted Beam-To-Column Flange Moment Connections, Welding Research Council Bulletin No. 218, August (1976).

8.   (With N. Snitbhan) "Plasticity Solutions for Slopes," Numerical Methods in Geomechanics (three-volume book), Edited by C.S. Desai, ASCE, New York, (1976) 731-743, (Vol. 2).

9.   (With N. Snitbhan) Finite Element Analysis of Large Deformation in Slopes, Numerical Methods in Geomechanics, (three-volume book), Edited by C.S. Desai, ASCE, New York, (1976) 744-756 (Vol. 2).

10.   (With H.L. Davidson) Nonlinear Analyses in Soil and Solid Mechanics, Numerical Methods in Geomechanics (three-volume book), Edited by C.S. Desai, ASCE, New York, (1976) 205-216 (Vol. 1).

11.   (With E. Dahl-Jorgensen) Stress-Strain Behavior of Polymer-Impregnated Concrete, New Horizons in Construction Materials, Edited by H.Y. Fang, Envo Publishing Co., Bethlehem, PA, November (1976) 303-326.

12.   (With J.A. Manson, H.C. Mehta and J.W. Vanderhoff) Use of Polymers in Highway Bridge Slabs, New Horizons in Construction Materials, Edited by H.Y. Fang, Envo Publishing Co., Bethlehem, PA, November (1976) 327-343.

13.   Foundation Stability-Theory and Applications, In Analysis and Design of Building Foundations, edited by H.Y. Fang, Envo Publishing Co., Bethlehem, PA, (1976) 37-102.

14.   (With H.C. Mehta and A.J. Pepe) Split-Cylinder Test and Double-Punch Test for Tensile Strength of Concrete, New Horizons in Construction Materials, edited by H.Y. Fang, Envo Publishing Co., Bethlehem, PA, November (1976) 625-642.

15.   (With G.P. Rentschler) Test and Analysis of Beam-To-Column Web Connections,

Proceedings of the National Structural Engineering Conference on Methods of Structural Analysis, ASCE, Edited by W.E. Saul and A.H. Peyrot, Madison, Wisconsin, Vol. II, August 22-25 (1976) 957-976.

1977 - Journal Articles (4)

1. (With H.C. Mehta, J.A. Manson, and J.W. Vanderhoff) Field Impregnation Techniques for Highway Concrete, Journal of Transportation Engineering, ASCE, Vol. 103, No. TE3, May (1977)355-368.
2. (With H.L. Davidson) Nonlinear Response of Undrained Clay to Footings, Computers and Structures, Vol. 7, (1977) 539-546.
3. (With D.A. Ross) The Axial Strength and Behavior of Cylindrical Columns, Journal of Petroleum Technology, AIME, March (1977) 239-241.
4. (With D.A. Ross) Tests of Fabricated Tubular Columns, Journal of the Structural Division, ASCE, Vol. 103, n. ST3, March (1977) 619-634 (Closure, Vol. 104, No. ST9, 1978, pp. 1536-1538).

1978 - Journal Articles (5)

1. (With H.L. Davidson) Nonlinear Response of Drained Clay to Footings, Computers and Structures, Vol. 8, (1978) 281-290.
2. (With N. Snitbhan) Elastic-Plastic Large Deformation Analysis of Soil Slopes, Computer and Structures, Vol. 9, (1978) 567-577.
3. (With T.Y.P. Chang) Plasticity Solutions for Concrete Splitting Tests, Journal of the Engineering Mechanics Division, ASCE, Vol. 104, No. EM3, Proc. Paper 13852, June (1978) 691-704. (Authors' Closure, Vol. 105, No. EM6, December (1979) 1064-1067.
4. (With E. Dahl-Jorgensen) Monomer Impregnated Through Case-In Perforated Pipes, American Concrete Institute Special Publication Volume SP58, "Polymers in Concrete," ACI, Detroit, (1978) 299-312.
5. (With H.C. Mehta, J.A. Manson and J.W. Vanderhoff) Stress-Strain Behavior of Polymer-Impregnated Concrete Beams, Columns, and Shells, American Concrete Institute Special Publication, Vol. SP58 - "Polymers in Concrete," ACI, Detroit, (1978) 161-186.

1979 - Journal Articles (4)

1. (With H. Suzuki and T.Y. Chang) Implosion Analysis of Concrete Cylindrical Vessels, Journal of Pressure Vessel Technology, ASME, Vol. 101, No. 1, February (1979) 98-102.
2. (With H. Mehta and T.Y.P. Chang) Experiments on Axially Loaded Concrete Shells, Journal of the Structural Division, ASCE, Vol. 105, No. ST8, Proc. Paper 14785, August (1979) 1673-1688.
3. (With S. Toma) Analysis of Axially Loaded Fabricated Tubular Columns, Journal of the Structural Division, ASCE, Vol. 105, No. ST11, November (1979) 2343-2366 (Errata, Vol. 106, No. ST11, (1980) 2357).
4. Constitutive Equations for Concrete, Introductory Report, Colloquium on Plasticity

in Reinforced Concrete, IABSE Publication, Copenhagen, Denmark, May 21-23, (1979) 11-34.

1980 - Journal Articles (13)

1. (With E. C. Ting) Constitutive Models for Concrete Structures, Journal of the Engineering Mechanics Division, ASCE, Vol. 106, No. EM1, February (1980) 1-19.
2. (With G.P. Rentschler and G.C. Driscoll) Tests of Beam-To-Column Web Moment Connections, Journal of the Structural Division, ASCE, Vol. 106, No. ST5, May (1980) 1005-1022.
3. (With R.L. Yuan) Tensile Strength of Concrete: Double-Punch Test, Journal of the Structural Division, ASCE, Vol. 106, No. ST8, August (1980) 1673-1693.
4. End Restraint and Column Stability, Journal of the Structural Division, ASCE, Vol. 106, No. ST11, November (1980) 2279-2296 (Closure, Vol. 108, No. ST8, ((1982) 1929-1933).
5. (With H. Suzuki) Constitutive Models for Concrete, Computers and Structures, Vol. 12, (1980) 23-32.
6. (With D.A. Ross and L. Tall) Fabricated Tubular Steel Columns, Journal of the Structural Division, ASCE, Vol. 106, No. ST1, January (1980) 265-282.
7. (With F. Cheong-Siat-Moy) Limit States Design of Steel Beam-Columns, A State-of-the-Art Review, Journal Solid Mechanics Archives, Vol. 5, Issue 1, February, Noordhoff, (1980) 29-73.
8. (With R.L. Yuan) Behavior of Sulfur-Infiltrated Concrete in Sodium Chloride Solution, ACI Special Publication Volume SP65-17, American Concrete Institute, Detroit, (1980) 291-307.
9. Plasticity in Soil Mechanics and Landslides, Journal of the Engineering Mechanics Division, ASCE, Vol. 106, No. EM3, June (1980) 443-464.
10. (With H. Suzuki and T.Y. Chang) Nonlinear Analysis of Concrete Cylinder Structures Under Hydrostatic Loading, Computers and Structures, Vol. 12, Pergamon Press, (1980) 559-570.
11. (With H. Suzuki and T.Y.P. Chang) End Effects of Pressure-Resistant Concrete Shells, Journal of the Structural Division, ASCE, Vol. 106, No. ST4, April (1980) 751-771.
12. (With S.L. Koh) Plasticity Approach to Landslide Problems, Engineering Geology, Vol. 16, No. 12, July (1980) 125-133.
13. (With S.S. Hsieh and E.C. Ting) A Plastic-Fracture Model for Concrete, ASCE Special Publication on "Fracture in Concrete", Edited by W.F. Chen and E.C. Ting, ASCE, October (1980) 50-64.

6

Academic Life at Purdue
普渡生涯

During my forty year academic career in American, twenty three years were spent in West Lafayette, Indiana where Purdue University was located. Although all my three boys were born in Bethlehem, Pennsylvania, they grew up and received most of their education in West Lafayette. So we considered West Lafayette our home town and Purdue our University. Whether at home watching college sports, or reading newspapers or US News & World Report on national ranking of the best universities or engineering schools, Purdue has always been in our mind and made us proud. Thus, we are a constant Purdue watcher; and have followed the changes and news events with great interest.

6.1 The Background 背景

I joined Purdue as Professor of Structural Engineering in 1976 after the retirement of Professor John E. Goldberg in 1975. John was an international known expert in structural engineering. He served as the Academic Head of the Structures Area in the School of Civil Engineering. He worked at Purdue from 1950 to 1975 and passed away in 1995. He was a legend in structural engineering at Purdue; and his stature was similar to Nathan Newmark to the University of Illinois, Lynn Beedle to Lehigh University, and George Winter to Cornell University. Most of the structures faculty members at Purdue at the time were his students. Their teaching styles and contents were very much influenced by him and they followed his footsteps. He was the role model for most of them.

Goldberg's impact on classical structural mechanics and structural engineering was profound. His analysis and design techniques were practical and elegant.   Even today, engineers apply his knowledge concerning the behavior and design of building structures.   After I became Head of Structures Area in 1980, we started to build and expand the core group of faculty in engineering and applied mechanics. In 1992, I was appointed the first George E. Goodwin Distinguished Professor of Civil Engineering.   We started several frontier areas of research including construction safety assessment, domain-specific software development environment, and advanced analysis for steel design, among others.

W.F. (惠發) was in Professor Goldberg's office when he retired in 1975.

6.2 My Students 學生

Over the 23 years at Purdue, I produced 46 doctoral students. The average graduation rate was two per year.   This was exactly twice the average production rate of doctoral students per faculty per year at Stanford University (see Appendix 13.1, Chapter 13). Most of them were research assistants; and they usually came in a cluster or group from different regions of the world depending on a specific period of time.

T. Atsuta was my first Japanese student when I began my teaching career at Lehigh University.   When he returned to Japan and through his recommendations, I accepted two of his colleagues as my research assistants: first S. Toma and later H. Sugimoto. Both of them worked as design engineers at the Kawasaki Heavy Industries Corporation. Their experiences expanded the breadth of my technical interests; and the timing was just right for new research direction in offshore structures. Three more Japanese students came later. They were E. Mizuno, Y. Ohtani, and E. Yamaguchi. They were recruited through their advisers at Nagoya, Kyoto and Tokyo Universities. All of them had a strong engineering science background. They were all very eager to learn and to hard work.

In the 1980's, Taiwan economy started to take off and a good number of students were sent abroad by government for graduate studies. They came from two sources: those from academic institutions, and those from military community. Unlike the University of California at Berkeley, Purdue campus was considered to be a conservative environment suitable for students with military background. The nine students from academic institutions were C.J. Chang, M.F Chang (張明芳), S.S. Hsieh (謝錫興), C. Cheng, T.K. Huang (黃添坤), J. L. Peng (彭瑞麟), Y.L. Huang (黃玉麟), C.H. Lai, and I.H. Chen (陳奕宏). Most worked as my research assistants. The three students from military institutions were F.H. Wu (吳福祥), W.S. King (金文森), and H.L Cheng (成曉琳). They were a highly selected group of Army officers.

From left: King W.S. (金文森), Huang T.K. (黃添坤), Hsieh P. (謝尚賢), Wu F.H. (吳福祥), Peng J.L. (彭瑞麟), Linlin (玲玲), Huang Y.L. (黃玉麟), W.F. (惠發), Cheng H.L. (誠曉林), Hwa K. (華根), Lee T.F. (李騰芳), and Wang Y.K (王永康).

In 1981, I was invited by the Nanjing Institute of Technology (南京工學院) (NIT) to give a lecture series in Nanjing, China for two weeks.

It was the first time for me to return to China since I escaped to Taiwan with my family in 1950. After I returned to Purdue, I started to accept a few Chinese students (see Sec.10.3, Chapter 10). The first two doctoral students were D.J. Han (韓大健) and X.L. Liu (劉希拉). Han was recommended to me by her department chair at the South East University of Technology in Guangzhou (廣州). I met Liu at NIT and knew him in Nanjing. They were characterized by their strong desire to catch up the missing years lost during the culture revolution. They were mature and appreciative the opportunity to learn. They were very hard workers, quite independent in carrying out research activities. In the years that followed, four more Chinese students were offered research assistantships under my guidance. They were L. Duan (段煉), H. Zhang (張宏), M.Z. Duan (段明珠), and W.H. Yang (楊卫红). L. Duan came from Taiyuan University directly; while Zhang was recruited from Peking University for his expertise in computer software development. M.Z. Duan and W.H. Yang transferred to Purdue from other universities in U.S.

In the 1980's, many Arab students from Middle East were encouraged and financially supported by their governments to come to U.S for graduate studies. They furnished the highlights of my academic career and expanded the breadth of my culture diversity and technical interests. Their activities were characterized by their desire to return to their countries to teach and to establish new business. They wrote well and spoke fluent English. They were generally well prepared in engineering fundamentals. The three students from Saudi Arabia were S.I. Al-Noury, F. Al-Mashary, and K.H. Mosallam. The six from Egypt and Jordan were A.F. Saleeb, S.E. El-Metwally, M.A. Barakat, M.M. El-Shiekh, M. M. Abdel Ghaffar, and A. M. El-Shahhat. Most of them were supported by their governments through the Peace Scholarship program for graduate studies. I had one student from the University of Morocco, M. Aboussalah.

The students from Korea seemed to be more interested in steel than in concrete structures. During my last few years at Purdue, I had the opportunity to supervise four Korea graduate students. They were S. E. Kim, C.B. Joh, Y.S. Kim, and C.S. Doo. They all worked on steel frames with rigid or semi rigid connections. Their work ethic and their devotion to hard work reflected truly the rise of Korea economic power after the devastate War.

6.3 On Offshore Structural Engineering 近海結構

In the early 1980's, the research on offshore structural engineering became an important topic. Our research was concerned with stability and strength of tubular members as used in offshore structures in static and dynamic loading; and during construction and installation. Four graduate students were assigned on various aspects of this research program. The four were S. Toma, A.F. Saleeb, H. Sugimoto, and I.S. Sohal. Toma worked on member behavior and design; Saleeb was on rapid installation of pipelines; Sugimoto was on post-buckling modeling; and Sohal was on local buckling and design. All five co-authored papers with me in professional journals, and reported their studies in specialty conferences when appropriate.

Toma and Sugimoto took a leave from the Kawasaki Heavy Industry Corporation for their graduate studies at Purdue. Back in Japan, Sugimoto in 1983 returned to his company, Kawasaki; but Toma in 1980 accepted a faculty position at the Hokkaigakuen University in Sapporo, Japan. Saleeb's M.S. thesis was on rapid installation of pipelines; but his doctoral thesis was on constitutive equations of soil medium. In 1981, he accepted a faculty position at the University of Akron in Akron, Ohio; and promoted steadily and recently was honored with a chair professorship. Sohal graduated in 1986; and accepted a faculty position at Rutgers University in New Jersey. I lost all contacts with Sohal; but kept a close contact with Toma in particular. Over the many years, I had coauthored the following three books with three of them. These are:

1. Advanced Analysis of Steel Frames, CRC Press (1994). (With S. Toma).
2. Analysis and Software of Cylindrical Members, CRC Press (1996). (With S. Toma).
3. Plastic Design and Second-Order Analysis of Steel Frames, Springer Verlag (1995). (With I. S. Sohal).
4. Constitutive Equations for Engineering Materials, Vol. 1, Elasticity and

Modeling, Wiley Inter-Science (1982). (With A.F. Saleeb).

In the summer of 1990, Toma nominated me and I received the JSPS Fellowship for Research in Japan sponsored by the Japan Society for the Promotion of Science. During the summer months, my son, Brian and I, visited most Japan's national universities and gave lectures. I met most of my Japanese students and visited their labs; and toured the famous motorcycle factory of the Kawasaki Heavy Industry near Tokyo. I was very impressed with the efficiency of the manufacture process with an in-time delivery, and minute by minute market information on sales, orders and inventory. The tour was arranged by Atsuta who, at the time, was Vice President of the company. We also had a mini-reunion with my former students at a resort guest house on the mountain. It was a quite memorable event.

6.4 On Constitutive Equations 本構方程

The application of solid mechanics to the design of geotechnical engineering problems requires the considerations of various approaches to the estimation of stress, strain, and displacement. These approaches include analytical, numerical, and physical techniques; but the finite element technique is certainly the most versatile. The central emphasis is placed on the development of constitutive equations to characterize the soil behavior realistically. In the early years, elastic analysis is used to estimate the settlement of footing, for example; but it cannot account for irreversible deformation.

In order to describe the behavior of soil medium beyond the elastic range, an elastic-plastic approach has been advocated in which yield is pressure-dependent. Classical plasticity theory using pressure dependent yield and associated flow rules enables one go beyond the elastic range in a time-independent but theoretically consistent way to develop the constitutive equations. Since plasticity theory has been well developed and readily available, it provides the foundation for uniqueness theorems for numerical solutions and limit theorems for simple engineering applications. However, for soil or rock-like materials like concrete, the assumption of "*normality*" and associated flow rules has never been confirmed in laboratory experiment. So, classical plasticity theory requires further study and refinement.

To summarize, the constitution equations are of central importance to soil and concrete mechanics analyses and the engineering design of these structural systems. Elastic behavior is well developed, but irreversible deformation is not. The process of establishing a niche in this specialty was initiated during the first few years of my arrival at Purdue. The National Science Foundation (NSF) recognized the importance of the subject area; and provided the funds for the proposed research. The four NSF grants supported the research on constitutive modeling and its numerical applications over a decade were:

1. Earthquake-Induced Landslides, 1979-81.
2. Three-Dimensional Elastic-Plastic-Fracture Analysis for Concrete Structures, 1982-84.
3. Strain Softening Modeling of Concrete in the Post-Fracture Range, 1985-88.
4. Characterization of the Inelastic Constitutive Behavior of Concrete Materials, 1993-97.

The other related studies on the applications of the constitutive equations were highway-related projects supported by the Federal Highway Administration (FHWA) in cooperation with the Indiana Department of Transportation (IDOT):

1. Design of Reinforced Embankments, 1984-86,
2. Embankment Widening and Grade Raising on Soft Foundation Soils, 1989-92.

The first four graduate research assistants were A.F. Saleeb from Cairo University, Egypt; and E. Mizuno from Nagoya University, Japan. Saleeb was assigned to work on elasticity-based modeling; while Mizuno was on plasticity-based modeling. Their goals were to characterize soil behavior in the geotechnical application in a realistic way. In the same year, M.F. Chang (張明芳) and S.S. Hsieh (謝錫興) from the National Cheng-Kung University (成功大學), Taiwan joined the research team. Chang was interested in the applications of limit analysis to earth pressure problems in soil mechanics; while Hsieh was assigned to work on concrete modeling starting with a crude delineation of zones stressed beyond the elastic range and considering the transition between brittle and ductile behavior.

The next group of two research assistants was D.J. Han（韓大健）from South China University of Technology（華南理工大學）, Guangzhou（廣州）, China; and W.O. McCarron from AMOCO Production Company, Tulsa, Oklahoma. Han was assigned to work on the loss of strength in the post-elastic range for concrete materials. The consequences of strength loss or more generally strain softening represented an important subject for concrete mechanics research. McCarron was interested in the investigation of embankment problems with different plasticity models including cap, nested and bounding surface models.

The last group of two research assistants was Y. Ohtani from Kyoto University, Japan and E. Yamaguchi from University of Tokyo, Japan. Ohtani was focused on constitutive modeling of concrete materials for engineering applications considering hardening and softening behavior. Yamaguchi began the micromechanics study by tracing the micro-crack propagation associated with softening behavior of concrete materials. As a result of this extended study on soil and concrete materials, a rough approximation to the complete stress-strain relation was achieved.

A two-volume book on constitutive equations for engineering materials was published in 1994 by Elsevier. Volume 1 dealt with the development of stress-strain models for metals, concrete, and soils based on the principles of elasticity and showed how these models could be applied to engineering practice. Volume 1 was co-authored with A. F. Saleeb in 1984 and was revised in 1994 with the publication of Volume 2. Volume 2 extended elastic-based stress-strain models to the plastic range and developed plasticity-based models for engineering applications. Volume 2 was in collaboration with W.O. McCarron and E. Yamaguchi. My other doctoral students worked on the concrete and soil plasticity related subject were M Aboussalah from University of Morocco on nonlinear analysis of concrete structures; C.J. Chang from Taiwan on seismic stability of marine slopes; and T.K. Huang（黃添坤）from the National Chung-Hsih University（中興大學）, Taiwan on design of embankments. Their contributions had been drawn on extensively for the preparation of the two-volume treatise on *Constitutive Equations for Engineering Materials*, Elsevier, 1994.

In addition, a sequence of four books on soil and concrete plasticity were published as a result of this research. These books brought, for the first time, this highly mathematical subject from applied mathematicians to civil engineering students and faculty in particular. The four books on plasticity were published during an eight-year period from 1982 to 1990.

1. Plasticity in Reinforced Concrete, McGraw-Hill (1982).
2. Soil Plasticity: Theory and Implementation, Elsevier (1985). (With G.Y. Baladi).
3. Plasticity for Structural Engineers, Springer-Verlag (1988). (With D.J. Han).
4. Structural Plasticity: Theory, Problems and CAE Software, Springer Verlag (1990). (With H. Zhang).
5. Limit Analysis in Soil Mechanics, Elsevier (1990). (With X.L. Liu).
6. Nonlinear Analysis in Soil Mechanics, Elsevier (1990). (With E. Mizuno).

6.5 On Safety of Temporary Structures 暫時結構

Building contractor because of economic factors wanted to move as quickly as possible in the construction of the building. The government, on the other hand, wanted to ensure safety; and thus developed guidelines and regulations for the temporary structures such as shoring and re-shoring for concrete building construction. The safety of construction speed depended on the maturity of concrete, the nature of shores and the timing of addition and removal of supports, called shores and re-shores. Contractors using traditional concrete knew from experience — some of it learned from construction accidents — the length of time to spend on each construction step of a new building to avoid its collapse. Traditional concrete for construction had strength of 4,000 to 5,000 pounds per square inch, but the development of high strength concrete in last decade can double that strength that was widely available on the market. Lacking of practical experience with high strength concrete, contractors might be misled by the extra strength of the new concrete and moved too quickly to remove supports from the building under construction.

I was interested in developing a computer program that could help contractors save money by cutting construction time. In the meantime, the program could help regulatory agencies to assess the safety issues with various proposed construction schedules. The program, which should be run on a personal computer, also could determine the long-term performance of a building based on a variety of construction scenarios, including the type and age of concrete used and the day's temperature. I assigned this research topic to X.L. Liu（劉希拉）from Tsinghua University（清華大學）, China, who had a wealth of experience on construction of reinforced concrete structures in China, especially his work assignment in China during the culture revolution period. This was the beginning of the next ten-year research on safety assessment of temporary structures during construction.

The safety assessment research continued for nearly a decade with six more graduate assistants from three countries: M.M. El-Shiekh and A.M. El-Shahhat from Cairo University, Egypt, K.H. Mosallam from Saudi Arabia, J.L. Peng（彭瑞麟）and Y.L. Huang（黃玉麟）from the National Chung-Hsih University, Taiwan, and M.Z. Duan（段明珠）from China. The faculty member joined the research team and participated in the study was D.V. Rosowsky. Rosowsky's expertise was on loads and reliability analysis.

Liu started the safety analysis of shoring and re-shoring of high-rise reinforced concrete buildings during construction; continued with the study on creep effect and probability distribution of loads; and completed with a computer program that could analyze construction factors and determined the safety of building procedures. The computer program in *Analysis of Construction Loads on Slabs and Shores by Personal Computer* was published in June 1988 in Concrete International: Design & Construction, a publication of the American Concrete Institute.

El-Sheikh focused on the effects of fast construction rate on deflections of reinforced concrete buildings. It could predict the building's long term performance. His program could simulate the performance of concrete slabs under a variety of construction scenarios.

El-Shahhat focused on an improved analysis of shore-slab interaction; and proposed a deflection-based analysis; and assessed further the construction safety of multistory concrete buildings.

Mosallam examined various aspects of design considerations for formwork in multistory concrete buildings; and proposed practical procedures for contractors. To this end, we published a comprehensive book on *Concrete Buildings: Analysis for Safe Construction*, CRC Press, 1991, 260 pp. This book served as a final report of our decade work on the design of formwork for safe construction of multistory concrete buildings.

With the addition of Rosowski, Peng and Huang, our research program was expanded to include two new frontier areas: field measurements of shore loads and safety study of steel scaffolds system. During a five-year period, we did the modeling of concrete placement loads, measuring the formwork and shoring loads, developing load monitoring and hazard warning systems, and investigating actual scaffolds failures during construction. For high clearance scaffolds systems during construction, we proposed structural modeling and their possible modes of failures; and provided analysis methods. At the end, we developed the design guidance for the regulatory agencies in U.S. and Taiwan, who sponsored the research and provided the site, for actual field measurements.

M. Z. Duan (段明珠) was my last graduate assistant worked on the shoring and re-shoring of formworks. He improved the simplified method for slab and shore load analysis. He also proposed design guidelines for safe concrete construction; and made the guidelines available to general public in Concrete International, ACI, October (1996). The computer program and the design guidelines marked the practical application of years of theoretical research at Purdue. It will be an interesting extension next in applying artificial-intelligence techniques to the computer program. This will allow the computer program to consider factors that cannot be calculated by formula, such as worker performance at various times of the day.

6.6 On Advanced Analysis for Frame Design 高等分析

There existed an increasing awareness of the need for practical second-order analysis (that included the inelastic behavior of structural members, connections and other components) for determining overall system response. Although the basic theory for such frame-design analysis was well established, there was still a real challenge in making it work in engineering practice using desktop computers and workstations. This new method for design of steel frames in a direct manner was called *Advanced Analysis for Frame Design.* The classical approach to structural design was to use the effective length factor K to relate the member strength to structural system behavior. It was an indirect design method with the well recognized fact that the actual failure mode of a structural system did not have any resemblance whatsoever to the elastic buckling mode of the structural system that was the basis for determination of the effective length factor K. To this end a major research program on advanced analysis was initiated at Purdue with a close coordination with Cornell team under the leadership of Professor William McGuire. Donald White, a doctoral student of McGuire, was recruited to our structural group at Purdue in early 1990's.

At the time Cornell team was focused on developing rigorous method using workstations and super-computers to solve thousands of degree of freedom for validation of several advanced analysis procedures. Simple calibration techniques and practical approaches were the focus and goal of Purdue team. The intermediate solutions at the time included plastic-zone, quasi-plastic hinge, elastic-plastic hinge methods, and various modifications thereof. All in some way account for residual stresses, geometric imperfections, and non-linearity and moment redistribution throughout a structure. The first graduate assistant joined the team was J.Y.R. Liew (劉德源) from the National University of Singapore. A group of four Korea graduate assistants were recruited in the subsequent years. They were S.E. Kim, Y.S. Kim, C.B. Joh and C.S. Doo. K. Wongkaew from Thailand was the last graduate student worked on the subject area at Purdue before I moved to the University of Hawaii in the fall of 1999.

Liew succeeded in the development of a second-order refined plastic hinge method for frame design; while S.E. Kim achieved a practical method for steel frame design with a calibration against the AISC LRFD specifications. Y.S. Kim developed practical design procedures for semi-rigid frame design and provided design aids for practitioners. Joh investigated the failure of the beam-to-column connection after the 1994 Northridge earthquake in Los Angles. It was the first time the theory of fracture mechanics was used to simulate the actual failure process with an actual field condition. Doo extended Joh's work and proposed a prediction model concerning the fracture strength of fully welded beam-to-column connections. Wongkaew took the challenging work by extending the advanced analysis method to three-dimensional space framework; and by also considering local buckling of structural members in the overall analysis.

As a result of this research, five books on the advanced analysis for frame design were published:

1. Plastic Design and Second-Order Analysis of Steel Frames, Springer-Verlag, (1995). (With I. Sohal).
2. Stability Design of Semi-Rigid Frames, John Wiley and Sons (1996). (With Y. Goto and J.Y.R. Liew).
3. LRFD Steel Design Using Advanced Analysis, CRC Press (1997). (With S.E. Kim).
4. Practical Analysis for Partially Restrained Frame Design, Structural Stability Research Council, Lehigh University, Bethlehem, PA (1998). (With Y.S. Kim).
5. Practical Analysis for Semi-Rigid Frame Design, World Scientific Publishing (2000).

6.7 On Structural Members and Frames 單元與框架

The Purdue team was also active on other aspects of steel research. The research was concerned with structural members and frames in steel and concrete. Starting with in-house presentations of the subjects discussed in the SSRC Task Group meetings and AISC specifications committee meetings, the subjects offered numerous opportunities for incoming students to select their thesis topics. These activities provided new technical information for possible updates of the *SSRC Guide to Stability Design Criteria of Metal Structures* as well as to the new editions of the AISC specifications over the last two decades.

For me the research assistants, who were also graduate students, furnished new ideas, created the breadth of my technical interests and provided solutions to some timely and important subjects at the time. They furnished the highlights of my professional career as documented in the list of publications I authored and coauthored with them (see Appendix 6.2). To introduce our research, especially on steel structures, into the ASCE Manual, SSRC Guide, and AISC Specifications, my speaking and committee schedule mushroomed into an almost full-time activity. One of the top priorities of these meetings was to develop design recommendations that were suitable for adoption either in codes or in guides. The work led to membership on the Committee on Specification formed by the AISC in 1961, as well as on the executive committees of SSRC and ASCE EMD, among others.

These graduate assistants were all hard workers. They had strong desire to learn and to create and to complete their doctoral degrees within three years after their M.S. degrees. Eric M. Lui (呂汶) originally from Hong Kong, received his undergraduate education in U.S; and worked under my guidance for both his M.S. and Ph.D. degrees. He started his research on the effect of end restraint on column strength; which helped me receive the AISC T.R Higgins Lecture Award. Later, he focused on his research on the effects of connection flexibility and panel zone deformation on the behavior of frames. This research made an important contribution to the adoption of the 1986 AISC LRFD specifications for frames with Partially Restraint construction.

As a result of his research, we published two books and one handbook over the ten-year period:
1. Structural Stability: Theory and Implementation, Elsevier (1987).
2. Stability Design of Steel Frames, CRC Press (1991).
3. The Handbook of Structural Engineering, CRC Press, 1997, 2nd Edition, (2005).

L. Duan (段鍊) came from Taiyuan University (太原大學), China, started his research on a variety of design interaction equations of beam-columns; and studied also the appropriateness of the beam-column moment amplification factor as used widely in AISC specifications. Later, he focused on the improvement of effective length factor for columns in frame design. His research and design recommendations also made important contributions to the revisions of the AISC LRFD specifications.

L. Duan（段鍊）was very active in helping California Department of Transportation to update and improve its highway specifications. Because of his expertise in bridge engineering, highway bridges in particular; we co-edited a handbook on bridge engineering with three separate volumes on special topics; and one introductory textbook on steel design for students in Taiwan:

1. Bridge Engineering Handbook, CRC Press (2000).
2. Structural Steel Design: LRFD Method, Science/Technology, Taiwan (1999). (With W.S. King, S.P. Zhou, in Chinese).
3. Bridge Engineering: Substructure Design, CRC Press (2003).
4. Bridge Engineering: Seismic Design, CRC Press (2003).
5. Bridge Engineering: Construction and Maintenance, CRC Press (2003).

The bridge engineering handbook won the Choice Magazine's Outstanding Academic Title award for 2000 - Choice Magazine, January 2001.

6.8 On Software Development Environment 軟件發展環境

Structural engineering, a branch of solid mechanics, was my profession at Purdue. Structural engineering was a computationally intensive field of civil engineering; and its intensity was just second only to physics at the time on campus. Most of my student thesis software was one-of-a-kind. The student designed and developed his software for a specific application. Code duplication was common within and across of discipline. The same piece of code might be developed again and again among different software components of same research topics because of difficult in reusing or sharing existing software. In my research group, it was not uncommon that students might devote an inordinate amount of their time to software development and maintenance. This time might otherwise be spent on more fundamental research issues if advanced software engineering approaches and tools were available to facilitate reuse. The critical issue - that was in my mind for a long time - was that efforts spent on software development by my students often could not be accumulated. This was a great waste or resources in a university research environment, especially for the same research group.

In 1986, I was invited by Professor Ren Wang (王仁), Chair of Mechanics Department at Peking University (北京大學), to visit his department. I met his graduate student, H. Zhang (張宏), who was a software guru at his department. I invited Zhang for a dinner and discussed my concern about the software reuse and how can we accumulate the development and facilitate the reuse. He was excited about the challenge and proposed the creation of a domain-specific software development environment specifically for structural engineering to enable the knowledge obtained from the solution of a particular problem to be accumulated and shared in the solution of other problems. If software components accumulated from previous software development can be utilized readily in the development of new applications, substantial applications can be built more efficiently.

In 1991, I proposed with my colleagues, D.W. White and H.E. Dunsmore, who was a professor of computer science, and received a two-year research grant from the NSF, entitled *"An Automated Environment for Engineering Software Development"*. We recruited Assistant Professor E.D. Sotelino from Brown University, whose specialty was on parallel computing. We also recruited graduate assistants H. Zhang (張宏) and June Lu from China to start to create the structural engineering domain-specific software development environment. Zhang's background was in mechanics and Lu was transferred from computer science department to structural engineering.

In 1993, we received a three-year major funding from the U.S. Department of Energy on *"Domain-Specific Object-Oriented Environment for Parallel Computing"*. The purpose of the research was on the development of a prototype domain-specific concurrent software development environment for large-scale engineering computing. This environment was being created by applying the object-oriented paradigm to parallel programming. As a test case the structural engineering domain was chosen, because it typified the utilization of the computer technology among many scientific and engineering areas. Because this environment was designed as an open environment, it would facilitate collaboration between researchers working in similar areas as well as the assessment and assimilation of computing advances in related areas.

The Structural Engineering Software Development Environment (SESDE) team was further expanded to include Patrick S.H. Hsieh (謝錫興) as a postdoctoral associate from Cornell University, and graduate assistants R. Gambheera, and S Modal, both from IIT, India, among others. A comprehensive description of the project with some key aspects of progress of the SESDE development at the time was summarized in the paper entitled "*Domain-Specific Object-Oriented Environment for Parallel Computing*", Steel Structures, Journal of Singapore Structural Steel Society, Vol. 3, No. 1, December 1992.

The lessons we learned from this experience were that it was much easier for us to "*send out*" our engineering students to learn computer programming and software development; than to "*recruit in*" non-engineering students to learn structural engineering. Simply put, it is a good policy to "*send out*" our students to learn, rather than to "*recruit in*" non-engineering students to help develop software for engineering applications.

6.9 Concluding Remarks 評論

As Professor Newmark commented to a group of his former and current students "*If you think you have done well in your studies, research, and analyses, remember Newton's comment: If I have seen far, it is because I stood on the shoulder of giants.*" Over my forty five years of academic career and life at Brown, Lehigh and Purdue, I had the privilege to stand on the shoulders of three giants – Drucker, my thesis adviser at Brown University (see Chapter 12); Beedle, my mentor at Lehigh University (see Chapter 5), and Goldberg, my colleague at Purdue – who helped us get up on those shoulders.

Appendix 6.1
Doctoral Students - Purdue University
普大博士生

1976 to 2000 (Total 46)

S. Toma (80)	S.I. Al-Noury (80)	K.V. Patel (80)
A.F. Saleeb (81)	E. Mizuno (81)	C.J. Chang (81)
M.F. Chang (81)	S.S. Hsieh (81)	H. Sugimoto (83)
D.J. Han (84)	J.S. Larralde-Muro (84)	E.M. Lui (85)
W.O. McCarron (85)	X.L. Liu (85)	C. Cheng (86)
I.S. Sohal (86)	S.E. El-Metwally (86)	Y. Ohtani (87)
E. Yamaguchi (87)	F.H. Wu (88)	M.A. Barakat (88)
M.M. El-Shiekh (89)	F. Al-Mashary (89)	M. Aboussalah (89)
W.S. King (90)	T.K. Huang (90)	L. Duan (90)
H. Zhang (91)	K.H. Mosallam (91)	J.Y.R. Liew (92)
M.M. Abdel-Ghaffar (92)	A.M. El-Shahhat (93)	J.L. Peng (94)
Jun Lu (94)	Y.L. Huang (95)	S.E. Kim (96)
M.Z. Duan (96)	R. Gambheera (97)	W.H. Yang (97)
C.B. Joh (98)	Y.S. Kim (98)	C.H. Lai (98)
I.H. Chen (99)	C.S. Doo (99)	K. Wongkaew (00)
H.L Cheng (00).		

Appendix 6.2
Publications
論文集
1981 - 2003

.

Dr. Chen is the author or co-author of more than 595 articles in various refereed Technical Journals (347), Conference Proceedings and Symposium Volumes (248). The following is the list of journal articles during his years at Purdue University (1976-99).

1981 - Journal Articles (6)
1. (With W.J. Graff) Bottom-Supported Concrete Platforms: Overview, Journal of the Structural Division, ASCE, Vol. 107, No. ST6, June (1981) 1059-1081.
2. (With K.V. Patel) Static Behavior of Beam-To-Column Moment Connections, Journal of the Structural Division, ASCE, Vol. 107, No. ST9, September (1981) 1815-1838.
3. (With M.F. Chang) Limit Analysis in Soil Mechanics and Its Applications to Lateral Earth Pressure Problems, Solid Mechanics Archives, Vol. 6, Issue 3, Sijthoff & Noordhoff International Publishers, Alphen aan denRijn, The Netherlands, July (1981) 331-399.
4. (With A.F. Saleeb) Elastic-Plastic Large Displacement Analysis of Pipe, Journal of the Structural Division, ASCE, Vol. 107, No. ST4, April (1981) 605-626.
5. (With E. Mizuno) Plasticity Models for Soils: Comparison and Discussion, 328-351, also "Plasticity Models for Soils," 553-591. Proceedings of the Workshop on Limit Equilibrium, Plasticity and Generalized Stress-Strain in Geotechnical Engineering, ASCE Publication, New York, (1981).
6. (With A.F. Saleeb) Nonlinear Hyperelastic (Green) Constitutive Models for Soils: Predictions and Comparisons, 265-285, also Nonlinear Hyperelastic (Green) Constitutive Models for Soils: Theory and Calibration," 492-538, Proceedings of the Workshop on Limit Equilibrium and Generalized Stress-Strain in Geotechnical Engineering, ASCE Publication, New York, (1981).

1982 - Journal Articles (11)
1. (With R.L. Yuan and G.R. McLelland) Experiments on Closing Reinforced Concrete Corners, Journal of the Structural Division, ASCE, Vol. 108, No. ST4, April, (1982) 771-779.
2. (With S.S. Hsieh, E.C. Ting) A Plastic-Fracture Model for Concrete, International Journal of Solids and Structures, Vol. 18, No. 3, U.K., (1982)181-197.
3. (With S.I. Al-Noury) Behavior and Design of Reinforced and Composite Concrete Sections, Journal of the Structural Division, ASCE, Vol. 108, No. ST6, June (1982)

1266-1284.
4. (With S. Toma) Cyclic Analysis of Fixed-Ended Steel Beam-Columns, Journal of the Structural Division, ASCE, Vol. 108, No. ST6, June (1982) 1385-1399.
5. (With S. Toma) External Pressure and Sectional Behavior of Fabricated Tubes, Journal of the Structural Division, ASCE, Vol. 108, No. ST1, January (1982) 177-194.
6. (With M.F. Chang) Lateral Earth Pressures on Rigid Retaining Walls Subjected to Earthquake Forces, Solid Mechanics Archives, Vol. 7, Issue 3, Martinus Nijhoff Publishers, The Hague, The Netherlands, September (1982) 315-362.
7. (With G.P. Rentschler and G.C. Driscoll) Beam-To-Column Web Connection Details, Journal of the Structural Division, ASCE, Vol. 108, No. ST2, February, (1982) 393-409.
8. (With H. Sugimoto) Small End Restraint Effects on Strength of H-Columns, Journal of the Structural Division, ASCE, Vol. 108, No. ST3, March (1982) 661-684 (Closure, Vol. 109, No. 4, 1983, pp. 1073-1077).
9. (With S.I. Al-Noury) Finite Segment Method for Biaxially Loaded R. C. Columns, Journal of the Structural Division, ASCE, Vol. 108, No. ST4, April (1982) 780-799.
10. (With S. Toma) Inelastic Cyclic Analysis of Pinned-Ended Tubes, Journal of the Structural Division, ASCE, Vol. 108, No. ST 10, October (1982) 2779-2294.
11. (With E. Mizuno) Analysis of Soil Response with Different Plasticity Models, Proceedings of the Symposium on Limit Equilibrium, Plasticity and Generalized Stress-Strain Applications in Geotechnical Engineering, R.N. Yong and E.T. Selig, Editors, ASCE, (1982) 115-138.

1983 - Journal Articles (18)
1. (With S. Toma) Design of Vertical Chords in Deepwater Platform, Journal of the Structural Engineering, ASCE, Vol. 109, No. 11, November (1983) 2733-2746.
2. (With D.J. Han) Buckling and Cyclic Inelastic Analysis of Steel Tubular Beam-Columns, Engineering Structures, Vol. 5, No. 2, (1983) 119-132.
3. (With T. Sawada) Earthquake-Induced Slope Failure in Non-Homogeneous, Anisotropic Soils, Soils and Foundations, Japanese Society of Soil Mechanics and Foundation Engineering, Vol. 23, No. 2, (1983) 125-139.
4. (With D.J. Han) Behavior of Portal and Strut Types of Beam-Columns, Engineering Structures, Vol. 5, No. 1, (1983) 15-25.
5. (With E.M. Lui) Strength of H-Columns with Small End Restraints, Journal of the Institution of Structural Engineers, Vol. 61B, No. 1, Part B Quarterly, The Institution of Structural Engineers, London, (1983) 17-26.
6. (With S. Toma) Analytical Models of Tubular Beam-Columns, International Association for Bridge and Structural Engineering, Proceedings P-67/83, IABSE Periodica 4/1983, Zurich, November (1983) 193-212.
7. (With E.M. Lui) End Restraint and Column Design Using LRFD, Engineering Journal, American Institute of Steel Construction, First Quarter, Vol. 20, No. 1, (1983) 29-39.
8. (With E. Mizuno) Plasticity Analysis of Slope with Different Flow Rules, Computers & Structures, Vol. 17, No. 3, (1983) 375-388.

9. (With E. Mizuno) Cap Models for Clay Strata to Footing Loads, Computers & Structures, Vol. 17, No. 4, (1983) 511-528.
10. (With S. Toma) Post-Buckling Behavior of Tubular Beam-Columns, Journal of the Structural Engineering, ASCE, Vol. 109, No. 8, August (1983) 1918-1932.
11. A Rapid Method of Computing Geometric Relations in Structural Analysis, Civil Engineering Magazine, ASCE, New York, May (1983).
12. (With H. Suzuki) Elastic-Plastic-Fracture Analysis of Concrete Structures, Computers & Structures, Vol. 16, No. 6, Pergamon Press, U.S., (1983) 697-705.
13. (With S. Toma) Cyclic Inelastic Analysis of Tubular Column Sections, Computers and Structures, Vol. 16, No. 6, Pergamon Press, London, U.K. (1983) 707-716.
14. Stability Design of Columns in North America, in the book "Design Limit States of Steel Structures" edited by J. Melcher, Technical University of BRNO, Czechoslovakia, Brno, (1983) 301-319.
15. (With R. Bjorhovde) Behavior of Columns - A Comprehensive Treatment, W.H. Munse Symposium Volume on Metal Structures - Research and Practice, W. J. Hall and M. P. Gaus, Editors, ASCE, May 17 (1983) 85-102.
16. Effective Length of Columns with Simple Connections, In the book "Developments in Tall Buildings", L.S. Beedle, Editor-in-Chief, Council on Tall Buildings and Urban Habitat, Hutchinson Ross Publishing Co., Stroudsburg, PA, (1983) 513-520.
17. (With E.M. Lui) Design of Beam-Columns in North America, Proceedings of the Third International Colloquium on Stability of Metal Structures, Structural Stability Research Council, Bethlehem, PA, May (1983) 253-291.
18. Soil Mechanics, Plasticity and Landslides, Daniel C. Drucker Symposium on the Mechanics of Material Behavior, University of Illinois at Urbana-Champaign, (1983). Also, In "Mechanics of Material Behavior," edited by G.J. Dvorak and R.T. Shield, Elsevier, Amsterdam, (1984) 31-58.

1984 - Journal Articles (10)

1. (With E. Mizuno), Plasticity Models for Seismic Analyses of Slopes, Journal of Soil Dynamics and Earthquake Engineering, Vol. 3, No. 1, CML Publications, (1984) 2-7.
2. (With E.M. Lui), Simplified Approach to the Analysis and Design of Columns with Imperfections, Engineering Journal, AISC, Vol. 21, No. 2, 2nd Quarter (1984) 99-117.
3. (With S. Toussi, J.T.P. Yao) A Damage Indicator for Reinforced Concrete Frames, ACI Journal, Title No. 81-25, May-June (1984) 260-267.
4. (With X.L. Liu) Reinforced Concrete Pipe Columns: Behavior and Design, Journal of Structural Engineering, ASCE, Vol. 110, No. 6, June (1984) 1356-1373.
5. (With X.L. Liu) Reinforced Concrete Centrifugal Pipe Columns, Journal of Structural Engineering, ASCE, Vol. 110, No. 7, July (1984) 1665-1678.
6. (With C.J. Chang and J.T.P. Yao) Seismic Displacements in Slopes by Limit Analysis, Journal of Geotechnical Engineering, ASCE, Vol. 110, No. 7, July (1984) 860-874.
7. (With I.S. Sohal) Moment-Curvature Expressions for Fabricated Tubes, Journal of Structural Engineering, ASCE, Vol. 110, No. 11, November (1984) 2738-2757.

8. (With M. Yener) On In-Place Strength of Concrete and Pullout Tests, Journal of Cement, Concrete and Aggregate, Vol. 6, No. 2, ASTM, Winter (1984) 90-99.
9. (With K.V. Patel) Nonlinear Analysis of Steel Moment Connections, Journal of Structural Engineering, ASCE, Vol. 110, No. 8, August (1984) 1861-1874.
10. (With A.F. Saleeb) Plasticity Modeling for Engineering Materials, Special Anniversary Volume (Verba Volant, Scripta Manent) for Professor Massonnet, Liege, Belgium (1984) 117-132.

1985 - Journal Articles (14)

1. (With E.M. Lui) Stability Design Criteria for Steel Members and Frames in the United States, Journal of Constructional Steel Research, Vol. 5, No. 1, (1985) 51-94.
2. (With X.L. Liu, and M.D. Bowman) Construction Load Analysis for Concrete Structures, Journal of Structural Engineering, ASCE, Vol. 111, No. 5, May (1985) 1019-1036.
3. (With H. Sugimoto) Moment Curvature Axial Compression Pressure Relationship of Structural Tubes, Journal of Constructional Steel Research, Vol. 5, No. 4, (1985) 247-264.
4. (With I.S. Sohal) Large Bending of Pipes, Engineering Structures, Vol. 7, No. 2, (1985) 121-130.
5. (With K.V. Patel) Analysis of a Fully Bolted Moment Connection Using NONSAP, Computers and Structures, Vol. 21, No. 3, (1985) 505-511.
6. (With H. Sugimoto) Inelastic Post-Buckling Behavior of Tubular Members, Journal of Structural Engineering, ASCE, Vol. 111, No. 9, (1985) 1965-1978.
7. (With S.P. Zhou) Design Criteria for Box-Columns under Biaxial Loading, Journal of Structural Engineering, ASCE, Vol. 111, No. 12, December (1985) 2643-2658.
8. (With E.M. Lui) Columns with End Restraint and Bending in Load and Resistance Factor Design, Engineering Journal, AISC, Vol. 22, No. 3, 3rd Quarter (1985) 105-132.
9. (With M. Yener) Evaluation of In-Place Flexural Strength of Concrete, ACI Journal, Proceedings Vol. 82, No. 6, November-December (1985) 788-796.
10. (With D.J. Han) A Nonuniform Hardening Plasticity Model for Concrete Materials, Journal of Mechanics of Materials, Vol. 4, No. 3, December (1985) 1-20.
11. (With X.L. Liu and M.D. Bowman) Construction Loads on Supporting Floors, Concrete International: Design and Construction, ACI, Vol. 7, No. 12, December (1985) 21-26.
12. (With T. Sawada and S.G. Nomachi) Stability of Slopes with Anisotropic Cohesion Strength against Earthquakes, Proceedings of the 33rd Japan National Congress for Applied Mechanics, 1983, Theoretical and Applied Mechanics, Vol. 33, University of Tokyo Press, (1985) 417-432.
13. (With D.J. Han) A Five-Parameter Mixed-Hardening Model for Concrete Materials, Topical Lecture, In: G. Bianchi and A. Sawczuk, eds., International Centre for Mechanical Sciences, Symposium on Plasticity Today - Modeling, Methods and Applications, Udine, Italy, (1983) Elsevier Applied Science Publishers, London, (1985) 587-602.
14. Constitutive Relations for Concrete Rock and Soils - Discusser's Report, Chapter 5,

IUTAM Prager Symposium Book on "Mechanics of Geomaterials: Rocks, Concrete and Soils", Z.P. Bazant, Editor, John Wiley & Sons, London, (1985) 65-86.

1986 - Journal Articles (12)

1. (With X.L. Liu, and M.D. Bowman) Shore-Slab Interaction in Concrete Buildings, Journal of Construction Engineering and Management, Vol. 112, No. 2, ASCE, June (1986) 227-244.
2. (With D.J. Han) Strain-Space Plasticity Formulation for Hardening-Softening Materials with Elastoplastic Coupling, Journal of Solids and Structures, Vol. 22, No. 8, (1986) 935-950.
3. (With E.M. Lui) Analysis and Behavior of Flexibly-Jointed Frames, Engineering Structures, Vol. 8, No. 2, April (1986) 107-118.
4. (With E.M. Lui) Steel Beam-To-Column Moment Connections - Part I: Flange Moment Connections, Solid Mechanics Archives, Vol. 11, Issue 4, Oxford University Press, Oxford, England, December (1986) 257-316.
5. (With E. M. Lui) Frame Analysis with Panel Zone Deformation, Solids and Structures, Vol. 22, No. 12, Pergamon Press, (1986) 1599-1627.
6. (With E. Yamaguchi) On Constitutive Modeling of Concrete Materials, Proceedings of the U.S./Japan Joint Seminar on Finite Element Analysis of Reinforced Concrete Structures, Tokyo, May 21-24, 1985, ASCE Special Publication, New York, (1986) 48-71.
7. (With W.O. McCarron) Modeling of Soils and Rocks Based on Concepts of Plasticity, AIT Symposium and Course on Laboratory & Field Tests and Analysis of Geotechnical Problems, A.A. Balkema, Rotterdam, The Netherlands, (1986) 467-510.
8. (With E. Mizuno) Plasticity Modeling and Its Application to Geomechanics, AIT Symposium and Course on Laboratory & Field Tests and Analysis of Geotechnical Problems, A.A. Balkema, Rotterdam, The Netherlands, (1986) 391-426.
9. (With E.M. Lui) Generalized Column Equation - A Physical Approach, In Advances in Tall Buildings, L.S. Beedle, Editor-In-Chief, Van Nostrand Reinhold, New York, (1986) 323-352.
10. (With E.M. Lui) Recent Developments in Structural Connections, In Advances in Tall Buildings, L.S. Beedle, Editor-In-Chief, Van Nostrand Reinhold, New York (1986) 353-365.
11. Semi-Rigid Connections in Steel Frames, In High-Rise Buildings: Recent Progress, L.S. Beedle, Editor, Lehigh University, Bethlehem, PA, (1986) 171-189.
12. (With N. Kishi) Data Base of Steel Beam-To-Column Connections, CE-STR-86-26, School of Civil Engineering, Purdue University, West Lafayette, IN, Two Volumes, (1986) 653 pages.

1987 - Journal Articles (23)

1. (With X.J. Zhang) Stability Analysis of Slopes with General Nonlinear Failure Criterion, Journal for Numerical and Analytical Methods in Geomechanics, Vol. 11, No. 1, (1987) 33-50.
2. (With D.J. Han) Constitutive Modeling in Analysis of Concrete Structures, Journal

of Engineering Mechanics, Vol. 113, No. 4, April,, ASCE, (1987) 577-593.

3. (W. O. McCarron) Application of a Bounding Surface Model to Boston Blue Clay, Computers and Structures, Vol. 26, No. 6, (1987) 887-897.

4. (With J. Larralde) Estimation of Mechanical Deterioration of Highway Rigid Pavements, Journal of Transportation Engineering, Vol. 113, No. 2, ASCE, New York, March (1987) 193-208.

5. (With X. L. Liu) Probability Distribution of Maximum Wooden Shore Loads in Multistory R.C. Buildings, Journal of Structural Safety, Vol. 4, (1987) 197-215.

6. (With X. L. Liu) Effect of Creep on Load Distribution in Multistory R.C. Buildings During Construction, ACI Structural Journal, Vol. 84, No. 3, May-June, Detroit, Michigan (1987), Vol. 2, 192-200.

7. (With S. P. Zhou) Inelastic Analysis of Steel Braced Frames with Flexible Joints, Journal of Solids and Structures, Vol. 23, No. 5, (1987) 631-649.

8. (With S. P. Zhou) Design of Beam-Columns using Allowable Stress Design and Load and Resistance Factor Design, Engineering Structures Vol. 9, July (1987) 201-209.

9. (With I. S. Sohal) Local Buckling and Sectional Behavior of Fabricated Tubes, Journal of Structural Engineering, ASCE, Vol. 113, No. 3, March (1987) 519-533.

10. (With E. M. Lui) Steel Frame Analysis with Flexible Joints, Journal of Constructional Steel Research, Vol. 8, (1987) 161-202.

11. (With Y. Goto) On the Computer-Based Design Analysis for Flexibly Jointed Frames, Journal of Constructional Steel Research, Vol. 8, (1987) 203-231.

12. (With E. M. Lui) Effects of Joint Flexibility on the Behavior of Steel Frames, Computers and Structures, Vol. 26, No. 5, (1987) 719-732.

13. (With E. M. Lui) Steel Beam-To-Column Connections Part II - Web Moment Connections, Solid Mechanics Archives, Vol. 12, Issue 1, March, Oxford University Press, Oxford, U.K., (1987) 327-378.

14. (With T.Y. Chang and H. Taniguchi) Nonlinear Finite Element Analysis of Reinforced Concrete Panels, Journal of Structural Engineering, ASCE, Vol. 113, No. 1, January (1987) 122-140.

15. (With S.P. Zhou) C_m Factor in Load and Resistance Factor Design, Journal of Structural Engineering, ASCE, Vol. 113, No. 8, August (1987) 1738-1754.

16. (With Y. Goto) Second-Order Elastic Analysis for Frame Design, Journal of Structural Engineering, ASCE, Vol. 113, No. 7, July (1987) 1505-1519.

17. With W.O. McCarron) A Capped Plasticity Model Applied to Boston Blue Clay, Canadian Geotechnical Journal, Vol. 24, No. 4, November (1987) 630-644 (Closure, Vol. 115, No. 2, 1989, pp. 503-506).

18. (With Y.C. Ohtani) Hypoelastic-Perfectly Plastic Model for Concrete Materials, Journal of Engineering Mechanics, ASCE, Vol. 113, No. 12, December (1987) 1840-1860.

19. (With I. S. Sohal) Local Buckling and Inelastic Cyclic Behavior of Tubular Members, Journal of Thin-Walled Structures, Vol. 5, (1987) 455-475.

20. (With H. Sugimoto) Analysis of Tubular Beam-Columns and Frames under Reversed Loading, Engineering Structures, Vol. 9, October (1987) 233-242.

21. (With E. Yamaguchi) On Micro-Mechanics of Fracture and Constitutive Modeling

of Concrete Materials, Proceedings of the 2nd International Conference on Constitutive Laws for Engineering Materials: Theory and Application, Tucson, Arizona, January 5-10,C.S. Desai et al., Editors, Elsevier, New York, (1987) 939-947.

22. (With W. O. McCarron) Application of a Bounding Surface Model, Proceedings of the 2nd International Conference and Short Course on Constitutive Laws for Engineering Materials: Theory and Applications, Editors, C. S. Desai, E. Krempl, P. D. Kiousis, and T. Kundu, Tucson, Arizona, January 5-10 (1987) 1257-1264.

23. (With S.W. Chan and S.L. Koh) Upper Bound Limit Analysis of Stability of a Seismic-Infirmed Earth slopes, Proceedings of the Symposium and Specialty Sessions on Geotechnical Aspects of Mass and Material Transportation, AIT, Bangkok, Thailand, December 3-14, 1984, A.A. Balkema Publishers, Rotterdam, The Netherlands, (1987) 373-428.

1988 - Journal Articles (22)

1. (With E. M. Lui) Static Flange Moment Connections, Journal of Constructional Steel Research, Vol. 10, Elsevier, (1988) 39-88.

2. (With E. M. Lui) Static Web Moment Connections, Journal of Constructional Steel Research, Vol. 10, Elsevier, (1988) 89-131.

3. (With D. A. Nethercot) Effects of Connections on Columns, Journal of Constructional Steel Research, Vol. 10, Elsevier, (1988) 201-239.

4. (With S. Sohal) Local and Post-Buckling Behavior of Tubular Beam-Columns, Journal of Structural Engineering, ASCE, Vol. 114, No. 5, May (1988) 1073-1090.

5. (With B. Kato and M. Nakao) Effects of Joint-Panel Shear Deformation on Frames, Journal of Constructional Steel Research, Vol. 10, (1988) 269-320.

6. (With X. L. Liu, H. M. Lee) Analysis of Construction Loads on Slabs and Shores by Personal Computer, Concrete International, American Concrete Institute, Vol. 10, No. 6, June (1988) 21-30.

7. Evaluation of Plasticity-Based Constitutive Models for Concrete Materials, Solid Mechanics Archives, Vol. 13, Issue 1, Oxford University Press, Oxford, England, (1988) 1-63.

8. (With S. S. Hsieh and E. C. Ting) Applications of a Plastic-Fracture Model to Concrete Structures, Computers & Structures, Vol. 28, No. 3, (1988) 373-393.

9. (With S. E. El-Metwally) Moment-Rotation Modeling of Reinforced Concrete Beam-Column Connections, ACI Structural Journal, American Concrete Institute, July-August, Detroit, (1988) 384-394.

10. (With I. S. Sohal) Local Buckling and Inelastic Cyclic Behavior of Tubular Section, Journal of Thin-Walled Structures, Vol. 6, No. 1, (1988) 63-80.

11. (With I. S. Sohal) Cylindrical Members in Offshore Structures, Journal of Thin-Walled Structures, Vol. 6, (1988) 153-285.

12. (With L. Duan) Effective Length Factor for Columns in Braced Frames, Journal of Structural Engineering, ASCE, Vol. 114, No. 10, October (1988) 2357-2370.

13. (With M. El-Sheikh) Effects of Fast Construction Rate on Deflections of R.C. Buildings, Journal of Structural Engineering, ASCE, Vol. 114, No. 10, October (1988) 2225-2238.

14. (With E.M. Lui) Behavior of Braced and Unbraced Semi-Rigid Frames, International Journal of Solids and Structures, Vol. 24, No. 9, (1988) 893-913.

15. (With L. Duan) Design Rules of Built-Up Members in Load and Resistance Factor Design, Journal of Structural Engineering, ASCE, Vol. 114, No. 11, November (1988) 2544-2554.

16. (With Y. Ohtani) Multiple Hardening Plasticity for Concrete Materials, Journal of Engineering Mechanics, ASCE, Vol. 114, No. 11, November (1988) 1890-1910.

17. (With N. Kishi, K. G. Matsuoka and S. G. Nomachi) Moment-Rotation Relation of Top- and Seat-Angle with Double Web-Angle Connections, Proceedings of the Workshop on Connections and the Behavior, Strength and Design of Steel Structures, Cachan, France, May 25-27, 1987, Elsevier, London, (1988) 121-134.

18. (With N. Kishi, K. G. Matsuoka and S. G. Nomachi) Moment Rotation Relation of Single/Double Web-Angle Connections, Proceedings of the Workshop on Connections and the Behavior, Strength and Design of Steel Structures, Cachan, France, May 25-27, 1987, Elsevier, (1988) 135-149.

19. Connection Flexibility in Steel Structures, In Second Century of the Skyscraper, L. S. Beedle, Editor-In-Chief, Council on Tall Buildings and Urban Habitat, Lehigh University, Bethlehem, PA, (1988) 857-884.

20. (With W.O. McCarron) An Elastic-Plastic Two-Surface Model for Non-Cohesive Soils, In Constitutive Equations for Granular Non-Cohesive Soils, A. Saada & G. Bianchini, Editors, A.A. Balkema Publishers, Rotterdam, The Netherlands, (1988) 427-446.

21. (With E. Mizuno) A Multi-Surface Model for Non-Cohesive Soils, In Constitutive Equations for Granular Non-Cohesive Soils, A. Saada & G. Bianchini, Editors, A.A. Balkema Publishers, Rotterdam, The Netherlands, (1988) 481-500.

22. Evaluation of Constitutive Models in Soil Mechanics, In Constitutive Equations for Granular Non-Cohesive Soils, A. Saada & G. Bianchini, Editors, A.A. Balkema Publishers, Rotterdam, The Netherlands, (1988) 687-693.

1989 - Journal Articles (20)

1. (With I.S. Sohal and L. Duan) Design Interaction Equations for Steel Members, Journal of Structural Engineering, ASCE, Vol. 115, No. 7, July (1989) 1650-1665 (Closure, Vol. 117, No. 7, July 1991).

2. (With I.S. Sohal and L. Duan) On Beam-Column Moment Amplification Factor, Engineering Journal, AISC, Vol. 26, No. 4, 4th Quarter (1989) 130-135.

3. (With M. El-Sheikh) Maximum Probabilistic Shore Load in Multistory R/C Buildings, Computers & Structures, Vol. 32, No. 6, (1989) 1347-1357.

4. (With L. Duan) Effective Length Factor for Columns in Unbraced Frames, Journal of Structural Engineering, ASCE, Vol. 115, No. 1, January (1989) 149-165 (Errata, Vol. 122, No. 2, February (1996) 224-225).

5. (With N. Kishi) Semi-Rigid Steel Beam-To-Column Connections: Data Base and Modeling, Journal of Structural Engineering, ASCE, Vol. 115, No. 1, January (1989) 105-119.

6. (With X.L. Liu, H.M. Lee) Shoring and Reshoring of High-Rise Buildings, Concrete International: Design & Construction, Vol. 11, No. 1, January (1989) 64-68.

7. (With Y. Goto) On the Validity of Wagner Hypothesis, International Journal of Solids and Structures, Vol. 25, No. 6, (1989) 621-634.

8. (With L. Duan) Design Interaction Equation for Steel Beam-Columns, Journal of Structural Engineering, ASCE, Vol. 115, No. 5, May (1989) 1225-1243 (Closure, Vol. 117, No. 8, (1991).

9. (With Van Wijk, J. Larralde, C.W. Lovell) Pumping Predicting Model for Highway Pavements, Journal of Transportation Engineering, ASCE, Vol. 115, No. 2, March (1989) 161-175.

10. (With L. Duan and F.M. Wang) Flexural Rigidity of Reinforced Concrete Members, ACI Structural Journal, Vol. 86, July-August (1989) 419-427 (Authors' Closure, Vol. 87, No. 3, May-June (1990) 364-365).

11. (With T.A. Bubenik, and I.S. Sohal) Effect of Pressure on Tubular Beam-Column Capacity, Journal of Constructional Steel Research, Vol. 13, No. 1 (1989) 23-42.

12. (With S.E. El-Metwally) Nonlinear Behavior of R/C Frames, Computers & Structures, Vol. 32, No. 6, (1989) 1203-1209.

13. (With Y. Ohtani) A Plastic-Softening Model for Concrete Materials, Computers & Structures, Vol. 33, No. 4, (1989) 1047-1055.

14. (With N. Kishi, K. Matsuoka and S. Nomachi) Formulation of Initial Connection Stiffness and Ultimate Moment Capacity of Steel Beam-To-Column Angle Connections, Journal of Structural Engineering, Japanese Society of Civil Engineers, Vol. 35A, March (1989) 97-105.

15. (With N. Kishi, K. Matsuoka and S. Nomachi) Data Base of Steel Beam-To-Column Connection Tests and Its Applications, Journal of Structural Engineering, Japanese Society of Civil Engineers, Vol. 35A, March (1989) 75-82.

16. (With X.L. Liu and Q.Y. Wang) Layered Analysis with Generalized Failure Criterion, Computers & Structures, Vol. 33, No. 5, (1989) 1117-1124.

17. (With S.E. El-Metwally) Load Deformation Relations for Reinforced Concrete Sections, ACI Structural Journal, Vol. 86, No. 2, March-April (1989) 163-167. (Authors' Closure, Vol. 87, No. 1, January-February (1990) 117-118).

18. (With N. Kishi) Moment-Rotation Relation of Top- and Seat-Angle Connections, Proceedings of the International Colloquium on Bolted and Special Connections, May 15-20, Moscow, (1989).

19. (With W.O. McCarron) Formulation and Implementation of Bounding Surface Model, Computer and Physical Modeling in Geotechnical Engineering, Balasubramaniam et al. (eds), Balkema, Rotterdam, The Netherlands, (1989) 439-474.

20. Design of Beam-Columns in Frames, Proceedings of SSRC 4th International Colloquium on Stability of Metal Structures: Code Difference Around the World, Lehigh University, Bethlehem, PA, (1989) 135-147.

1990 - Journal Articles (16)

1. (With T.K. Huang) Simple Procedure for Determining Cap-Plasticity-Model Parameters, Journal of Geotechnical Engineering, ASCE, Vol. 116, No. 3, March (1990) 492-513.

2. (With N. Kishi) Moment-Rotation Relations of Semi-Rigid Connections with

Angles, Journal of Structural Engineering, ASCE, Vol. 116, No. 7, July (1990) 1813-1834.

3. (With L. Duan) Design Interaction Equations for Cylindrical Tubular Beam-Columns, Journal of Structural Engineering, ASCE, Vol. 116, No. 7, July (1990) 1794-1812.

4. (With M. Barakat) Practical Analysis of Semi-Rigid Frames, Engineering Journal, AISC, Vol. 27, No.2, 2nd Quarter, Chicago, (1990) 54-68.

5. (With S.P. Zhou and L. Duan) Comparison of Design Equations for Steel Beam-Columns, Structural Engineering Review, Edinburgh, Scotland, Vol. 2, No. 1, (1990) 45-53.

6. (With F. Al-Mashary) Elastic Second-Order Analysis for Frame Design, Journal of Constructional Steel Research, Vol. 15, No. 4, (1990) 303-322.

7. (With K.H. Mosallam) Design Considerations for Formwork in Multistory Concrete Buildings, Engineering Structures, Vol. 12, No. 3, July (1990) 163-172.See also Construction and Building Materials, Butterworth Heinemann, Oxford, U.K., Vol. 6, No. 1, (1992) 23-30, reprinted from Engineering Structures.

8. (With E. Yamaguchi) Cracking Model for Finite Element Analysis of Concrete Materials, Journal of Engineering Mechanics, ASCE, Vol. 116, No. 6, June (1990) 1242-1260.

9. (With F.H. Wu) A Design Model for Semi-Rigid Connections, Engineering Structures, Vol. 12, April (1990) 88-97.

10. (With S.E. El-Metwally and A.F. Ashour) Instability Analysis of Eccentrically Loaded Concrete Walls, Journal of Structural Engineering, ASCE, Vol. 116, No. 10, October (1990) 2853-2872.

11. (With L. Duan) A Yield Surface Equation for Doubly Symmetrical Sections, Engineering Structures, Vol. 12, April (1990) 114-119.

12. (With E. Yamaguchi) Post-Failure Behavior of Concrete in Compression, Journal of Engineering Fracture Mechanics, Vol. 37, No. 5-6, (1990) 1011-1024.

13. (With S.P. Zhou, and L. Duan) Second-Order Inelastic Analysis of Braced Portal Frames: Evaluation of Design Formulae in LRFD and GBJ Specifications, Journal of Singapore Structural Steel Society, Singapore, Vol. 1, No. 1, December (1990) 5-15.

14. (With S.E. El-Metwally and A.M. El-Shahhat) Three-Dimensional Nonlinear Analysis of R/C Slender Columns, Computers and Structures, Vol. 37, No. 5, (1990) 863-872.

15. (With D.W. White and H. Zhang) Preparing Structural Engineering Research and Education for the 21 Century, Invited Theme Article for Journal of the Chinese Institute of Civil and Hydraulic Engineering, Vol. 2, No. 2, Taipei, Taiwan, (1990) 95-106.

16. (With D.W. White) Second-Order Inelastic Analysis for Frame Design, Keynote Lecture, Proceedings of National Symposium on Advances in Steel Structures, India Institute of Technology, Madras, India, February 7-9, Tata McGraw-Hill, New Delhi, (1990) 461-474.

1991 - Journal Articles (18)

1. (With K.H. Mosallam) Determining Shoring Loads for Reinforced Concrete Construction, ACI Structural Journal, Vol. 88, No. 3, May-June (1991) 340-350.
2. (With Y. Goto and S. Suzuki) Bowing Effect on Elastic Stability of Frames under Primary Bending Moments, Journal of Structural Engineering, ASCE, Vol. 117, No. 1, (1991) 111-127.
3. (With E. Yamaguchi) Microcrack Propagation Study of Concrete under Compression, Journal of Engineering Mechanics, ASCE, Vol. 117, No. 3, March (1991) 653-673.
4. (With C. S. Cai and X.L. Liu) Further Verifications of Beam-Column Strength Equations, Journal of Structural Engineering, ASCE, Vol. 117, No. 2, February (1991) 501-513.
5. (With E. Yamaguchi and H. Zhang) On the Loading Criteria in the Theory of Plasticity, Journal of Computers and Structures, Vol. 39, No. 6, (1991) 679-683.
6. (With M. Barakat) Design Analysis of Semi-Rigid Frames: Evaluation and Implementation, Engineering Journal, AISC, Vol. 28, No. 2, 2nd Quarter, Chicago, (1991) 55-64.
7. (With S.E. El-Metwally) Behavior and Strength of Concrete Masonry Walls, ACI Structural Journal, American Concrete Institute, Vol. 88, No. 1, January/February (1991) 42-48.
8. (With J.Y.R. Liew and D.W. White) Beam-Column Design in Steel Framework - Insights and Current Methods and Trends, Journal of Constructional Steel Research, Vol. 18, No. 4, (1991) 269-308.
9. (With Y. Goto and S. Suzuki) Analysis of Critical Behavior of Semi-Rigid Frames With or Without Load History in Connections, Solids and Structures, Vol. 27, No. 4, (1991) 467-484.
10. (With H.M. Lee, X.L. Liu) Creep Analysis of Concrete Buildings During Construction, Journal of Structural Engineering, ASCE, Vol. 117, No. 10, October (1991) 3135-3148.
11. (With F. Al-Mashary) Simplified Second-Order Inelastic Analysis for Steel Frames, The Structural Engineer, Journal of the Institution of Structural Engineers, No. 69, No. 23, London, December (1991) 395-399.
12. New Developments in Stability Design of Steel Structures, Annual Lecture, The Structural Steel Society Annual Meeting, Steel News & Notes, Singapore Structural Steel Society, Vol. 6, No. 1, June/July, Singapore, (1991) 4-6.
13. (With J.Y.R. Liew) Refining the Plastic Hinge Concept for Advanced Analysis/Design of Steel Frames, Steel Structures, Journal of Singapore Structural Steel Society, Vol. 2, No. 1, Singapore, (1991) 13-28.
14. (With T.K. Huang) Cap Plasticity Model for Embankment: From Theory to Practice - Keynote Lecture, Proceedings of 2nd International Symposium on Environmental Geotechnology, Tongji University, Shanghai, China, May 15-17, 1989, Vol. 2, Editors - H.Y. Fang and S. Pamukcu, Envo Publishing Co., Bethlehem, PA, (1991) 1-40.
15. (With E.M. Lui) Analysis and Design of Steel Frames in the U.S.A.-90's and Beyond, Keynote Lecture, Proceedings of International Conference on Steel &

Aluminum Structures, ICSAS 91, May 22-24, Singapore 1991, Vol. 1 - Steel Structures, S.L. Lee and N.E. Shanmugam, Editors, Elsevier Applied Science, London, (1991) 1-19.

16. (With A. D. Pan) Finite Element and Finite Block Methods in Geomechanics, Proceedings of the 3rd International Conference on Constitutive Laws for Engineering Materials: Theory and Applications and Workshop on Innovative Use of Materials in Industrial and Infrastructure Design and Manufacturing, Tucson, January 7-11 (1991) 669-675.

17. Design Analysis of Semi-Rigid Frames with LRFD, Proceedings of the Second International Workshop on Connections in Steel Structures, R. Bjorhovde, A. Colson, G. Haaijer and J.W.B. Stark, Editors, Pittsburgh, April 10-12, AISC, Chicago, (1991) 370-379

18. (With D. W. White and J. Y. R. Liew) Considerations of Inelastic Stability in Frame Design, Proceedings of the William McGuire Symposium, Cornell University, Ithaca, New York (1991) 27-41.

1992 - Journal Articles (12)

1. (With J.M. Zeng, L. Duan and F.M. Wang) Flexural Rigidity of Reinforced Concrete Columns, ACI Structural Journal, American Concrete Institute, Vol. 89, No. 2, March-April (1992) 150-158.

2. (With W.S. King and D.W. White) On Second-Order Inelastic Analysis Methods for Steel Frame Design, Journal of Structural Engineering, ASCE, Vol. 118, No. 2, February (1992) 408-428.

3. (With H.M. Lee and X.L. Liu) Construction Live Load Caused by Powered Buggies, Concrete International: Design and Construction, American Concrete Institute, Vol. 14, No. 1, January (1992) 47-51.

4. Design of Beam-Columns in Steel Frames in the United States, Thin-Walled Structures, Vol. 13, Nos. 1 & 2, (1992) 1-83.

5. (With W.S. King and D.W. White) A Modified Plastic Hinge Method for Second-Order Inelastic Analysis of Rigid Frames, Structural Engineering Review, Vol. 4, No. 1, Chapman and Hall, London, (1992) 31-41.

6. (With S. Toma) European Calibration Frames for Second-Order Inelastic Analysis, Engineering Structures, Butterworth-Heinemann, Vol. 14, No. 1 (1992) 35-48.

7. (With A.M. El-Shahhat) Improved Analysis of Shore-Slab Interaction, ACI Structural Journal, American Concrete Institute, Vol. 89, No. 5, September-October (1992) 528-537.

8. (With T.K. Huang and J.L. Chameau) Application of Cap-Plasticity Model to Embankment Problems, Journal of Computers & Structures, Vol. 44, No. 6, (1992) 1349-1370.

9. (With K. Mosallam) Construction Load Distributions for Laterally Braced Formwork, ACI Structural Journal, American Concrete Institute, Vol. 89, No. 4, July-August (1992) 415-424.

10. (With E.D. Sotelino and D.W. White) Domain-Specific Object-Oriented Environment for Parallel Computing, Steel Structures, Journal of Singapore Structural Steel Society, Vol. 3, No. 1, (1992) 47-60.

11.　(With R.J.Y. Liew) Seismic Resistant Design of Steel Moment Resisting Frames Considering Panel-Zone Deformation, Proceedings of the US-Japan Seminar on "Stability and Ductility of Steel Structures under Cyclic Loading," Osaka, Japan, July 1-8, (1991), Editors, Y. Fukumoto and G. Lee, CRC Press, Boca Raton, Florida, (1992) 323-334.

12.　(With S.G. Nomachi, T. Kida and T. Sawada) On Slope Displacement by a Logarithmic Spiral Failure Slide During Earthquake, Proceedings of Sixth International Symposium on Landslides, Christchurch, New Zealand, February 10-14 (1992) 1193-1198.

1993 - Journal Articles (19)

1.　(With L. Duan and J.T. Loh) Analysis of Dented Tubular Members Using Moment Curvature Approach, Journal of Thin-Walled Structures, Vol. 15, (1993) 15-41.

2.　(With M. Abdel-Ghaffar and D. W. White) An Error Estimate and Step Size Control Method for Nonlinear Solution Techniques, Journal of Finite Elements in Analysis and Design, Vol. 13, (1993) 137-148.

3.　(With W.S. King, L. Duan, R.G. Zhou and Y.X. Hu) K-Factors of Framed Columns Restrained by Tapered Girders in US Codes, Engineering Structures, Vol. 15, No. 5, (1993) 369-378.

4.　(With R.G. Zhou, Y.X. Hu and L. Duan) An Approximate Solution of Creep of Orthogonal Anisotropic Concrete Thin Plates, Engineering Structures, Vol. 15, No. 1, January (1993) 61-66.

5.　(With L. Duan and J.T. Loh) Moment-Curvature Relationships for Dented Tubular Sections, Journal of Structural Engineering, ASCE, Vol. 119, No. 3, March (1993) 809-830, Discussion and Closure, Vol. 120, No. 7, July (1994) 2255-2257.

6.　(With Y. Goto and S. Suzuki) Stability Behavior of Semi-Rigid Sway Frames, Engineering Structures, Vol. 15, No. 3, (1993) 209-219.

7.　(With T. Sawada and S.G. Nomachi) Assessment of Seismic Displacements of Slopes, Soil Dynamics and Earthquake Engineering, Vol. 12, No. 6, (1993) 357-362.

8.　(With Y.X. Hu, R.G. Zhou, W.S. King, L. Duan) On Effective Length Factor of Frame Columns in ACI Code, ACI Structural Journal, Vol. 90, No. 2, March-April (1993) 135-143.

9.　(With N. Kishi, Y. Goto, and K.G. Matsuoka) Design Aid of Semi-Rigid Connections for Frame Analysis, Engineering Journal, Vol. 30, No. 3, 3rd Quarter (1993) 90-107.

10.　(With J.Y.R. Liew and D.W. White) Beam-Columns, Constructional Steel Design: An International Guide, P.J. Dowling, J.E. Harding, and R. Bjorhovde, Editors, Elsevier Applied Science, London, (1993) 105-132.

11.　Concrete Plasticity:　Macro and Micro Approaches, International Journal of Mechanical Sciences, Vol. 35, No. 12, December (1993) 1097-1109.

12.　(With L. Duan and W.S. King) K-Factor Equation to Alignment Charts for Column Design, ACI Structural Journal, Vol. 90, No. 3, May-June (1993) 242-248.

13.　(With A. El-Shahhat) Deflection-Based Analysis for Concrete Buildings during Construction, Structural Engineering Review, Vol. 5, No. 4, November (1993) 285-300.

14. (With A. El-Shahhat, D.V. Rosowsky) Construction Safety of Multistory Concrete Buildings, ACI Structural Journal, Vol. 90, No. 4, July-August (1993) 335-341.
15. (With J.Y.R. Liew and D.W. White) Second-Order Refined Plastic-Hinge Analysis for Frame Design. Journal of Structural Engineering, Vol. 119, No. 11, November (1993), Part I 3196-3216, Part II, 3217-3237.
16. (With W.S. King) LRFD Analysis for Semi-Rigid Frame Design, Engineering Journal, AISC, Vol. 30, No. 4, 4th Quarter (1993) 130-140.
17. (With N. Kishi, Y. Goto and K. Matsuoka) Analysis Program for the Design of Flexibly Jointed Frames, Computers and Structures, Vol. 49. No. 4 (1993) 705-713.
18. (With J.Y.R. Liew and D.W. White) Limit States Design of Semi-Rigid Frames Using Advanced Analysis: Part 1: Connection Modeling and Classification, Part 2: Analysis and Design, Journal of Constructional Steel Research, Vol. 26, No. 1, (1993) 1-58.
19. Preparing Civil Engineering Research and Education for the 21st Century, Proceedings of the Workshop on Transportation Development and Construction Research, National Cheng-Kung University, Tainan, Taiwan, October 12-14 (1993) 163-184.

1994 - Journal Articles (19)

1. (With M. Resheidat, M. Ghanma and C. Sutton) Improved "EI" Estimate for Reinforced Concrete Circular Columns, Materials & Structures, Vol. 27, (1994) 515-526.
2. (With W.S. King) Practical Second-Order Inelastic Analysis of Semi-Rigid Frames, Journal of Structural Engineering, ASCE, Vol. 120, No. 7, July (1994) 2156-2175.
3. (With S. Toma) Calibration Frames for Second-Order Inelastic Analysis in Japan, Journal of Constructional Steel Research, Vol. 28, No. 1, (1994) 51-77.
4. (With S.L. Chan) Second-Order Inelastic Analysis of Steel Frames by Personal Computers, Journal of Structural Engineering, Vol. 21, No. 2, Madras, India, (1994) 99-106.
5. (With N. Kishi, Y. Goto and M. Komuro) Sway Analysis of Tall Building Frames with Mixed Use of Rigid and Semi-Rigid Connections, Journal of Constructional Steel, JSSC, Vol. 2, November (1994) 1-8.
6. (With J.Y.R. Liew and D.W. White) Notational Load Plastic-Hinge Method for Frame Design, Journal of Structural Engineering, ASCE, Vol. 120, No. 5, May (1994) 1434-1454.
7. (With J.Y.R. Liew) Implications of Using Refined Plastic Hinge Analysis for Load and Resistance Factor Design, Journal of Thin-Walled Structures, Vol. 20, (1994) 17-47.
8. (With L. Duan and J. Loh) Ultimate Strength of Damaged Members, Journal of Offshore and Polar Engineering, Vol. 4, No. 2, June (1994) 127-133.
9. (With D.V. Rosowsky, Y.L. Huang, and T. Yen) Modeling Concrete Placement Loads during Construction, Structural Engineering Review, Vol. 6, No. 2, (1994) 71-84.
10. (With D.V. Rosowsky, D. Huston, P. Fuhr) Measuring Formwork Loads During Construction, ACI Concrete International, Vol. 16, No. 11, November (1994) 21-25.

11. (With M.D. Cohen and A. Goldman) The Role of Silica Fume in Mortar: Transition Zone versus Bulk Paste Modification, Cement and Concrete Research, Vol. 24, No. 1, (1994) 95-98.

12. (With T. Sawada, S.G. Nomachi) Seismic Bearing Capacity of a Mounded Foundation near a Down-Hill Slope by Pseudo-Static Analysis, Soils and Foundations, Vol. 34, No. 1, March (1994) 11-17.

13. Engineering Mechanics in Structural Engineering Research and Education, Steel Structures, Journal of Singapore Structural Steel Society, Vol. 5, No. 1, December (1994) 85-93.

14. (With N. Kishi, R. Hasan, and Y. Goto) Power Model for Semi-Rigid Connections, Steel Structures, Journal of Singapore Structural Steel Society, Vol. 5, No. 1, December (1994) 37-48.

15. (With A.M. El-Shahhat, D.V. Rosowsky) Partial Factor Design for Reinforced Concrete Buildings During Construction, ACI Structural Journal, Vol. 91, No. 4, July-August (1994) 475-485.

16. (With J. Lu, D.W. White, H.E. Dunsmore and E. Sotelino) F++: An Object-Oriented Application Framework for Finite Element Programming, The 2nd Annual Object-Oriented Numerics Conference, Sunriver, Oregon, April 24-27 (1994) 438-447.

17. (With T.K. Huang) Plasticity Analysis in Geotechnical Engineering: From Theory to Practice, Developments in Geotechnical Engineering, Balasubramaniam et al. Balkema, Rotterdam, (1994) 49-79.

18. Second-Order Inelastic Analysis for Frame Design, Proceedings of the 50th Anniversary Conference, Structural Stability Research Council, Bethlehem, PA, (1994) 197-209.

19. (With T. Sawada and S.G. Nomachi) Stability Analysis and Model Test for Seismic Displacement of a Slope, Proceedings of the International Symposium on Pre-Failure Deformation Characteristics of Geomaterials, Sapporo, Japan, September 12-14, S. Shibuya, T. Mitachi and S. Miura, Editors, A.A. Balkema, Rotterdam, (1994) 665-671.

1995 - Journal Articles (14)

1. (With S. Toma and D.W. White) A Selection of Calibration Frames in North America for Second-Order Inelastic Analysis, Engineering Structures, Vol. 17, No. 2, February (1995) 104-112.

2. (With K.M. Abdalla) Expanded Database of Semi-Rigid Steel Connections, Computers and Structures, Vol. 56, No. 4, August (1995) 553-564.

3. (With S.L. Chan) Second-Order Inelastic Analysis of Steel Frames Using Element with Midspan and End Springs, Journal of Structural Engineering, ASCE, Vol. 121, No. 3, March (1995) 530-541.

4. (With N.E. Shanmugam) An Assessment of K Factor Formulas, AISC Engineering Journal, First Quarter, Vol. 32, No. 1 (1995) 3-11.

5. (With M. Resheidat, M. Ghanma and C. Sutton) Flexural Rigidity of Biaxially Loaded Reinforced Concrete Rectangular Column Sections, Computers and Structures, Vol. 55, No. 4 ,(1995) 601-614.

6. (With J. Lu, D.W. White and H.E. Dunsmore) A Matrix Class Library in C++ for

Structural Engineering Computing, Computers & Structures, Vol. 55, No. 1, April (1995) 95-112.

7. (With J.Y.R. Liew) Analysis and Design of Steel Frames Considering Panel Joint Deformation, Journal of Structural Engineering, ASCE, Vol. 121, No. 10, October (1995) 1531-1540. Closure, Vol. 123, No. 3, March (1997) 381-383.

8. (With S.L. Chan, Z.H. Zhou, J.L. Peng and A.D. Pan) Stability Analysis of Semi-Rigid Steel Scaffolding, Engineering Structures, Vol. 17, No. 8, (1995) 568-574.

9. (With S. E. Kim) Practical Advanced Analysis for Frame Design-Case Study, Journal of Steel Structures, Singapore Structural Steel Society, Vol. 6, No. 1, (1995) 61-73.

10. (With M. Luan, G. Lin, Y. Guo, A.D. Pan and Y.K. Wang) Generalized Sliding-Wedge Method and Its Application to Stability Analysis in Soil Mechanics, Chinese Journal of Geotechnical Engineering, Vol. 17, No. 4, July (1995) 1-9.

11. (With A.M. El-Shahhat) Toward a Life Cycle Analysis of Concrete Structures, International Journal for Engineering Analysis and Design, Vol. 2, No. 3, New Delhi, India, (1995) 35-54.

12. (With S.H. Hsieh, E.D. Sotelino, and D.W. White) Toward Object-Oriented Parallel Finite Element Computations, High Performance Computing - ASIA'95, September, Taiwan September 16-18 (1995) 10 pp.

13. (With A.E. Maleck and D.W. White) Practical Applications of Advanced Analysis in Steel Design, Proceedings of the 4th Pacific Structural Steel Conference, Vol. 1, Steel Structures, Singapore, October (1995) 119-126.

14. (With T. Yen, J.L. Peng, I.C. Lin and A.D. Pan) Why Frequent Scaffold Failures During Construction? Construction News Record, Taiwan Construction Technology Research Center, December (1995) 32-43 (In Chinese).

1996 - Journal Articles (20)

1. (With J.L. Peng, D.V. Rosowsky, A. Pan, S. L. Chan and T. Yen) Analysis of Concrete Placement Load Effects Using Influence Surfaces, ACI Structural Journal, Vol. 93, No. 2, (1996) 180-187.

2. (With X.H. Zhao) Stress Analysis of a Sand Particle with Interface in Cement Paste under Uniaxial Loading, International Journal for Numerical and Analytical Methods in Geomechanics, Vol. 20, No. 4, (1996) pp. 275-285.

3. (With Y.L. Huang, T. Yen and Y.C. Lin) Loading Behavior of Form Supports during Concrete Placing, Journal of the Chinese Institute of Civil and Hydraulic Engineering, Vol. 8, No. 2, (1996) pp. 273-279.

4. (With E. Yamaguchi) Spotlight on Steel Moment Frames, Civil Engineering Magazine, ASCE, March (1996) 44-46.

5. (With X.H. Zhao) The Influence of Interface Layer on Microstructural Stresses in Mortar, International Journal for Numerical and Analytical Methods in Geomechanics, Vol. 20, No. 3, March (1996) 215-228.

6. (With G. Ramesh and E.D. Sotelino) Effect of Transition Zone on Elastic Moduli of Concrete Materials, Cement and Concrete Research, Vol. 26, No. 4 (1996) 611-622.

7. (With M. Resheidat, K. Numayr and C. Sutton) Improved EI Estimation for RC Rectangular Columns, Journal of Structural Engineering, Madras, India, Vol. 23,

No. 3, October (1996) 151-157.

8. (With K.M. Abdalla and A. Alshegeir) Analysis and Design of Mushroom Slabs with a Strut-Tie Model, Computer & Structures, Vol. 58, No. 2, (1996) 429-434.

9. (With S.E. Kim) Practical Advanced Analysis for Semi-Rigid Frame Design, AISC Engineering Journal, Fourth Quarter, Vol. 33, No. 4, (1996) 129-141.

10. (With N. Kishi, Y. Goto and R. Hasan) Behavior of Tall Buildings with Mixed Use of Rigid and Semi-Rigid Connections, Computers and Structures, Vol. 61, No. 6 ,(1996) 1193-1206.

11. (With M. Resheidat, M. Ghanma, C.D. Sutton) Flexural Rigidity of Biaxially Loaded RC Rectangular Column Sections, Journal of Structural Engineering, Structural Engineering Research Center, Madras, India, Vol. 22, No. 4, (1996) 201-210.

12. (With T. Yen, Y.L. Huang, Y.C. Lin and R.C. Chi) On the Resistance Capacity of Steel Scaffolds, Journal of the Chinese Institute of Civil and Hydraulic Engineering, Vol. 8, No. 1, (1996) 33-43 (in Chinese).

13. (With J.L. Peng, A.D. Pan, D.V. Rosowsky, T. Yen and S.L. Chan) High Clearance Scaffold Systems During Construction, Part I - Structural Modeling and Modes of Failure, and Part II - Structural Analysis and Development of Design Guidelines, Engineering Structures, Vol. 18, No. 3, March (1996) 247-267.

14. (With M.Z. Duan) Design Guidelines for Safe Concrete Construction, Concrete International, ACI, October (1996) 44-49.

15. (With S.E. Kim) Practical Advanced Analysis for Braced Steel Frame Design, Journal of Structural Engineering, ASCE, Vol. 122, No. 11, November (1996) 1266-1274.

16. (With S.E. Kim) Practical Advanced Analysis for Unbraced Steel Frame Design, Journal of Structural Engineering, ASCE, Vol. 122, No. 11, November (1996) 1259-1265.

17. (With Y.M. Lan) Applications of Advanced Composite Materials to Civil Substructures, Journal of Civil Engineering Technology, Taiwan Professional Civil Engineer Association, Vol. 6, September (1996) 107-118.

18. (With S.E. Kim) Reduced Tangent Modulus Plastic-Hinge Method for Steel Structure Design, Korea Society of Structural Steel, Vol. 8, No. 1, March (1996) 145-154.

19. (With T. Yen, J.L. Peng, A.D. Pan and S.L. Chan) Applications of Steel Scaffolds and Investigation of Failure Modes of Formwork Supports, Construction News Record, Taiwan Construction Technology Research Center, (1996) 13-22   (In Chinese).

20. (With T. Yen, J.L. Peng, D.V. Rosowsky and A.D. Pan) Safety of High-Clearance Structures During Construction - Pattern Load Effects of Concrete Placement, Construction News Record, Taiwan Construction Technology Research Center, No. 161, June (1996) 18-32.

1997 - Journal Articles (10)

1. (With N. Kishi, Y. Goto and R. Hasan) Study of Eurocode 3 Steel Connection Classification, Engineering Structures, Vol. 19, No. 9, (1997) 772-779.

2. (With N. Kishi and Y. Goto) Effective Length Factor of Columns in Semi-Rigid and Unbraced Frames, Journal of Structural Engineering, ASCE, Vol. 123, No. 3, March (1997) 313-320. Closure/Discussion, Vol. 124, No. 10, October (1998) 1230-1231.

3. (With J.L. Peng, T. Yen, Y. Lin and K.L. Wu) Performance of Scaffold Frame Shoring under Pattern Loads and Load Paths, Journal of Construction Engineering and Management, ASCE, Vol. 123, No. 2, June (1997).

4. (With C. Joh) Application of Fracture Mechanics to Steel Connections in Moment Frames under Seismic Loading Advances in Structural Engineering, Vol. 1, No. 1, (1997) 23-37.

5. (With S.E. Kim) Further Studies of Practical Advanced Analysis for Weak-Axis Bending, Engineering Structures, Vol. 19, No. 6, June (1997) 407-416.

6. (With J.L. Peng, A.D. Pan, T. Yen and S.L. Chan) Structural Modeling and Analysis of Modular Falsework Systems, Journal of Structural Engineering, ASCE, Vol. 123, No. 9, September (1997) 1245-1251.

7. (With R. Hasan, N. Kishi and M. Komuro) Evaluation of Rigidity Extended End-Plate Connections, Journal of Structural Engineering, ASCE, Vol. 123, No. 12, December (1997).

8. (With W.H. Yang and M.D. Bowman) The Behavior and Load-Carrying Capacity of Seated-Beam Connections, Engineering Journal, AISC, Vol. 34, No. 31, 3rd Quarter (1997) 89-103.

9. Structural Stability: From Theory to Practice, Proceedings of Innovations in Structural Design: Strength, Stability, Reliability, A Symposium Honoring T. V. Galambos, Minneapolis, Minnesota, June 6-7, J.F Hajjar and R.T. Leon, Editors, Structural Stability Research Council, Bethlehem, PA, (1997) 29-37.

10. Design of Steel Structures with LRFD Using Advanced Analysis, Proceedings of the 5th International Colloquium on Stability and Ductility of Steel Structures, Nagoya, Japan, July 29-31 Edited by T. Usami, 2 volumes (1997) 21-32.

1998 - Journal Articles (15)

1. (With S.E. Kim) A Sensitivity Study on Number of Elements in Plastic-Hinge Refined Analysis, Computers & Structures, Vol.66, No. 5 (1998) 665-673.

2. (With X.H. Zhao) Solutions of Multi-Layer Inclusion Problems under Uniform Field, Journal of Engineering Mechanics, ASCE, Vol. 124, No. 2, February (1998) 209-216.

3. (With X.H. Zhao) Effective Elastic Moduli of Concrete with Interface Layer, Computers and Structures, Vol. 66, Nos. 2-3, (1998) 275-288.

4. (With X.H. Zhao) The Effective Elastic Moduli of Concrete and Composite Materials, Journal of Composites Engineering Part B, 29B, (1998) 31-40.

5. Implementing Advanced Analysis for Steel Frame Design, Progress in Structural Engineering & Materials, Vol. 1, Issue 3, April (1998) 323-328.

6. (With Y.S. Kim) Practical Analysis for Partially Restrained Frame Design, Journal of Structural Engineering, ASCE, Vol. 124, No. 7, July (1998) 736-749.

7. (With Y.S. Kim) Design Tables for Top- and Seat-Angle with Double Web-Angle Connections, Engineering Journal, AISC, Vol. 35, No. 2, Second Quarter (1998) 50-75.

8. (With K. Takanashi, M. Nakashima and D. White) Innovations in Stability Concepts and Methods for Seismic Design in Structural Steel-Forward, Engineering Structures, Vol. 20, No. 4-6, June (1998) 239-241.

9. (With J.L. Peng, D.V. Rosowsky, A.D. Pan and S.L. Chan) Simplified Modeling and Analysis of Pattern Loading Effects on Shoring Systems During Construction, Advances in Structural Engineering, Vol. 1, No. 3, (1998) 203-218.

10. (With N. Kishi, Y. Goto and M. Komuro) Effective Length Factor of Columns in Flexibly Jointed and Braced Frames, Journal of Constructional Steel Research, Vol. 47, No. ½, July/August (1998) 93-118.

11. (With R. Hasan and N. Kishi) A New Nonlinear Connection Classification System, Journal of Constructional Steel Research, Vol. 47, No. ½, July/August (1998) 119-140.

12. (With S. E. Kim) Design of Steel Structures with LRFD Using Advanced Analysis, In Stability and Ductility of Steel Structures edited by T. Usami and Y. Itoh, Pergamon/Elsevier, (1998) 153-166.

13. (With G. Ramesh and E.D. Sotelino) Effect of Transition Zone on Elastic Stresses in Concrete Materials, Journal of Materials in Civil Engineering, ASCE, Vol. 10, No. 4, (1998) 275-282.

14. Concrete Plasticity: Past, Present and Future, Proceedings of the International Symposium on Strength Theory: Application, Development & Prospects for 21st Century, Edited by Mao-Hong Yu and Sau-Cheong Fan, Science Press, Beijing, Conference held in Xian, China, September 9-11 (1998) 7-49.

15. Structural Engineering: Past, Present and Future, Proceedings of the Fourth National Conference on Structural Engineering, Taipei, Taiwan, Vol. 1/3, September 9-11 (1998) 1-16.

1999 - Journal Articles (9)

1. (With G. Li, R.G. Zhou and L. Duan) Multiobjective and Multilevel Optimization for Steel Frames, Engineering Structures, Vol. 21, No. 6, June (1999) 519-529.

2. (With S.E. Kim) Design Guide for Steel Frames Using Advanced Analysis Program, Engineering Structures, Vol. 21, No. 4, April (1999) 352-364.

3. (With C.B. Joh) Fracture Strength of Welded Flange-Bolted Web Connection, Journal of Structural Engineering, ASCE, Vol. 125, No. 5, May (1999) 565-571.

4. (With D.R. Zeng, S.P. Zhou) The Strength and Stiffness Analysis of the Bird-Shape Plates, Industrial Construction, Vol. 29, No. 7, July (1999) 4-6.

5. (With S.P. Zhou) Local Stability Analysis of Bird-Shape Plates, Industrial Construction, Vol. 29, No. 7, July (1999) 7-11.

6. (With S.P. Zhou, Y.C. Cheng) Torsional Buckling of Bird-Shape Plates, Industrial Construction, Vol. 29, No. 7, July (1999) 12-16.

7. (With W.H. Yang and M.D. Bowman) Experimental Study on Bolted Unstiffened Seat Angle Connections, Journal of Structural Engineering, ASCE, Vol. 125, No. 11, November (1999) 1224-1231.

8. (With S. E. Kim) Guidelines to Unbraced Frame Design with LRFD, The Structural Design of Tall Buildings, Vol. 8, (1999) 273-288.

9. (With E. D. Sotelino and D. W. White) Future Challenges for Simulation in

Structural Engineering, 4th World Computational Mechanics Congress, Buenos Aires, Argentina, June 29 – July 2 (1998) 18pp.

2000 – Journal Articles (9)

1. (With S.E. Kim, M.K. Kim) Improved Refined Plastic Hinge Analysis Accounting for Strain Reversal, Engineering Structures, Vol. 22, No. 1, January (2000) 15-25.
2. Structural Stability: From Theory to Practice, Engineering Structures, Vol. 22, No. 2, February (2000) 116-122.
3. Plasticity, Limit Analysis and Structural Design, International Journal of Solids and Structures, Vol. 37, (2000) 81-92.
4. (With S. A. S. Youakim, S. E. E. El-Metewally) Nonlinear Analysis of Tunnels in Clayey/Sandy Soil with a Concrete Lining, Engineering Structures, Vol. 22, No. 6, June (2000) 707-722.
5. (With J. Y. R. Liew, H. Chen) Advanced Inelastic Analysis of Frame Structures, Journal of Constructional Steel Research, Vol. 52, Nos. 1-3, July-September (2000) 245-265.
6. (With I. H. Chen) Major Design Impact of 1997 LRFD Steel Seismic Code Revision in USA, Journal of Structural Engineering, Madras, India, Vol. 27, No. 1, April (2000) 1-16.
7. (With J.Y. Richard Liew, H. Chen, and N.E. Shanmugam) Improved Nonlinear Plastic Hinge Analysis of Space Frame Structures, Engineering Structures, Vol. 22, No. 10, October (2000) 1324-1338.
8. (With Y. L. Huang, H. J. Chen, T. Yen, Y. G. Kao, and C. Q. Lin) A Monitoring Method for Scaffold-Frame Shoring System for Elevated Concrete Formwork, Computers and Structures, Vol. 78, (2000) 681-690.
9. (With C. S. Doo and E. D. Sotelino) Practical Approach to Cracking of Welded Moment-Resisting Frames, Proceedings of the Symposium Honoring Professor Alfredo H-S. Ang's Career, University of Illinois at Urbana-Champaign, May 19-20, (2000). (In CD-ROM).

2001 - Journal Articles (5)

1. (With S. P. Zhou, L. Duan, Y. C. Cheng) Global Stability Analysis for Bird-Shaped Girders, Journal of Structural Engineering, ASCE, Vol.127, No.3, March (2001) 306-313.
2. (With J. L. Peng, A. D. E. Pan) Approximate Analysis Method for Modular Tubular Falsework, Journal of Structural Engineering, ASCE, Vol. 127, No.3, March (2001) 256-263.
3. (With N. Kishi, A. Ahmed, N. Yabuki) Nonlinear Finite Element Analysis of Top- and Seat-Angle with Double Wed-Angle Connections, Structural Engineering and Mechanics, Vol.12, No. 2, Techno-Press, February (2001) 201-214.
4. (With S. E. Kim and S. H. Choi) Practical Second-Order Inelastic Analysis for Three-Dimensional Steel Frames, Steel Structures, Vol. 1 (2001) 213-223.
5. (With C. Joh) Seismic Behavior of Steel Moment Connections with Composite Slab, Steel Structures, Vol. 1 (2001) 175-183.

2002 - Journal Articles (2)

1. (With H. L. Cheng, E. D. Sotelino) Strength Estimation for FRP Wrapped Reinforced Concrete Columns, Steel and Composite Structures, Vol. 2, No. 1, February (2002) 1-20.
2. (With K. Wongkaew) Consideration of Out-of-Plane Buckling in Advanced Analysis for Planar Steel Frame Design, Journal of Constructional Steel Research, Vol. 58, No. 5-8, (2002) 943-965.

2003 - Journal Articles (3)

1. (With Y. M. Lan, E. D. Sotelino) The Strain-Space Consistent Tangent Operator and Return Mapping Algorithm for Constitutive Modeling of Confined Concrete, International Journal of Applied Science and Engineering, Vol.1, No.1 (2003) 17-29.
2. (With M. H. Teng, E. D. Sotelino) Performance Evaluation of RC Bridge Columns Wrapped with FRP, ASCE Journal of Composite for Construction, Vol.7, No. 2 (2003) 83-92.
3. (With L. Duan) Basic Design Criteria for Bridge Crossing Open Sea and Bay Area, Marine Georesources and Geotechnology, Vol. 21, (2003) 289-305.

A regular monthly gathering of Purdue's Ping-Pong Club members in West Lafayette. Front row from left: Emily Lin (王若雪), Mrs. P. T. Yeh (葉唐慶祥), K. C. and Jean Chao (趙廣緒, 蘇君塋,), Y. L. Tong (唐裕林).

Colleagues and friends at Purdue Chinese Ping-Pong club in West Lafayette, Indiana in 2005.

First row: Jean and K. C. Chao (趙廣緒, 蘇君瑩).

Second row: C. L. and Josephine Chen (陳清林, 宋靜峰), Lucy Shieh (倪鐵柳), C. K. Lee (楊宗琪), Y. C. Wang (陳英), Louise Lin (李紹雲), Emily Lin (王若雪).

Third row: P. M. Lin (林本銘), Y. N. Shieh (謝越寧), Michael Lin (林政賢), John Wang (王叔平), Daniel Lee (李修步), C. J. Song (宋正吉).

7

My Life Story
我的一生

I stepped down as dean of engineering at the University of Hawaii in June 2006 after seven years service. I was part of the University's top management team since my arrival in 1999. This article was prepared for my high school 50th class reunion Yearbook（*台灣師大附中高卅三班畢業五十週年紀念*）with the title "*半世紀的友情*". In this article, I offer some perspectives on my personal life, career and struggle over the last 50 years. I trace the history of my schooling, career, honors and recognitions, family, and lessons about life. The real motivation for me to write this life-long story for the Yearbook is to share our collective wisdom with each other, our spouse, and our next generation in particular.

7.1 Step down as Dean 辭院長職

The engineering faculty at the University of Hawaii sat in a large auditorium as I gave a speech typical of a college dean during a budget briefing. I recounted the college's accomplishments, citing record high enrollment, research funding and donations. I expounded on the college's creation of centers of excellence in wireless communication and corrosion research, and the establishment of the program in the new field of biomedical engineering. Then, I reached the conclusion. I announced that I would be stepping down as the dean of the college as soon as the new dean was recruited. As audience members sat up in their seats, gasped, and tried to make sense of the unexpected announcement, I explained how the college can best be served to realize its long term vision with a new dean at this stage of progress.

I am not burned out, but I know some strategies to prevent burnout.

The dean's job is like a corporate CEO position; you are responsible for everything in the college, and your signature is everywhere. You must keep things in perspective, be patient, diplomatic, and tolerate ambiguity and frustration. You have few friends and you must leave work at work and delegate. There are times to step up and there are times to step down. I never planned to be a professor but ended up as an academician. I never planned to take an administrative job in a university but ended up as a dean in a large public school. Life is unpredictable. I would like to share with you how I got from there to here.

7.2 My Schooling 我的學業

My earliest memory of schooling began in Taiwan in 1950, Class 27（附中初 27 班）for junior high and Class 33（附中高 33 班）for senior high. I still vividly remember the Korean War breaking out when we arrived at Taipei. The way, we, the three brothers got to Taiwan was similar to the boat people escaping Vietnam after the war ended in the1970's (see Chapter 1). The other earliest memories during my childhood were about the Second World War against Japan followed by the Civil War between the Nationalists and Communists. It is just war, war, war; and run, run, run. No schooling, period.

My formal schooling began in junior high Class 27. My twin brother, Wai-Kai（惠開）, the twin classmates Henry Mee（米明琳）, John Mee（米明瑯）and the Han brothers, Robert and Charles（韓純武，韓祖武）, sort of twin-like, come to mind immediately. Our friendship with the twins can be traced back more than 50 years. Even more amazing, John and Grace（張諱寶）now live less than 10 miles from our condo in downtown Honolulu. We are fortunate to have frequent gatherings with the Mee's family as well as his elder brother, Old Mee（老米）. As twins, my brother and I hated to be dressed alike and treated as the center of entertainment. So, we decided to attend different schools; Wai-Kai transferred to Taipei Institute of Technology（台北工專）; and I entered Class 33 after graduating from junior high. Many of Wai-Kai's schoolmates frequently thought I was him when we attended two different but nearby schools. I finally gave up the endless explanation and just pretended to be him, continuing any conversation whenever I bumped into his friends on the road.

7.3 My Career Path 我的事業

I chose civil engineering as my career for the wrong reason; the sole reason is because my two elder brothers were already in electrical engineering. I never liked chemical engineering. By process of elimination, civil engineering seemed to be a good choice. The same thing happened to my youngest son, Brian（中宇）. Like father like son. What a coincidence, isn't it?

The year I took the national college entrance examination, I was required to list, for the first time, only the priority of our selected fields, instead of universities. I was admitted to my first choice in my selected field, but the university was my second choice, the National Cheng-Kung University in Southern Taiwan（台南成功大學）. As a result, I moved away from home and lived in a dormitory for four years and Paul Sun（孫長貴）was one of my roommates. Later, I had two more years of ROTC training and service（預官訓練與服役）, then I went abroad. At the time, everyone was dreaming of going to U.S.A. in particular, for advanced studies, since there was not much future in Taiwan. It seems that my entire life was already planned by my family's high expectations: a Ph.D. degree, an engineering job, a Chinese wife, and immigration for family members.

My dream universities were Cornell and Stanford. I received a scholarship from Stanford for my graduate studies but the stipend was not enough to survive. So, I had no choice but to take an assistantship from Lehigh University in Bethlehem, PA. It was a painful decision but it turned out to be a blessing. Again, the same thing happened to Brian. He also received a scholarship from Stanford and decided to attend University of Texas at Austin instead. Another coincidence!

At the time, Lehigh's structural engineering group was well focused on the development of a new steel design method called "*plastic design*", replacing the century-old method of "*allowable stress design*". It is a revolutionary concept pioneered at Cambridge University in UK during

WWII. Lehigh's Fritz laboratory had the world's largest testing machine and I participated in the full scale testing of steel structures. The new method was later adopted by the American Specifications for steel construction and quickly became a worldwide standard. The structural group became world famous in steel construction. Finally, I received my M.S. degree there.

Linlin (玲玲) and W.F. (惠發) visited Yale campus in July 1965.

As my study at Lehigh came to an end, I received an offer from Cornell University to continue my steel research for my Ph. D. study. It was a dream-come-true, but for some reason I was attracted to Brown University in Providence, RI for its highly theoretical work in solid mechanics. Tired of experimental work at Lehigh? Maybe. I was particularly interested in the mathematical theory of plasticity, a branch of applied mathematics describing the inelastic behavior of materials. The subject "*solid mechanics*" sounded sexy, mathematical, and intellectually challenging. It was a very hot subject at the time. It may sound strange now but that's the way it was at the time. In today's terminology, it is equivalent to the topic of "nano-material" or "bio-material". However, as it turned out, the exotic plasticity theory and its daring applications frequently complicated things. Design simplifications were always necessary for its practical implementation. Nothing can be more practical than a simple theory. Consequently, simple limit analysis methods were developed and widely applied to steel structures at Brown. My Ph. D. thesis was focused on the application of this new method, not on steel, but on concrete structures for the first time ever.

The Lehigh "gang" at one of the Structural Stability Research Council gatherings.

7.4 My Celebrity Status 我的名聲

When I returned to Lehigh to teach as an Assistant Professor, I started to apply and extend this simple limit analysis method to soil mechanics. I developed simple techniques for obtaining real world solutions. I wrote books on this method and also made the mathematical theory easily understandable to civil engineers. It became instantly popular and made me a celebrity of sorts.

I noticed this status change when I attended conferences or visited universities. Students lined up for my signature and attendees wanted to take pictures with me. They started to ask specific questions in the books written by me many, many years ago. I really did not know what they were talking about. They thought I still remembered every detail of my work! Even nowadays, my signed books can still command a high price on Amazon.com. Often, I could find my books, newspaper articles or photos from a Google search much faster than I could dig them out from my own files. I am often surprised to learn that I am so highly regarded by my international peers or students.

In a recent email I received from a graduate student in Brazil, the student was so excited to find my 1974 book on limit analysis and soil plasticity at a book store in a remote area. The book had been out of print for many years. It was quite moving to read his letter. The graduate student expressed his and his professor's view on my work:

"Next Wednesday, I will show this book to my professor and I am sure he will wish for a copy of this book and I will give him this copy. He is very fond of you, him and all the classmates. In his classes, he always worships you as God."

Recently, I was truly flattered by a request from a UK author, asking for a color photo of me for his forthcoming book entitled *"Giants of Engineering Science"*. The book covers the contributions of ten of the most prolific US engineering scientists of the last forty years. According to him, I will be featured in Chapter 9 in his book for my contributions to structural and geo-mechanics engineering science (see Sec. 3.6, Chapter 3). What a compliment! A Chinese Proverb says *"man fears of choosing a wrong profession, while woman fears of marrying a wrong guy"* （男怕選錯行，女怕嫁錯郎）. I may have chosen the right profession, but I did so for the wrong reason. This is life. This may be fate. It is indeed a risky business to plan our own future.

In my career over the last 40 years, I have discovered one basic principle in engineering – keep things simple. I have built a reputation for developing simple solutions. It has been a wonderful experience and rewarding career for me. If the solutions are complicated, we have not found the right solutions yet. Consider how simple Newton's law of motion is. Einstein's theory of relativity is simple and elegant. Laws of nature have to be simple. This may also be true for our daily life. A life of good fulfillment should also be simple.

7.5 My Honors and Recognitions 我的榮譽與讚譽

The celebrity status comes with some benefits and opportunities. When I was elected to the US National Academy of Engineering, the executive search firms started to contact me more aggressively on a regular basis. With the status, the Chief Executive Officer (CEO) level positions were more available and easily accessible. Seeking a higher position is admirable but the reality of one's true interest and work-family balance is something that cannot be avoided. At the end of the day, a decision has to be made and ambition has to be curtailed at some point.

Meet with friends Durkee and Beedle at the National Academy of Engineering annual meeting in Washington D.C.

Retirement is looming ahead and the inner urge of spending more time with family and kids is calling and rising. A Chinese Proverb says *"there are people resigning from high office and returning to hometown with family; and there are people studying late to secure a high office"* (有人辭官歸故里，有人漏夜趕功名). It is ironic to see our kids burning candles on both ends in order to make it. Work and family rest on a very delicate balance. Life, career, family and struggle for the younger generation will continue as we, the older generation, fade away.

When I was elected to Taiwan's National Academy of Science, called Academia Sinica (中央研究院), I was contacted or nominated more frequently for university CEO positions in both Taiwan and Hong Kong. It seems that a membership in Academia Sinica is a necessary union card for a high academic position in the Far East. This seems to follow exactly the old Chinese saying *"for a high achiever in academia, the next move is to seek a high office"* (學而優則仕).

Very few civil engineers are elected to the Academia Sinica. Engineering is usually considered to be a second class work in the world of science and mathematics. Civil engineers deal with earth and wood, and are not considered intellectually comparable to physicists and mathematicians. That is true. There is no Nobel Prize in engineering. Thus, I was very fortunate to be recognized and accepted by this very elite group of intellectuals as a colleague. The biannual meeting and some special treatments such as an escort service from the airplane gate directly to the special limo at the airport, and a formal dinner with Taiwan's President were the kind of perks that impressed my wife Linlin (宣玲玲, Lily) and my family.

Richard Chu (朱兆凡) and his wife, and Linlin (玲玲) and W.F. (惠發) at the banquet hosted by President Chen Shui-bian (陳水扁) of Taiwan in July 1992 for Academicians.

In engineering practice, it is not the intellectual challenge that counts. Rather, it is the impact of major engineering achievements. The highest honor a professional society can bestow to its member, second only to the title of Society President, is to award the individual with an Honorary Membership. I was fortunate to be awarded this membership by the American Society of Civil Engineers. There is only a handful of such membership awarded yearly among the nearly 140,000 members.

To be recognized by your colleagues in the same school is even more of a challenge, especially at Purdue with its nationally top-ranked engineering programs. There were already several NAE members on the faculty list at Purdue's civil engineering department. Per an old Chinese saying, *"jealousy always exists in the same professional group"* (同行相嫉). I was the first one to be named and assumed the prestigious title *"Distinguished Professorship"* in the history of the School of Civil Engineering. It was indeed a great honor to be recognized by my colleagues and peers. This is why it was so difficult for me to resign a few years later and accept the deanship at UH.

Receiving the ASCE Honorary Membership award held in Minneapolis, October 1997.

7.6 My Family 我的家庭

I met Linlin at my twin brother Wai-Kai's (惠開) engagement party. I began to know her when she was still in high school. Her major was in mathematics at the University of Illinois. She continued her graduate study in mathematics at Lehigh University. We married in June 1966, right after my graduation from Brown University. She is a tall, out-going, sociable, sweet, good looking, and hard-working girl, and can easily make friends. She loves ballroom dance. She taught a class at a community college near our home at Lehigh, and gave a cooking class at Purdue while raising our three boys (see Chapter 4).

The members of the Ping-Pong Club at Purdue, West Lafayette, Indiana, 2004.

Front from left: K.Y. Chung (李楊宗琪), C.S. Yeh (葉唐慶祥), Michael Lin (林政賢), Emily Lin (林王若雪), Jean Chao (趙蘇君瑩), Lucy Shieh (謝倪鐵柳), Sophie Yeh (葉). Second from left: H.M. Loh (唐陸曉明), Y.L. Tong (唐裕林), Daniel Lee (李修步), K.C. Chao (趙廣緒), C.J. Song (宋正吉), Y.N. Shieh (謝越寧), Josephine Chen (陳宋靜峰).

I worked at Lehigh University for exactly 10 years, and then moved westward to Purdue University and stayed there for nearly 23 years. Our three boys, Eric (中傑), Arnold (中毅), and Brian (中宇), were all born in Bethlehem, PA. But they grew up in West Lafayette, IN. We considered West Lafayette our hometown. West Lafayette is a college town, an ideal place for raising a family. Our life was simple and routine. I was busy with my work and Linlin (玲玲) was busy with the three boys. The years in West Lafayette were the most productive years in my career and the most rewarding years for Linlin in raising our three wonderful boys. We had a lot of fun with the boys during those early years.

Linlin (玲玲) is the key figure in my life for what I have been able to achieve, especially in bringing about the closeness of our family group with our three wonderful boys. To me, she is the central figure in the family. I took a sabbatical leave in 1999 and lived in Honolulu for six months. Linlin and I both loved the weather and Hawaii's aloha spirit. I decided to apply for the Dean's position at the University of Hawaii when the opportunity arose in the summer of 1999. It was a big decision for all of us in a very short time, just two weeks. It was a big change for me. It was a happy choice for Linlin. It had some depressing effect for the three boys, especially for Eric（中傑）, who had more sentimental attachment to our self-built self-designed house in West Lafayette. We had to say goodbye to our long-time friends there. The boys had to say goodbye to their long-time schoolmates there. We had many farewell parties. It was an emotional period for me to convince my close colleagues and administrators to let me go. UH paid Purdue my sabbatical costs for not staying at least one year after returning from my leave.

Our Lehigh home at "*1949 Markham Drive*" in Bethlehem: a visit from Auntie Chao (趙姑媽) family in 1967. 趙天星, 趙慶珍 (second row, middle and right) 姑媽 (front, right)

Time flies. Eric（中傑）, my eldest son, an electrical engineer with a Ph. D. from the University of Illinois, worked nearly five years at MIT's Lincoln Lab in Lexington, MA. He has talent in writing and is a good communicator. He loves chess and is good at it. During the Internet bubble, he decided to go to University of Chicago's Business School for an MBA degree. He is now working at Northrop Grumman as Director of Finance in Redondo Beach, CA. He is quite happy with his new career in finance. He loves the warm weather of Los Angeles and lives in a condo close to the beach. He is the first one in our family tree to break out of the traditional path of engineering, entering the business world. He is doing quite well.

Family reunion at Avalon home, *3214 Belmont Terrace*, Fremont, California.
From left: Eric (中傑), W.F. (惠發), Brian (中宇), Linlin (玲玲), and
Arnold (中毅), dog Kaffee.

Arnold (中毅), my second son, also an electrical engineer with a Ph. D. from the University of Illinois, joined a start up company in the Bay Area during the Internet bubble. His specialty is in optical switches and amplifiers. He has a quiet personality like me and loves gardening. He was the first son in the family to marry and had his wedding in Hawaii. His wife, Lin Ng (黃慧琳), worked with an investment bank at the time. She has natural talents in linguistics and is fluent in more than four languages. They both love Bay Area living. Arnold is now working at the optical switch company called Infinera in Sunnyvale, and Lin is working with Stanford's business school. We have a granddaughter, Chloe（陳秀敏）born on December 10, 2006.

Left: Arnold dressed up for his senior graduation dance.
Right: Eric（中傑）, Arnold（中毅）, and Brian（中宇）at Kassel during W.F.（惠發）1984 visit under the US Senior Scientist Award, Alexander von Humboldt Foundation, Germany.

W.F. (惠發) and Linlin（玲玲）at Brian Wedding in Austin, Texas, April 2005.

Brian（中宇）, like me, a structural engineer with a Ph. D. from the University of Texas at Austin, taught one year at Bucknell University, and then joined a consulting firm in Dallas. His specialty is in failure investigation. He, like Linlin（玲玲）, has the most outgoing personality in the family. He also loves dancing, swing dancing in particular. He met his wife, Christine de Asis, while swing dancing in Austin, and married there last April. She, following her father's footsteps, is studying for her medical degree in Dallas. They lived in Dallas during Christine's medical school time but love Austin living. They are now living in San Antonio for Christine's medical residency and Brian's consulting work. We have a grandson, Emerson Alexander born on March 10, 2007.

7.7 50-Year Friendship 50 年的友情

The real motivation for me to write this life-long story for the Yearbook is to share our collective wisdom with each other, our spouse, and our next generation in particular. Steve Jobs, the CEO of Apple Computer, told three stories from his life in his recent commencement speech at Stanford. These true life stories have a real meaning to me as I look back at my career, life, family and struggle. Similar stories were told again and again by our classmates in our previous gatherings over the last 20 years. Let's use Job's three stories as a way to reflect on our collective wisdom, career, life, family and struggle.

His first story was about how dropping out of college affected him. The bottom line was that he had found his lifework in the work that he loved. Stay put, you will be successful. His second story was about getting fired from Apple. It turned out that getting fired was the best thing that ever happened to him. The humiliation kept him humble and focused and allowed him the freedom to enter one of the most creative periods of his life. Otherwise, complacency leads to mediocrity. Always seize the opportunity and make the best out of it. His third story is about being diagnosed with cancer and facing death. It was a wake-up call for him. Life is fragile and health is important. The 50th anniversary reunion of our classmates is a good reminder for all of us. Life is short. Life should be fun, and what fun is life if you don't have good health.

W.F. (惠發) and Linlin (玲玲) in Los Angeles Class 33 High School 50th Reunion, June 2005.

台灣師大附中高卅三班畢業五十週年紀念
二 00 五年六月
(The High School of Taiwan Normal University)

半世紀的友情

Family reunion at Ja Ja Hsuan's wedding.: Brian (中宇), Christine, W.F. (惠發), Linlin (玲玲), Eric (中傑)， Lin (慧琳), and Arnold (中毅) in 2005 in Fremont, CA.

Left: from left—Brian (中宇), W.F. (惠發), Eric (中傑), Linlin (玲玲) (seat middle), and Arnold (中毅).
Right: standing—Brian (中宇), Eric (中傑), Jerome (中堯), Arnold (中毅), and Hollis Jr. (濤濤), Seated—Desiree (蕊), Betsy, and Melissa (美珊).

Eric（中傑）featured in Raytheon Company's diversity panel.

8

Academic Contributions
學術貢獻
From Theory to Practice
從理論到實踐

In my career spanning over 40 years, I have built a reputation for developing simpler solutions based on rigorous theories of engineering science. From my first work on bi-axially loaded columns, to limit analysis and soil plasticity, up to my current work on advanced analysis for structural design, I have operated on one basic principle – keep things simple with a rigorous theoretical basis.

My academic life may be divided into three stages:

1) Before the 30 years old, it is an educational stage. It is a period to learn the basic science and engineering preparing for a life-long engineering career.

2) From 30 to 70 years old, apply the fundamental theories of engineering science to engineering practice.

3) After 70 years old, it is time to look back, examine the big picture and share the experiences with others.

This is the purpose of the present chapter – a self assessment on my academic career in the areas of structural engineering and geo-mechanics over the last 40 years.

8.1 Background 背景

My life began on December 23, 1936 in Nanjing (南京), China but grow up in Taiwan from junior high to college. As evidenced by my career and college degrees, education played a very important part in my life. My early schooling took place in Taipei (台北), Taiwan. I later attended National Cheng-Kung University (成功大學) in Tainan (台南),

Taiwan（台灣）where I graduated in civil engineering with B.S. (1959). I continued my graduate education at Lehigh University at Bethlehem, Pennsylvania with M.S. in structural engineering (1963) and at Brown University at Providence, Rhode Island with a Ph.D. in solid mechanics (1966).

Among my life accomplishments, I was honored with distinguished alumni awards given by the National Cheng-Kung University (1988) and Brown University (1999). I received several major awards given by the American Society of Civil Engineers (ASCE). In 2003, I received the Lifetime Achievement Award given by the American Institute of Steel Construction (AISC).

One of my most prestigious honors was being elected to the US National Academy of Engineering (NAE) in 1995. Two year later, I was awarded honorary membership by the ASCE. The following year I was elected to the prestigious Taiwan's National Academy of Science formally called the Academia Sinica（中央研究院）.

In a 2003 UK publication, I was pleasantly surprised and honored to be featured in the biographical monograph *"Giants of Engineering Science"* as one of ten of the world leading engineering scientists in American (see Sec. 3.6, Chapter 3).

8.2 Solid Mechanics 固體力學

My specialty is in the areas of structural engineering and geo-mechanics. My formal engineering education is in structural engineering with a M.S. degree from Lehigh University; and a Ph.D. degree in solid mechanics from Brown University. My academic career has been with the traditional field of structural engineering in the department of civil engineering at Lehigh University (10 years), Purdue University (23 years), and most recently the University of Hawaii (more than 7 years). These universities provide traditional engineering technology educations for engineers working in engineering practice, or simply put, they are trained as professional engineers.

Structural engineering and geo-mechanics are branches of solid mechanics. Solid mechanics is a branch of applied mechanics involving a mathematical formulation of basic differential equations and their solutions. The field of applied mechanics is closely related to the field of applied mathematics which is under the division of engineering at Brown University. Brown, as Harvard and Cal Tech, provides engineering science education for engineers working in the area of research and development. Simply put, their education and training are more suitable for academic and research institutions. They are really engineering scientists, rather than professional engineers. So, I have a mixture of both engineering technology and engineering science education.

The mathematical formulation of any solid mechanics problems involves three basic conditions: physics, materials, and geometry. The basic physical condition is the equilibrium equations or motion; the basic material condition is the constitutive equations or stress-strain relations; while the basic geometric condition is the compatibility equations or kinematical assumptions.

Solutions for a rigorous mathematical formulation of these differential equations are seldom solvable, and drastic simplifications and idealizations are necessary for obtaining engineering solutions. The development of structural engineering and geo-mechanics over the last 40 years follows closely the advancement in material modeling and computing environment: ranging from slide rule to mainframe computing in earlier years to the evolution from PC computing to Grid computing in recent years.

In this article, I shall trace the history of the progress in simplifications and idealizations of these three basic conditions as used in any mathematical formulation in order to achieve better and more useful solutions for engineering applications. Mathematical formulations and their numerical calculations have advanced hand-in-hand as the computing environment and material modeling have evolved.

My life-long career involves the interaction of the advancement of these three areas: mechanics, materials and computing. I have been benefited much from the Brown University type of integrative engineering science education involving mechanics, materials, and computing. This will be described in the following.

8.3 My Research at a Glance 研究一覽

Over the last 40 years, remarkable developments have occurred in computer hardware and software. Advancement in computer technology has spurred the development of structural calculations ranging from the simple strength of materials approach in early years; to the finite element type of structural analysis for design in recent years; and to the modern development of scientific simulation and visualization for structural engineering problems in the years to come.

Such capabilities and advancements are allowing the solution of problems before thought *"unsolvable"*, and consequently, are now driving progress in a number of areas in civil engineering. This article traces the development in the areas of structural engineering and geo-mechanics. The topics covered in the following are those that I have been involved and have made significant contributions.

8.3.1. Steel Connections 鋼結點

Large full-scale tests on welded and bolted steel beam-to-column connections were carried out at Fritz Engineering Laboratory, Lehigh University under the leadership of late Professor L.S. Beedle. The doctoral students involved included D. J. Fielding (1971), J.S. Huang (1973) and G.P. Rentschler (1979). The test information provided, for the first time, a rational basis for safe and economical designs of steel building structures as coded in the AISC specifications. This knowledge has lessened building cost without sacrificing safety.

8.3.2. Beam-Columns 梁柱

The comprehensive research program was aimed to solving problems associated with the analysis and design of columns under biaxial loading. The program was under the leadership of Professor L.W. Lu (呂烈武) at Lehigh University. The doctoral students involved included S. Santathadaporn (1971), and T. Atsuta (1972).

The analysis methods developed were relatively simple and powerful emphasizing physical aspects of the problem while utilizing traditional structural engineering analysis concept. The solutions were accomplished by using high speed mainframe computers. The theoretical results provided, for the first time, a rational basis for safe and economical designs of wide flange and box columns in high rise steel buildings by the AISC-LRFD specifications.

The results were also formed as the basis for my two-volume classical book entitled *"Theory of Beam-Columns"* co-authored with T. Atsuta, McGraw-Hill, 1976 and 1977. We brought together and presented systematically for the first time the most comprehensive theory of beam-columns in elastic or plastic, in-plane or in-space, short or long, of steel or concrete.

Photo with Nanking Institute of Technology (南京工學院) faculty and staff members in Nanking, China for the 1981 lectures on beam-columns and limit analysis. Professor Ding Dajun (丁大鈞) (third from right) and W.F. (惠發) (sixth from right).

8.3.3. Limit Analysis 極限分析

The development of simple limit analysis techniques to soil mechanics was aimed for engineering practitioner concerned directly with soil and foundation problems. The graduate students involved in this work in the later years at Purdue University included doctoral students M.F. Chang (張明芳) (1981) and C.J. Chang (1981).

My first book on "*Limit Analysis and Soil Plasticity*" was published by Elsevier, 1976 while I was with Lehigh University. It brought together in a unified manner for the first time so much of what by then was only known only to a few in the field. It soon became a very popular one and also become a classic work on the subject area. The doctoral students contributed to this work included A.C.T. Chen (陳啓宗) (concrete plasticity, 1973), H.L. Davidson (critical state soil mechanics, 1974).

8.3.4. Soil and Concrete Plasticity 塑性力學

The development of nonlinear analysis methods were aimed to solving problems associated with large deformation inelastic media with weak tensile strength. The research began by developing powerful yet simple soil and concrete plasticity models for finite element analysis.

The doctoral students involved in this work included A.F. Saleeb (hypo- and hyper-elastic models, 1981), E. Mizuno (seismic modeling, 1981), S.S. Hsieh (謝錫興) (concrete modeling, 1981), D.J. Han (韓大健) (hardening and softening plasticity, 1984), W.O. McCarron (cap models, 1985), Y. Ohtani (hypo-elastic models, 1987), E. Yamaguchi (micro-crack modeling, 1987), and T.K. Huang (黃添坤) (soil plasticity, 1990).

They all contributed greatly to the subject areas by developing simple constitutive models and analysis methods for embankments, slope stability, foundations, retaining structures, concrete indentation problems, and micro-crack propagation of concrete materials. The research on micro-cracks was a breakthrough in understanding concrete material behavior used as a composite material consisting of coarse aggregates, sands, cements and cement-sand interface, among others.

The information on modeling, analysis methods, and numerical results also provided, for the first time, a rational basis for developing my two-volume treatise on *"Constitutive Equations for Engineering Materials"*, Volume One *"Elasticity and Modeling"* co-authored with A.F. Saleeb, John Wiley, 1982, and Volume Two *"Plasticity and Modeling"* in collaboration with W.O. McCarron, and E. Yamaguchi, Elsevier, 1994.

Photo with faculty and staff members for a series of seminars on plasticity, structural mechanics and engineering at Tsinghua University (清華), Beijing. Xila Liu (劉希拉) (fourth from left, W.F. (惠發) sixth, D. J. Han (韓大健), seventh, and S. P. Zhou (周綏平) eighth).

8.3.5. Advanced Analysis 高等分析

With today's computer technology, two aspects, the stability of separate members and the stability of the structure as a whole, can be treated rigorously for the determination of the maximum strength of the structures. This design approach uses directly the second-order inelastic analysis methods for steel design is called advanced analysis. The aim of the research is to develop practical theory and procedures that enable us to incorporate advanced analysis into current indirect AISC-LRFD analysis and design practice suitable for adoption in engineering practice.

The doctoral students contributed to this development included S. Toma (1980), I.S. Sohal (1986), W. S. King（金文森）(1990), L. Duan （段鍊）(1990), J.Y.R. Liew（劉德源）(1992), S.E. Kim (1996), I.H. Chen (1999), K. Wongkaew (2000) and K. Hwa（華根）(2003). My co-worker, Professor D. W. White at Purdue University helped conceive the research idea, planned the work, and found financial support. This research program has already been revolutionizing present method in designing steel building frames as reflected in the 2005 version of the newly released AISC-LRFD specifications.

The development of a practical theory that requires only simple modifications of an in-house elastic program with a modified tangent modulus together with a refined plastic hinge concept formed the basis of my book entitled "*LRFD Steel Design Using Advanced Analysis*" co-authored with S.E. Kim, CRC Press, 1997. A more traditional approach with second-order analysis was presented in the book "*Plastic Design and Second-Order Analysis of Steel Frames*" co-authored with I.S. Sohal, Springer-Verlag, 1995. The early companion book entitled "*Advanced Analysis of Steel Frames*" co-authored with S. Toma, CRC Press, 1994, provided a comprehensive background coverage of the second-order analysis: from elastic to inelastic, rigid to semi-rigid connections, and the simple plastic hinge method to the sophisticated plastic-zone method.

8.3.6. Semi-Rigid Frames 柔性框架

The steel framework is one of the most commonly used structural systems in modern construction. The analysis of such systems is governed by the assumptions adopted in the modeling of their structural elements, especially those concerning the behavior of beam-to-beam column connections. Conventional methods of steel frame analysis use two idealized connection models: the rigid-joint model and pinned-joint model. Because the actual frame connections always falls in between these two extremes, much attention has been focused in recent years toward a more accurate modeling of such connections. The recent research on structural connections has resulted in considerable progress and understanding of the subject that prompted changes in design provisions in the 1986 AISC LRFD specifications: Type PR (Partially Restrained) construction and Type FR (Fully Restrained) construction.

The aim of our research was to provide practical methods for the stability design of semi-rigid frames. The work began by collecting and developing connection data base, then by developing simple yet accurate connection models for analysis, and finally by developing practical analysis methods suitable for engineering practice. The doctoral students contributed to this work included E.M. Lui (呂汶) (1985), F.H. Wu（吳福祥）(1988), W.H. Yang（楊卫红）(1997), C.B. Joh (1998), and Y.S. Kim (1998). My co-workers, Professor Y. Goto and N. Kishi from Japan contributed greatly to several breakthroughs in achieving the goal.

The information on connection data basis, simple connection models, practical design procedures, and software development provided a rational basis for safe and economical designs of semi-rigid frames. They were documented systemically in the books entitled "Stability Design of Semi-Rigid Frames" co-authored with Y. Goto, J.Y.R. Liew（劉德源）, John Wiley, 1996, and more recently *Practical Analysis for Semi-Rigid Frame Design*" edited by W.F.Chen, World Scientific Publishing, 2000. The early companion book "*Stability Design of Steel Frames*" co-authored with E.M. Lui（呂汶）, CRC Press, 1992 provided a state-of-the-art summary on the theory and design of structural stability.

8.4 Slide Rule and Calculator Environment 計算尺

8.4.1 General Trend in Structural Engineering in the 1960-70's 趨勢

The basic approach during the slide rule and calculator period in structural engineering was known as the "*strength of materials approach*". The basic material model used was linear elasticity, while the basic kinematical or geometric condition used was the powerful assumption *"plane sections remain plane after bending"* for bar problem including columns, beams, and beam-columns. Mostly, only axial forces and bending moments could be considered. Similar concepts were later extended to plates and shells.

As a result of these simplifications, the equations of equilibrium were expressed in the simple form of generalized stress (stress resultant) and generalized strain (strain resultant), or simply put, in the form of one-dimensional moment and curvature relationships instead of a three-dimensional stress and strain relationships. The generalized stress was related to stress through the equilibrium condition, while the generalized strain was related to strain through the kinematical conditions. The generalized stress was then related to the generalized strain through a constitutive relation between basic internal forces and basic displacements.

Typical solutions for most of the strength of materials problems were solved either in closed form by series expansion, numerically by the finite-difference method, or approximately by the successive approximation method. Many of these well known classical solutions for beams, columns, beam-columns, plates and shells were well documented in a series of famous books by Timoshenko and his associates.

For high rise buildings, simple relations between generalized stresses and generalized strains were developed for structural members in the form of *"slope-deflection equations"* for structural system analysis. In engineering practice, a more powerful approximate method known as the *"moment distribution method"* was developed by Hardy Cross at the University of Illinois. It was substantively an iterative hand method based on the St. Venant principle where the moment distribution in a particular structural member of a high-rise building frame is predominately affected by the surrounding members adjacent to it.

For low rise buildings, a simple plastic theory was developed in which the material model used was elastic perfectly plastic. The kinematical assumption used was the concept of a *"plastic hinge"*. Upper- and lower-bound solutions were obtained by the mechanism method bounded above and by the equilibrium method with moment check bounded below.

Based on these simple and practical solution techniques, the traditional *"Allowable Stress Design Method"* and the newly developed *"Plastic Design Method"* were widely used in engineering practice in those early years. The environment for basic computing at the time was essentially based on slide rule and calculator. Simplicity was the key for a rapid and successful implementation of these methods for design of real world structural engineering problems.

8.4.2 Contributions to Limit Analysis in Soil Mechanics 貢獻

In the early 1960's, computers were in their infancy while the theory of plasticity was in its golden time. We had a beautiful theory but few practical solutions. We need simple methods for practical solutions. Nothing can be more practical than a simple theory. As a result the limit analysis methods were developed and widely applied to steel structures. When I returned to Lehigh University to teach, I started to apply these limit analysis techniques to soil mechanics and developed simple methods for obtaining practical solutions and wrote a book on *"Limit Analysis and Soil Plasticity"*. It became a very popular one and also a classical one.

In the following, I have selected some highlights of book reviews around the world to reflect on the excitement generated on the publication of this book.

APPLIED MECHANICS REVIEW by W. PRAGER, Switzerland, VOL. 29, MAY 1976

The exceedingly well-written treatise can serve not only as a valuable reference work on the use of limit analysis in soil mechanics, but also as an excellent text on limit analysis in general, even though the vast majority of examples is taken from soil mechanics.

APPLIED MECHANICS REVIEW by A. KEZDI, Hungary, VOL. 32, JANUARY 1979

It is gratifying to see how the author commands even the most difficult problems; he produced a clear and didactic treatment. The book will certainly serve to increase the application of limit analysis in practical work because the book, starting with the basic elements of a problem, shows the method of solution and ends with practical applications. It can be recommended to every civil engineer who is more or less engaged with soil mechanics.

THE CIVIL ENGINEER IN SOUTH AFRICA

In spite of attending and lecturing at three post-graduate soil mechanics courses I think it is fair to say that on reading this new volume I perceived for the first time what may well be termed an underlying unity of approach or a basic philosophy of understanding for analysis applicable to the whole sphere of the mechanics of soil *"structures"*.

Not only is the method of analysis simple, it is also the best way yet, to my knowledge, of visualizing and appreciating how failure can take place in a soil mass. And to this volume must go the credit that it succeeds admirably well in presenting and elucidating both by strict theory and many worked examples the whole methods.

This volume is likely to become a *classic* on the subject and standard required reading for all who take soil mechanics and engineering seriously enough to desire to keep abreast of current developments, curiously enough to desire to have freshly presented old, sometime tortuous derivations of familiar formulae. Indeed, this volume is definitely of interest to all engineers, not only geotechnical, who are seeking a simple and practical analytical method for solutions to problems in the mechanics of solids, but also to those who try to understand and perceive the philosophy of engineering and the essential unity of all physical processes.

GEODERMA by A.J. KOOLEN, Wageningen, VOL. 15, 1976

From reading this book it appears that the author has a near encyclopedic knowledge of the huge amount of scientific results on soil mechanics problems that have been published during the last two centuries. Many relations with these results are shown where the author explains the latest developments which have been taken place during the last 20 years. Only a few people have contributed to these developments and the aim of the book is to present in a logical and systematic manner the most important recent results to a large public.

This book may be thought of as being a "*milestone*" in the growing tendency by soil mechanics practitioners to adopt limit analysis, because it is the first single reference book on the subject. It can be recommended to anybody who is interested in the calculation of collapse loads. The book, being a good combination of "handbook" dealing with applications, and "textbook", will surely find its way into the various collections of soil mechanics books.

GEOTECHNIQUE by R.G.J., MARCH 1976

This book is a most welcome addition to the soil mechanics literature. It contains a wealth of material which is of great interest to all engineers concerned with the calculations of limit loads in civil engineering.

This book, although somewhat expensive for 630 pages, should nevertheless be enthusiastically received by the university teacher, the researcher, and the civil engineer who practice limit analysis.

CIVIL ENGINEERING, Canada, FEBRUARY 1977

Soil mechanics theory has been brought up to date with this fine reference book and it is highly recommended, despite the cost, for those enthusiasts amongst us. An enthusiast is perhaps a person who keeps a copy of Civil Engineering under the bed…. But maybe the definition of limit analysis is much simpler!

This book is a *landmark* in soil mechanics. ….. The more familiar the engineer gets with this reference book, the greater the value – it's worth ordering. Perhaps the housekeeping will stand the price!

8.5 Mainframe Computing Environment 主机

8.5.1 General Trend in Structural Engineering in the 1970-80's 趨勢

The basic approach used in mainframe computing period was known as the "*finite-element method*". The basic material model used was the extension from linear elasticity to inelasticity, or plasticity in particular. The basic kinematical or compatibility condition used for a finite-element formulation was known as the "*shape function*". The equilibrium condition was achieved through a weak format *"equation of virtual work"* instead of the usual free body equilibrium formulation.

As a result of these simplifications, the force displacement relation for a finite element was expressed in the form of the generalized stress and generalized strain relationship. This basic relationship for an element in a discrete continuum of a structural system is known as the *"nodal force and nodal displacement equation"*. The stresses in elements were related to the generalized stresses or nodal forces through the virtual work equation. Elemental strains were related to the generalized strains or nodal displacements through the kinematical assumption, or shape function. The incremental generalized stress and generalized strain relation for a finite element was then obtained through the constitutive equation of a particular material.

Photo with faculty and staff members at a Chungking (重慶) University during one of W.F. (惠發)'s lecture tours in China with D. J. Han (韓大健) (fourth from left), S.P. Zhou (周綏平), Linlin (玲玲) (sixth from left), Arnold (seventh) and W. F. (eighth) and Xila Liu (劉希拉) (second from right).

Structural engineering problems were thereby reduced to the solution of a set of simultaneous incremental equations for a structural system, with the primary unknowns usually being nodal displacements. Since the solution included the inelastic behavior of materials, which was load path dependent, the numerical scheme used was an incremental and iterative process. Many numerical procedures were developed during this period, most notably the tangent stiffness method.

Combining powerful mainframe computing with an effective numerical procedure, many previously unsolvable inelasticity problems were solved using new formulations, including among others the classical elastic-plastic solutions of bars, plates, and shells.

With these extensive numerical solutions so generated, simple design equations were then developed. Some were adapted and adopted by various code-writing bodies around the world. Typical literature on the collections of these solutions for structural members and frames were summarized in the "*Theory of Beam-Columns*" by Chen and Atsuta (1976, 1977) and "*Structural Stability Research Council Guide*" edited by T.V. Galambos, (1988).

With a large amount of numerical data so generated, it became necessary for engineers to use probability theory and reliability analysis to analyze the data and develop design procedures for practical implementation. As a result of this development, a new generation of codes based on reliability analysis, sometimes with an extensive computer simulation, was developed and adopted around the world.

For the first time in engineering practice, the load effect and the structural resistance effect were treated independently in design, each with its own safety or load factor. In the US, the "*Load and Resistance Factor Design Specifications for Steel Buildings*" was adopted by the American Institute of Steel Construction in 1986. It was a revolutionary development.

8.5.2 Contributions to Constitutive Modeling of Materials 貢獻

Constitutive equations are of central importance to structural and engineering mechanics analyses and the engineering design of structural system. Elastic behavior is well understood, but irreversible deformation is not. At best, only crude delineation of zones stressed beyond the elastic range is within the present state of the art. Constitutive equations for concrete-like materials beyond the elastic range represent an area of great important for finite element analysis.

With Dean Rodney Clifton after receiving Brown University's Distinguished Alumnus Medal in Providence, Rhode Island in 1999.

Classical plasticity theory using pressure dependent yield and associated flow rules enables one goes beyond the elastic range in a time-independent but theoretically consistent way. However, the classical theory requires the assumption of *"normality"* and associated flow rules for these types of engineering materials. Most of these materials are probably frictional to some degree, so that the classical plasticity theory must be modified in order to calculate the distribution of stress and the progress of post-elastic deformation which is micro-cracked and damaged.

To this end, we began by developing various expressions as one-through five- parameter failure criteria models for concrete instead of the traditional Mohr-Coulomb failure or yield criteria. In the elastic range, the range of reversible deformation, the constitutive equations are embodied in Hooke's law or some nonlinear elastic models. Combining the failure criteria with elastic models, prediction of acceptable accuracy can be made for some engineering applications in reinforced concrete analysis. This development also provided the basis for my first book on *"Plasticity in Reinforced Concrete"* McGraw-Hill, 1982.

Once a mathematically and physically attractive failure criterion has been established, the next step is to use the work-hardening theory to establish the stress-strain relation in the plastic range. To this end, Han and I developed a relatively sophisticated work-hardening model called *"a nonuniform hardening plasticity model for concrete materials"*. The important features of inelastic behavior of concrete including: brittle failure in tension, ductile behavior in compression, hydrostatic pressure sensitivity and volumetric dilation under compressive loading can all be represented by this refined constitutive model.

Since engineering materials such as concrete, rock and soil exhibit a strong-softening behavior in the post-peak stress range, showing a significant elastic-plastic coupling for the degradation of elastic modulus with increasing plastic deformation. Stress-space formulation of plasticity based on Drucker's stability postulate for these materials encounters difficulties in modeling the softening/elastic-plastic coupling behavior. Strain-space formulation is therefore necessary for further progress. In 1986, Han and I presented a consistent form of the constitutive relation for an elastic-plastic material with stiffness degradation in the range of work-hardening as well as strain-softening. This work was a breakthrough in understanding the concrete-like materials and provided the framework for modeling materials for finite element analysis.

Various analysis approaches to the estimate of stress, strain and displacement including analytical, numerical, physical and analog techniques are available, but the finite element technique is certainly the most versatile. In all cases, we must place special emphasis on constitutive equations in computing. To this end, the two-volume treatise on *"Constitutive Equations for Engineering Materials"* Elsevier, 1994 was published in collaboration with my former students Saleeb, McCarran and Yamaguchi to help engineers select reliable constitutive equations concerning post-elastic deformation and failure criteria with guidelines for the selection of proper parameters for their constitutive equations.

In the following, I have selected a few book reviews on the subject areas reflecting the timely publications of these books.

SOIL PLASTICITY: THEORY AND IMPLEMENTATION

BY W.F. CHEN AND G.Y. BALADI, ELSEVIER, 1985
APPLIED MECHANICS REVIEW, VOL.39, NO.9, SEPTEMBER 1986

Computers are becoming more accessible daily. The need for assistance from computer analysis techniques in designing geotechnical problems is a fact of life. The publication of this volume will greatly facilitate the appropriate use of the techniques.

As both an educator and a practicing soil engineer, this reviewer is….strongly recommends it to anyone interested in learning the basic of computer methods in soil plasticity.

PLASTICITY FOR STRUCTURAL ENGINEERS

BY W.F. CHEN AND D.J. HAN, SPRINGER, 1988
APPLIED MECHANICS REVIEW, VOL.42, NO.2, FEBRUARY 1989
Professors Chen and Han are to be congratulated for their quite unique contribution to the frequently intimidating field of plasticity for structural engineers. The book is very well written. It presents clearly the theoretical and computational aspects, treating metallic and non-metallic materials. The coverage of materials is very through without being burdensome to read.

The book must be a treasure in the bookshelf of those who seriously commit themselves in the study of structural plasticity. The book is worth the price right down to the last penny.

STRUCTURAL PLASTICITY: THEORY, PROBLEMS, AND CAE SOFTWARE

BY W.F. CHEN AND H. ZHANG, SPRINGER, 1991
APPLIED MECHANICS REVIEW, VOL.44, NO.8, AUGUST 1991
The book is written in a concise and logical manner. It can be recommended as an excellent teaching aid in various structural engineering courses, as well, and it will be of immediate practical assistance to engineers and researchers dealing with the problems of structural analysis and design.

8.6 Desktop Computing Environment 書桌机

8.6.1 Current Trend in Structural Engineering 趨勢

In recent years, we have entered into the desktop or free computing environment. This environment has fostered the development of large-scale simulations of structural systems over their life-cycle and has enabled computer simulation to join theory and experimentation as a third path for engineering design and performance evaluation. The development of model-based simulation for any civil engineering structures or facilities must involve the following four steps:

(1) *Modeling of Materials*

The constitutive equations for materials are now moving from time-independent to time-dependent behaviors such as creep, relaxation, temperature variation, and deterioration or aging. These equations must be developed by engineers on the basis of mechanics, physics, and materials science. In a numerical analysis of these materials in a structural system, the proper modeling of discontinuity and fracture or crack for tension-weak materials becomes increasing important.

(2) *Solution Algorithm*

For a realistic life-cycle simulation of constructed facilities, it is not uncommon for engineers to deal with the mathematical modeling of radically different scales, in time and/or space. The computational effort for different parts of a large structural system may be drastically different. For example, in the analysis of reinforced concrete bridge system under seismic loading, a macro scale is necessary to model the overall behavior of the structure-soil interaction. Yet, a micro scale is needed at a local level to trace crack initiation and propagation.

Computation efficiency can be achieved in this case by using parallel finite-element analyses for the structural system. Parallel macro and micro analyses can be performed by multiple machines, such as PC cluster systems. This computational method requires repartitioning of the domain during the course of the analysis, making the development of suitable interfaces, data communication tools, or central databases with different levels for different scales in time and in space is of critical importance.

As another example, the finite-element method is preferred in structural engineering and solid mechanics, while the finite-difference method is more commonly used in fluid mechanics. When dealing with structure-fluid interaction problems, as frequently encountered in offshore structural engineering, the development of suitable data translators or data communication tools is necessary in order to use existing codes, which are based on two different methodologies.

(3) *Software Development*

There are hundreds of software systems on the market to support software development. A software support system is a compatible set of tools, usually based on a specific software development methodology, which can be employed for several phases of software and operation.

The key to a domain-specific software development environment is software reuse. Software reuse enables the knowledge obtained from the solution of a particular problem to be accumulated and shared in the solution of other problems. If software components accumulated from previous software development can be utilized readily in the development of new applications, substantial applications can be built more efficiently. This is an ideal environment for university research and education. This idea was carried out and implemented, for the first time, at Purdue University with my colleagues, D.W. White and E. Sotelino, and doctoral students, H. Zhang（張宏）and J. Lu, among others (1994) with major financial supports from the National Science Foundation (NSF).

At present, the key to software development is software integration. Since most commercial software has its own particular function and input/output formats, it may prohibit direct data access. It seems very necessary to unify the documentation from different software and to make the newest and largest efforts in the development of standard models, such as Industry Foundation Class (IFC). Following the development of grid computing, the interoperability of facilities and software at different location in the network can be realized.

(4) *Visualization and Verification*

Modeling is science, simulation is computing, and computing requires solution algorithms and software development. Visualization is a necessary step to aid in the interpretation of the simulated results. Validation of the simulation of an engineering problem must be verified by experimental work.

Model-based simulation is inherently interdisciplinary in science and engineering, where computation plays the key role. The entire process of model-based simulation involves the following seven steps:

1. Experimental measurements as the basis for the development of relevant constitutive equations for a physical system;

2. Design of a proper algorithm for its numerical solutions;

3. Implementation of the procedures with necessary documentation and software interface development;

4. Selection of appropriate hardware to run the computer simulation of the physical system;

5. Validation of the computer model with physical testing;

6. Graphical visualization of the simulated results; and finally

7. Sharing of the simulation model with others through high speed network communication.

8.6.2 Contributions to Advanced Analysis 貢獻

With the desktop computing environment, we are now able to combine the theory of stability and the theory of plasticity and apply them directly to simulate a more realistic behavior of structural members and frames with great confidence. As a result, the structural system approach to design has been advanced and the new 2005 AISC LRFD specifications were issued to codify this new analysis procedure.

This marks the beginning of the modern development of structural design based on the second-order inelastic analysis, known as advanced analysis. The advanced analysis method has been proposed as an alternative design method for the conventional prescriptive design method, where member capacity equations are tied to effective length factors to estimate system strength. Performance-based design using direct simulation of structural system behavior under ultimate loads can now be achieved by advanced analysis.

Computer programs that perform second-order elastic analyses are readily available in engineering practice. In order to make the second-order inelastic analyses for frame design usable in engineering practice, simple modifications of the key factors in the elastic analysis can be made. These modifications include the following key behavioral effects of steel members: gradual yielding associated with residual stresses and flexure; and geometric imperfections. To meet the current AISC LRFD requirements, these modifications can be calibrated against the current AISC LRFD specifications. This was described in my book *"LRFD Design Using Advanced Analysis"* co-authored with S.E. Kim (1997).

Once these simple modifications are made, the complex second-order inelastic analysis problem is now reduced to a second-order elastic analysis problem. Thus, the existing computer programs for a second-order elastic analysis can be used for the structural system design without the use of the effective length factor K as required in the current design practice. This process has made the advanced analysis method work in engineering practice, since it only requires the use of existing second-order elastic analysis software.

The practical advanced analysis methods as codified by the 2005 AISC Specifications will encourage designers to use more accurate analysis methods and computer programs. The advanced analysis can be extended to include semi-rigid connections as separate structural elements in a steel frame design.

Advanced analysis can more accurately predict possible failure modes of a structure, exhibit a more uniform level of safety, and provide a better long-term serviceability and maintainability. It is the state-of-the-art design method for the structural engineering profession for years to come.

This advancement is well reflected in the following book review.

STABILITY DESIGN OF STEEL FRAMES

BY W.F. CHEN AND E.M. LUI, CRC PRESS, 2000
CANDIAN JOURNAL OF CIVIL ENGINEERING, VOL.21, NO.171, 1994

This book is to provide a concise encapsulation of the extensive amount of research that has been conducted over the last 20 years on the analysis and design of steel members and frames. During that time, there have been prodigious advances in computer hardware and in structural analysis and design software, and the limit states (or load and resistance factor) design philosophy has been adopted in North America. These advances are reflected in the content.

....the AISC specifications are referenced almost exclusively. Nevertheless, because of the fundamental nature of much of the subject matter, the book should be a valuable reference in structural stability courses and the senior structural steel designers.

8.7 Future Trends in Structural Computing 前瞻

Advancements in computer technology in recent years have spurred the development of scientific simulation and visualization for problems, which traditionally have been addressed via experimentation and theoretical models. Such capacities are allowing the solution of problems previously thought "*unsolvable*", and consequently, are now driving progress in a number of areas in civil engineering.

Simulation plays an increasingly critical role in all areas of science and engineering. Exciting examples of these simulations are occurring in areas such as automotive crash-worthiness for component design in the auto industry; and system design and manufacturing in aerospace, among others. The benefits of the development of the advanced computing capabilities in simulating real world structural engineering problems include:

o　Enabling engineers to more thoroughly explore the design space at a much lower cost
o　Providing an efficient tool for engineers to provide adequate safety while reducing project life-cycle costs
o　Helping engineers maximize the value of their physical tests with computer simulations
o　Providing engineers with greater opportunity for creativity and innovation

With the current rapid progress in computing, it is possible now to realistically simulate the structural behavior over the life span of a structural system, including planning, design, construction, service, deterioration, rehabilitation, and demolition. The only limitations are the lack of similar progress in theory to describe the deterioration of structural materials under various environmental conditions. There is an urgent need to develop the fundamental science of deterioration and application thereof to develop practical constitutive equations for engineering materials. Good progress has been made in recent years, but much more remains to be done.

8.8 Rules for Good Engineering Practice 金科玉律

I would like to share with you, the reader, my three simple cardinal rules for a good practice of structural engineering:

Cardinal Rule One: Ductility can be forgiving of one's mistakes.

Be wary when you deal with a new environment with a familiar material but with possibly a much less ductility due to size effect, temperature change, damage due to cold work or the new way of making steels. For example, structural steel is considered to be a homogeneous, isotropic, and ductile material, but none of these properties is true at a fully welded beam-to-column joint with thick flange plates.

Cardinal Rule Two: Connection detailing is everything.

Most structural failures initiated from the connections as a result of some poor detailing in joints. Structural members seldom fail in a sudden manner but joints fail frequently in brittle fracture. Extensive computation and modeling may not be helpful because in-situ material properties and residual stresses are not known at these joints. Simple design rules developed on the basis of full scale tests are more relevant and result in safer joints with lower labor and material costs.

Cardinal Rule Three: Redundancies are the best defense against unexpected failures.

Good structural engineers need to look ahead and anticipate unexpected events to guard against such progressive failures as witnessed in the World Trade Center collapse. Redundancies can also compensate for some undetected quality control problems that may occur during construction.

In summary, the true fulfillment for engineering is to see one's work find its way into engineering practice. That is to be "*a place in practice*". To achieve this goal, one must always keep in mind: *make things simple*.

References
參考文

1. Chen W.F. (1975). Limit Analysis and Soil Plasticity, Elsevier, Amsterdam, 638 pp.
2. Chen W.F. (1982, 1994). Constitutive Equations for Engineering Materials, Vol. 1 - Elasticity and Modeling, Wiley Inter-science, New York, 580 pp. (with A.F. Saleeb). Vol. 2 -Plasticity and Modeling, Elsevier, Amsterdam, 1096 pp.
3. Chen W.F. (1982). Plasticity in Reinforced Concrete, McGraw-Hill, New York, 474 pp.
4. Chen W.F. and Baladi G.Y. (1985). Soil Plasticity: Theory and Implementation, Elsevier, Amsterdam, 231 pp.
5. Chen W.F. and Lui E.M. (1987). Structural Stability: Theory and Implementation, Elsevier, New York, 486 pp.
6. Chen W.F. and Han D.J. (1988). Plasticity for Structural Engineers, Springer-Verlag, New York, 600 pp.
7. Chen W.F. and Zhang H. (1990). Structural Plasticity: Theory, Problems and CAE Software, Springer-Verlag, New York,. 250 pp.
8. Chen W.F. and Liu X.L. (1990). Limit Analysis in Soil Mechanics, Elsevier, Amsterdam, 477 pp.
9. Chen W.F. and Mizuno E. (1990). Nonlinear Analysis in Soil Mechanics, Elsevier, Amsterdam, 661 pp.
10. Chen W.F. and Lui E.M. (1991). Stability Design of Steel Frames, CRC Press, Boca Raton, Florida, 1991, 380 pp.
11. ASCE (1971). American Society of Civil Engineers Manual 41, Plastic Design in Steel: A Guide and Commentary, ASCE, New York, 336 pp.
12. AISC (1986, 2005). The Load and Resistance Factor Design Specification for Structural Steel Buildings, American Institute of Steel Construction, Chicago.
13. Chen W. F. and Atsuta T. (1976). Theory of Beam-Columns, Vol. 1 - In-Plane Behavior and Design, McGraw-Hill, New York, 513 pp.
14. Chen W. F. and Atsuta T. (1977). Theory of Beam-Columns, Vol. 2 - Space Behavior and Design, McGraw-Hill, New York, 732 pp.
15. Chen W. F., Editor (1993). Semi-Rigid Connections in Steel Frames, Council on Tall Buildings and Urban Habitat, McGraw-Hill, New York, 318 pp.

16. Chen W. F. and Toma S. (1994). Advanced Analysis of Steel Frames, CRC Press, Boca Raton, Florida, 384 pp.

17. Chen W. F. and Sohal I. (1995). Plastic Design and Second-Order Analysis of Steel Frames, Springer-Verlag, New York, 509 pp.

18. Chen W. F. and Kim S. E. (1997). LRFD Steel Design Using Advanced Analysis, CRC Press, Boca Raton, Florida, 448 pp.

19. Chen W. F., Goto Y., and Liew J. Y. R. (1996). Stability Design of Semi-Rigid Frames, John Wiley & Sons, New York, 468 pp.

20. Chen W. F. and Kim Y. S. (1998). Practical Analysis for Partially Restrained Frame Design, Structural Stability Research Council, Lehigh University, Bethlehem, Pennsylvania, 82 pp.

21. Chen W. F., Editor (2000). Practical Analysis for Semi-Rigid Frame Design, World Scientific Publishing Co., Singapore, 465 pp.

22. Chen W. F. and Han D. J. (1988). Plasticity for Structural Engineers, Springer-Verlag, New York, 606 pp.

23. Chong K. P., Saigal S., Thynell S., and Morgan H., Editors (2002). Modeling and Simulation-Based Life-Cycle Engineering, Spon Press, London, UK, 348 pp.

24. Galambos T. V., Editor (1988). Guide to Stability Design Criteria for Metal Structures, 4th edition John Wiley & Sons, New York.

25. Gere J. M. (1962). Moment Distribution, Van Nostrand, Princeton, New Jersey.

26. Montero P., Chong K. P., Larsen-Basse J. and Komvopoulos K. (2001). Long-Term Durability of Structural Materials, Elsevier, Netherlands, 296 pp.

27. Timoshenko S. P. (1953). History of the Strength of Materials, McGraw-Hill, New York.

28. Timoshenko S. P. and Gere J. M. (1961). Theory of Elastic Stability, McGraw-Hill, New York.

29. Timoshenko S. P. and Goodier J. N. (1951). Theory of Elasticity, McGraw-Hill, New York.

30. Timoshenko S. P. and Young D. H. (1962). Elements of Strength of Materials, Van Nostrand, Princeton, New Jersey.

31. White D. W. and Chen W. F. (1993). Plastic Hinge Based Methods for Advanced Analysis and Design of Steel Frames: An Assessment of the State-of-the-Art, Structural Stability Research Council, Bethlehem, Pennsylvania, 299 pp.

32. Tebedge N. and Chen W.F. (1974). Design Criteria for H-Columns under Biaxial Loading, Journal of the Structural Division, ASCE, Vol. 100, No. ST3, March, 579-598.

33. Zhou S.P. and Chen W.F. (1985). Design Criteria for Box-Columns under Biaxial Loading, Journal of Structural Engineering, ASCE, Vol. 111, No. 12, December, 2643-2658.

Appendix 8.1
Books and Contributed Chapters
書篇專集

Dr. Chen was the author or co-author of 19 books, edited 23 books, and contributed 41 chapters to other books and special publications.

Books (Total 19)

1. Limit Analysis and Soil Plasticity, Elsevier, Amsterdam, 1975, 638 pp.
2. Theory of Beam-Columns, Vol. 1 - In-Plane Behavior and Design, McGraw-Hill, New York, 1976, 513 pp. (with T. Atsuta).
3. Theory of Beam-Columns, Vol. 2 - Space Behavior and Design, McGraw-Hill, New York, 1977, 732 pp. (with T. Atsuta).
4. Plasticity in Reinforced Concrete, McGraw-Hill, New York, 1982, 474pp.
5. Constitutive Equations for Engineering Materials, Vol. 1 - Elasticity and Modeling, Wiley Inter-Science, New York, 1982, 580 pp. (with A.F. Saleeb). Vol. 2 - Plasticity and Modeling, Elsevier, Amsterdam, 1994, 1096 pp. (in 2 Volumes).
6. Tubular Members in Offshore Structures, Pitman, London, 1985, 271 pp. (with D.J. Han).
7. Soil Plasticity: Theory and Implementation, Elsevier, Amsterdam, 1985, 231 pp. (with G.Y. Baladi).
8. Structural Stability: Theory and Implementation, Elsevier, New York, 1987. 486 pp. (with E.M. Lui).
9. Plasticity for Structural Engineers, Springer-Verlag, New York, 1988. 600 pp. (with D.J. Han).
10. Structural Plasticity: Theory, Problems and CAE Software, Springer-Verlag, New York, 1990. 250 pp. (with H. Zhang).
11. Limit Analysis in Soil Mechanics, Elsevier, Amsterdam, 1990. 477 pp. (with X.L. Liu).
12. Nonlinear Analysis in Soil Mechanics, Elsevier, Amsterdam, 1990. 661 pp. (with E. Mizuno).
13. Stability Design of Steel Frames, CRC Press, Boca Raton, Florida, 1991, 380 pp. (with E. M. Lui).
14. Concrete Buildings: Analysis for Safe Construction, CRC Press, Boca Raton, Florida, 1991, 260 pp. (with K. H. Mosallam).
15. Plastic Design and Second-Order Analysis of Steel Frames, Springer-Verlag, New York, 1995, 509 pp. (with I. Sohal).
16. Stability Design of Semi-Rigid Frames, John Wiley and Sons, New York, 1996, 468 pp. (with Y. Goto and J.Y.R. Liew).
17. Analysis and Software of Cylindrical Members, CRC Press, Boca Raton, Florida, 1996, 309 pp. (with S. Toma).

18. LRFD Steel Design Using Advanced Analysis, CRC Press, Boca Raton, Florida, 1997, 448 pp. (with S.E. Kim).
19. Practical Analysis for Partially Restrained Frame Design, Structural Stability Research Council, Lehigh University, Bethlehem, PA, 1998, 82 pp. (with Y.S. Kim).

Edited Books (Total 23)

1. Beam-To-Column Building Connections: State-of-the-Art, ASCE Portland Convention, April 14-18, Preprint Volume No. 80-179, ASCE Publication, New York, 1980, 260 pp.
2. Fracture in Concrete, Proceedings of a Session at the ASCE Hollywood Convention, October 27-31, Florida, ASCE, New York, 1980, 110 pp. (with E.C. Ting).
3. Recent Advances in Engineering Mechanics and Their Impact on Civil Engineering Practice, Two Volumes, ASCE, New York, 1983, 1326 pp. (with A.D.M. Lewis).
4. Connection Flexibility and Steel Frames, Proceedings of a Session at the ASCE Detroit Convention, October 21-25, ASCE Publication, No. 482, New York, 1985, 122 pp.
5. Joint Flexibility in Steel Frames, Journal of Constructional Steel Research, Special Issue, Vol. 8, Elsevier Applied Science, London, 1987, 290 pp.
6. Steel Beam-To-Column Building Connections, Journal of Constructional Steel Research, Special Issue, Vol. 10, Elsevier Applied Science, London, 1988, 482 pp.
7. Cylindrical Members in Offshore Structures, Journal of Thin-Walled Structures, Special Issue, Vol. 7, Elsevier Applied Science, London, 1988, pp. 153-285, 132 pp. (with I. S. Sohal).
8. Semi-Rigid Connections in Steel Frames, Council on Tall Buildings and Urban Habitat, McGraw-Hill, New York, 1993, 318 pp.
9. Plastic Hinge Based Methods for Advanced Analysis and Design of Steel Frames: An Assessment of the State-of-the-Art, Structural Stability Research Council, Bethlehem, PA, 1993, 299 pp. (with D.W. White).
10. Advanced Analysis of Steel Frames, CRC Press, Boca Raton, Florida, 1994, 384 pp. (with S. Toma).
11. The Civil Engineering Handbook, CRC Press, Boca Raton, Florida, 1995, 2609 pp. Second Edition, August 2002, 2800pp. (with J. Y. Richard Liew).
12. The Handbook of Structural Engineering, CRC Press, Boca Raton, Florida, 1997, 1600 pp. Second Edition, February 2005, 1786 pp. (with Eric Lui).
13. Innovations in Stability Concepts and Methods for Seismic Design in Structural Steel, Proceedings, US – Japan Seminar, Special Issues of Engineering Structures, Vol. 20, Nos. 4-6, April-June 1998, D.W. White and W. F Chen, editors, 569 pp.
14. Bridge Engineering Handbook, CRC Press, Boca Raton, Florida, October, 2000, 1600pp. (with L. Duan).
15. Structural Steel Design: LRFD Method, Science/Technology Book Company, Taiwan, 1999, 268pp. (with W.S. King, S.P. Zhou and L. Duan). (In Chinese).
16. The Earthquake Engineering Handbook, CRC Press, Boca Raton, Florida, September 2002, 1376pp. (with C. Scawthorn).
17. Practical Analysis for Semi-Rigid Frame Design, World Scientific Publishing Co., Singapore, 2000, 465pp.

18. High Performance Computing, Guest Editors, W. F. Chen and J. Y. Richard Liew, Special Issue of Computer-Aided Civil and Infrastructure Engineering, Vol. 16, No. 5, September, 2001, 78 pp.
19. Bridge Engineering: Substructure Design, CRC Press, Boca Raton, Florida, 2003, 186pp. (with L Duan).
20. Bridge Engineering: Seismic Design, CRC Press, Boca Raton, Florida, 2003, 456pp. (with L. Duan).
21. Bridge Engineering: Construction and Maintenance, CRC Press, Boca Raton, Florida, 2003, 236pp. (with L. Duan).
22. Earthquake Engineering for Structural Design, CRC Press & Taylor & Francis Group, Baca Raton, Florida, January 2006, 264pp. (with E.M. Lui).
23. Principles of Structural Design, CRC Press & Taylor & Francis Group, Boca Raton, Florida, January 2006, 528pp. (with Eric M. Lui).

Contributing Chapters (Total 41)

1. Plastic Design in Steel, Chapter 6, ASCE Manual 41, ASCE, New York, 1971, pp. 61-84.
2. Connections, Chapter 7 in ASCE Monograph on Structural Design of Tall Steel Buildings, L. S. Beedle, Editor-In-Chief, ASCE, New York, 1978, pp. 485-575.
3. Beam-Columns, Chapter 8 in SSRC Guide to Stability Design Criteria for Metal Structures, 3rd Edition, B.G. Johnston, Editor, Wiley-Interscience, New York, 1976, pp. 189-226.
4. Constitutive Relations and Failure Theories, Chapter 2, In Finite Element Analysis of Reinforced Concrete Structures," ASCE Special Publication, New York, 1981, pp. 34-148. (with Z.P. Bazant, D. Darwin, O. Buyukozturk, K.J. Willam et al.).
5. Box and Cylindrical Columns Under Biaxial Bending, Chapter 2 in Vol. 1 - Axially Compressed Structures, in Developments in the Stability and Strength of Structures, R. Narayanan, Editor, Applied Science Publishers, London, 1982, pp. 83-127.
6. The Continuum Theory of Rock Mechanics, Chapter 3 in Mechanics of Oil Shale, K.P. Chong and J.W. Smith, Co-Editors, Applied Science Publishers, London, 1984, pp. 71-126.
7. Constitutive Modeling in Soil Mechanics, Chapter 5 in the book on Mechanics of Engineering Materials, C.S. Desai and R.H. Gallagher, Co-Editors, John Wiley & Sons, London, 1984, pp. 91-120.
8. Soil Mechanics, Plasticity and Landslides, A Chapter in the Special Anniversary Volume on Mechanics of Material Behavior to Honor Dean Daniel C. Drucker, G. J. Dvorak, and R. T. Shield Co-Editors, 1984, Elsevier Science Publishers, B.V., Amsterdam, pp. 31-58.
9. Buckling and Cyclic Inelastic Analysis of Steel Tubular Beam-Columns, A Special Paper in Honor of Professor Dr. Bruno Thurlimann in the Special Issue of the Journal of the Engineering Structures, Vol. 5, IPC Science and Technology Press, April, London, 1982, pp. 119-132. (with D.J. Han).
10. Beam-To-Column Moment-Resisting Connections, Chapter 6 in Vol. 4 -Steel Framed Structures (Stability and Strength), in Developments in the Stability and Strength of Structures, R. Narayanan, Editor, Applied Science Publishers, London,

1985, pp. 153-203. (with E.M. Lui).

11. Fabricated Tubular Columns Used in Offshore Structures, Chapter 5 in Vol. 5 - Shell Structures (Stability and Strength), in Developments in the Stability and Strength of Structures, R. Narayanan, Editor, Applied Science Publishers, London, 1985, pp. 137-184. (with H. Sugimoto).

12. Constitutive Relations for Concrete, Rock and Soils: Discusser's Report Chapter 5, In Mechanics of Geomaterials, Z. Bazant, Editor, John Wiley & Sons, New York, 1985, pp. 65-86.

13. Plasticity Modeling and Its Applications to Geomechanics, In Recent Developments in Laboratory and Field Tests and Analysis of Geotechnical Problems, A. S. Balasubramaniam/S. Chandra/D.T. Bergado, Editors, A.A. Balkema, Rotterdam, 1986, pp. 391-426. (with E. Mizuno).

14. Modeling of Soils and Rocks Based on Concepts of Plasticity, In Recent Developments in Laboratory and Field Tests and Analysis of Geotechnical Problems, A. S. Balasubramaniam/S. Chandra/D.T. Bergado, Editors, A.A. Balkema, Rotterdam, 1986, pp. 467-510. (with W.O. McCarron).

15. Analysis of Steel Frames with Flexible Joints, Chapter 10, In Structural Connections: Stability and Strength, R. Narayanan, Editor, Applied Science Publishers, London, 1989, pp. 335-444.

16. Bearing Capacity of Shallow Foundations, Chapter 4, In Foundation Engineering Handbook, H.Y. Fang, Editor, 2nd Edition, Van Nostrand Reinhold, New York, 1991, pp. 144-165. (with W.O. McCarron).

17. Beam-Columns, North America Chapter Editor, Chapter 5, In Stability of Metal Structures - A World View, L. S. Beedle, Editor-in-Chief, Structural Stability Research Council, Lehigh University, Bethlehem, PA, 1991, pp. 318-330, 358, and 370-376.

18. Beam-Columns, Chapter 2.5, In Constructional Steel Design: An International Guide, P.J. Dowling, J.G. Harding and R. Bjorhovde, Editors, Elsevier Science Publishers, London, 1992, pp. 105-132. (with D.W. White and R. Liew).

19. Introduction, Plastic Hinge Based Methods for Advanced Analysis and Design of Steel Frames: An Assessment of the State-of-the-Art, Structural Stability Research Council, Lehigh University, Bethlehem, PA, 1993, pp. 1-22. (with D. W. White).

20. Toward Advanced Analysis in LRFD, Plastic Hinge Based Methods for Advanced Analysis and Design of Steel Frames: An Assessment of the State-of-the-Art, Structural Stability Research Council, Lehigh University, 1993, pp. 95-173. (with D. W. White and J. Y. R. Liew).

21. Design of Beam-Columns in Steel Frames in the United States, Special Issue, Thin-Walled Structures, Vol. 13, Nos. 1 & 2, 1992, pp. 1-83.

22. Analysis of Members in Semi-Rigid Unbraced (Sway) Frames, Chapter 4, In Semi-Rigid Connections in Steel Frames, L. S. Beedle, Editor-in-Chief, McGraw-Hill, New York, 1993, pp. 73-100. (with E. M. Lui).

23. Creating Design Application Models from Historical Experimental Database, Appendix A, In Semi-Rigid Connections in Steel Frames, L. S. Beedle, Editor-in-Chief, McGraw-Hill, New York, 1993, pp. 211-232. (with N. Kishi).

24. Moment-Rotation Relationship of Semi-Rigid Steel Beam-to-Column Connections,

Appendix B, In Semi-Rigid Connections in Steel Frames, L. S. Beedle, Editor-in-Chief, McGraw-Hill, New York, 1993, pp. 233-268. (with F. H. Wu).

25. Constitutive Models, Chapter 2, In Finite Element Analysis of Reinforced Concrete Structures II, J. Isenberg, Editor, ASCE Proceeding Publication, New York, 1993, pp. 36-117. (with E. Yamaguchi, M. D. Kotsovos and A.D. Pan).

26. Trends Toward Advanced Analysis, Chapter 1, In Advanced Analysis of Steel Frames, W. F. Chen and S. Toma, Editors, CRC Press, Boca Raton, Florida, 1994, pp.1-45. (with J. Y. R. Liew).

27. Second-Order Plastic Hinge Analysis of Frames, Chapter 4, In Advanced Analysis of Steel Frames, W. F. Chen and S. Toma, Editors, CRC Press, Boca Raton, Florida, 1994, pp. 139-194. (with J. Y. R. Liew).

28. Plastic-Zone Analysis of Beam-Columns and Portal Frames, Chapter 5, In Advanced Analysis of Steel Frames, W. F. Chen and S. Toma, Editors, CRC Press, Boca Raton, Florida, 1994, pp.195-258. (with S. P. Zhou).

29. Benchmark Problems and Solutions, Chapter 7, In Advanced Analysis of Steel Frames, W. F. Chen and S. Toma, Editors, CRC Press, Boca Raton, Florida, 1994, pp. 321-371. (with S. Toma).

30. LRFD-Limit Design of Frames, Chapter 6, In Steel Design Handbook - LRFD Method, A. R. Tamboli, Editor, McGraw-Hill, New York, 1997, pp. 6-1 to 6-83. (with J.Y.R. Liew).

31. Bridge Structures, Chapter 10, In Handbook of Structural Engineering, W.F. Chen, Editor-in-Chief, CRC Press, Boca Raton, Florida, 1997, pp. 10-1 to 10-103. (with S. Toma and L. Duan).

32. Multistory Frame Structures, Chapter 12, In Handbook of Structural Engineering, W.F. Chen, Editor-in-Chief, CRC Press, Boca Raton, Florida, 1997, pp. 12-1 to 12-73. (with J.Y.R. Liew and T. Balendra).

33. Effective Length Factors of Compression Members, Chapter 17, In Handbook of Structural Engineering, W.F. Chen, Editor-in-Chief, CRC Press, Boca Raton, Florida, 1997, pp. 17-1 to 17-52. (with L. Duan).

34. An Innovative Design for Steel Frame Using Advanced Analysis, Chapter 28, In Handbook on Structural Engineering, W.F. Chen, Editor-in-Chief, CRC Press, Boca Raton, Florida, 1997, 28-1 to 28-56. (with S.E. Kim).

35. Effective Length of Compression Members, Chapter 52, In Bridge Engineering Handbook, W. F. Chen and L. Duan, Editors, CRC Press, Boca Raton, Florida, 2000. (with L. Duan).

36. Impact Effect of Moving Vehicles, Chapter 56, In Bridge Engineering Handbook, W. F. Chen and L. Duan, Editors, CRC Press, Boca Raton, Florida, 2000. (with L. Duan).

37. Bridges, Chapter 18, In Earthquake Engineering Handbook, W. F. Chen and Charles Scawthorn, CRC Press, Boca Raton, Florida, 2002. (with L. Duan).

38. Steel Frame Design Using Advanced Analysis, Chapter 5, In Handbook of Structural Engineering, Second Edition, W.F.Chen and Eric M. Lui, Editors, CRC Press, Boca Raton, Florida, 2005. (with S.E. Kim).

39. Seismic Design of Bridges, Chapter 20, In Handbook of Structural Engineering, Second Edition, W.F.Chen and Eric M. Lui, Editors, CRC Press, Boca Raton,

Florida, 2005. (with L. Duan, M. Reno, S. Unjoh).

40. Bridge Structures, Chapter 25, In Handbook of Structural Engineering, Second Edition, W.F.Chen and Eric M. Lui, Editors, CRC Press, Boca Raton, Florida, 2005. (S. Toma and L. Duan).

41. Effective Length Factors for Compression Members, Chapter 31, In Handbook of Structural Engineering, Second Edition, W.F.Chen and Eric M. Lui, Editors, CRC Press, Boca Raton, Florida, 2005. (with L. Duan)

9

Life in Rapidly Changing Times
生活在急變中的世界
Engineers in Transition
工程師日新月異

As a dean, I was in charge of every convocation ceremony for the College of Engineering. To the graduates, I always started with congratulations on successfully completing their course of study. They were happy to hear that their long study sessions, many homework problems, laboratories, and reports had been completed. But then I reminded them that they were not quite through yet. I told them that in order to compete successfully in world markets, they would need to learn continually in order to keep up.

Since 21st century is the engineered century and we are at a particularly exciting time. We are fortunate because at this time the industrial world is making the transition from the industrial age to the high technology age. The high technology age will be just that – an age of technology. It is a time of change and indeed it is a time for change. This chapter, based on several speeches I gave at major conferences and lectures, provided my perspective on high tech, new economy and career flexibility in rapidly changing times.

9.1 High Tech and New Economy 高科技與新經濟

We use computers and communications gear to work smarter, more efficiently through the economy to increase our *"productivity"*. After World War II, according to available statistics, productivity increased 2.7 percent a year for 25 years. From 1995, it has been growing at 3 percent a year. This increase has to do with information technology. Will the current productivity growth hold up? Yes, based on just current technology, we have only implemented 20 to 30 percent level what the technology can do for business, if workers and consumers are willing to change their ways of doing business. This is known as *"new economy"*. In the new economy, we know that technology changes rapidly, but people do not. The most important thing is not technology. It is the thinking and redesigning of one's business.

The reasons for rapid changes are due to computers and communications. There are three key developments for communications toward the 21st century:

1. Internet: connect more people together and around the world.
2. Wireless: free people from local physical infrastructure—from wireless local network, over mobile phones, to satellite communications.
3. Multimedia: provide broadband contents—graphics, video, music, and voice.

As a result, science and engineering are in transition with two major elements:

1. Increasing globalization of R&D—contribute to stronger economy.
2. Prolific growth of Information Technology (IT)—impact on all facets of society.

On a recent survey made by the American Society of Engineering Education to its members, the first question was *"What are the three most significant issues facing engineers today?"* The second question was *"What will be the three most significant issues facing engineers in the next ten years?"* The majority of members identified the following three most significant issues for both questions:

1. Engineering education in the 21st century.
2. Engineering and public policy.
3. Managing complexity/information explosion.

As for the engineering education in the 21st century, four key issues were identified:

1. Lifelong learning.
2. Increasing interdisciplinary nature of engineering.
3. Preparing at the K-12 level for engineering careers.
4. Teaching and learning engineering fundamentals.

As for the engineering and public policy, three issues were identified:

1. Communications between engineers and the public.
2. Involved in public policy decisions.
3. Support for R&D.

As for the third issue on managing complexity information explosion, four issues were identified:

1. Implication for education.
2. Need for systems/multidisciplinary approach in engineering.
3. Effect of globalization.
4. Industry/university partnerships.

In summary, new economy is a knowledge-based economy, education is the key element. Buying computers and software is easy, but rethinking and redesigning the way we do business is not. Technology changes rapidly, but people do not, and we must properly educate and prepare our workforce for the new economy. We must address to our fundamental issue of K-12 educational problem, and build a first-rate engineering and science program in the universities. We should not worry about our job and career since we are part of the exciting and rapid changing times. We should enjoy the ride and tell the future generation about the once-in-a-lifetime experience. Let's make the most of it.

9.2 Career Flexibility in Rapidly Changing Times 事業的柔軟性

Lecture delivered at Taiwan Bureau of High Speed Rail（台灣高速鐵路局）, Taipei, Taiwan, June 28, 2006.
The full length speech was published in the *Magazine of the Chinese Institute of Civil and Hydraulic Engineering*, Vol. 33, No.4, August 2006 Taiwan.（土木水利雜誌，33 卷 第 4 期）

Title: Academic Achievements and Contributions of Academician Chen

9.2.1 Introduction 介紹

Thank you very much Director Wu（吳局長）.
I'm honored to be here today to deliver my lecture. Before I go any further, I want to assure you that I fully understand that the purpose of

my lecture is inspiration and not to promote myself.

About a few months ago, I was contacted by Director Wu to speak to you today. At that time, he said that I could choose the topic and to make it easy on myself. No pressure. Then on Thursday, June 15, he sent me an email in which he said and I quote:

"The topic will be: The academic achievements and contributions of Academician Chen. The lecture will last for 2 hours in a conference hall for nearly two hundred.... I wonder if this arrangement is OK for you." Wow!!! "…..My academic achievements and contributions to engineering….." for two hours!

I recall that a famous Chinese scholar, Mr. Lin Yu-Tung (林語堂), once said that *"a good speech is like a girl's skirt, the shorter the better."* I promise that I will make my lecture short, to the point, and enjoyable; but not too short beyond what Director Wu would allow me to do.

I'm not quite sure how much of an impact my work may have on our civil engineering profession in general and engineering education in particular. However, I will provide you with a few of my personal thoughts, experiences, and perspectives. I hope some of these may help you in your professional career. I'll leave it to each of you to determine what sort of impact my comments may have on you.

And…just so that you can keep track of what are my key points and where I am in my presentation, There are four lessons that I have learned in my career and life. I would like to share these lessons with you. These lessons are:

1. If you do what you love to do, you will be successful.
2. You must be a life-long learner in this information technology age.
3. You must be flexible in your career choice to be successful.
4. Keep things simple for your engineering work to be useful.

These lessons are interwoven into the story that I am going to tell you. I'll point them out along the way rather than address each lesson individually.

My research basically involved three topics of interaction - *mechanics, materials, and computing.* In my research and teaching career over the last 40 years, I have witnessed the tremendous growth and interaction in these three topics. Each played a key role at different periods in time.

1

Together, they have made major contributions to structural and geotechnical engineering applications and practice. This is the story I want to share with you today.

A group photo for members of Academia Sinica (中研院) during the biannual meeting held in Taipei, Taiwan in July 2004.
(Center seated-President Chen Shui-Bian 陳水扁)

9.2.2 Information Technology Age 資訊時代

My first thoughts are about what we refer to as the high technology age. Some also refer to this time as the information age. I think, perhaps, it should be called the *Information Technology Age.* The foundation of this information technology age is the rapid development of computers, personal computers in particular.

We can simply refer to this as *"computing"* power in the terminology of engineering science. Computing power, in my opinion, is one of the greatest accomplishments thus far in the Information Technology Age and it has greatly affected my research.

I think that it's important to reflect on all of the dramatic changes that we have experienced in the Information Technology Age and how this has influenced me and will surely affect you in your future.

Let me use as examples a few glimpses of what a typical engineering office and my research environment looked like 40 years ago and what some of the tools were at that time. Let's compare them very briefly with what we use today. I will show you in this process how these technology changes affected my research and our engineering practice over the years.

In 1960's

- o I was allowed to use FORTRAN language as one of the two foreign language requirements for my Ph. D. degree. Can you image that? A computer language as a substitute for French, German or other important scientific languages.
- o I taught the first FORTRAN course for our Civil Engineering students at Lehigh University in 1966. That was an experience I won't forget. I recalled that I told my department chairman at the time that I never took the course before but he simply said "you are young, right, you can learn."
- o I used slide rule to do my homework and HP calculator for my research work. How many of you even know what a slide rule is and what it looks like?
- o Specification formulas were simple. They were limited to no more than taking a square root.
- o The materials model that was used exclusively was linear elasticity.
- o Mathematical solutions were limited to series expansions with closed form.
- o Timoshenko's books were the bible of the time. Do any of you know of these books?
- o Method of superposition was used extensively.
- o Allowable stress design was used exclusively.
- o We had a beautiful mathematical theory of plasticity but few solutions.
- o Newmark's numerical methods with calculators were popular.
- o Hardy Cross's moment distribution method was widely used in engineering practice for building design.

These were the tools and theories from the 1960's. Many of these are probably no longer used and you may have not heard of them. They were, however, the basis that I used at the beginning of my career.

In 1970's

- o In the next decade, the 1970's, we used main frame computing and waited in computer centers for solutions. This was the time of punch cards and key punch machines.
- o The finite element method was used extensively for numerical solutions.
- o The material model used was plasticity with no work-hardening.
- o Simple plastic theory was used to design steel frames.
- o Ultimate strength method was used for reinforced concrete design.

o We had IBM typewriters to produce manuscripts with equations.
o I wrote my first structural engineering books on *"Theory of Beam-Columns"* extending Timoshenko's work to plastic behavior and design.
o I wrote my first soil mechanics book on *"Limit Analysis and Soil Plasticity"* to provide theoretical justifications for the famous Terzaghi solutions. Terzaghi, in case you don't know, was the father of soil mechanics.
o We had our first energy crisis and long lines at gas stations.
o Offshore structural engineering emerged as a new area of research and development.
o I wrote the first book on *"Plasticity in Reinforced Concrete"* extending the plasticity theory to concrete materials for offshore structural applications in particular.

In 1980's
o Along came the 1980's, the third decade of my research. We had computers on our desks with possibly 2MB of disk space.
o Probably laptops were just beginning to be developed. The few that existed weighed 20 lbs.
o Cell phones were slightly larger than the size of a brick and weighed just as much.
o Fax machines were just being developed and used with thermal paper where the ink faded over time.
o Communications were done via hardcopy, US Mail, overnight delivery, fax, phones.
o We have analog phones.
o Cameras for instant photos were Polaroid where you had to pull out the film hold it for 30 to 60 seconds and then peel them open. That was "instant" at that time.
o Drawings were just that - they were drawn manually. CAD just started.
o Offices included drafting machines or drawing tables, computers with 12-inch screens, fax machines, hand-held calculators, drawing files and racks, hardcopy libraries, file cabinets etc.
o We had Load Resistance Factor Design Code for Steel Structures. It is a maximum strength-based code for member design that uses plastic theory.
o I wrote my first book on *"Constitutive Equations of Engineering Materials"* to meet the need of finite element analysis. This was the first time that the capability of computing power and finite element analysis method exceeded the advancement of material modeling.
o I wrote the first book on *"Soil Plasticity: Theory and Implementation"* which extended the plasticity theory to geotechnical materials for offshore foundation applications in particular.
o I wrote the popular textbook on *"Plasticity for Structural Engineers"* to

help civil engineers understand the highly mathematical theory of plasticity. Plasticity theory became a basic course for civil engineers.

o   I wrote the first book on "*Tubular Members in Offshore Structures*" extending the well established wide-flange section members for building design to circular-section tubular members for offshore structural applications.

Present Day

o   Today, I challenge you to find more than 1 or 2 typewriters in your office.
o   We have desktop computers with several MB RAM and disk space in Gigabytes.
o   Laptops weight under 3 pounds.
o   Cell phones are smaller than your fist; and they can take photos, send email, give you your calendars, surf the internet, and so on.
o   PDAs are everywhere.
o   FTP sites, project websites, Email and IM are the preferred mode of communication.
o   Digital cameras provide instant photos.
o   Drawings are electronic.
o   Offices are mostly desks with computers. We work on projects thousands of miles away but we communicate and transmit data in real time between people and offices.
o   Your personal Email comes to your PDA, even when you travel to most destinations in US, Europe or China. When you are traveling, you use your IP phone on your laptop. Everyone you talk to thinks that you are in your office and talking to them.
o   I work with my secretary next door via the Internet and find most information via Google searches.
o   We're connected 24/7.
o   I wrote a series of books on "*Advanced Analysis for Steel Design*" moving away from traditional member design toward system design leading to performance-based codes for structural engineering.
o   Material models include fracture mechanics, composite FRP, and durability, from time independent to time dependent behavior.
o   Modeling (physics), simulation (computing), visualization (virtual reality) and verification (experimentation) are the rapidly developing fields.
o   I edited a series of Handbooks on *Civil Engineering,* on *Structural Engineering,* on *Bridge Engineering* and on *Earthquake Engineering* via the Internet communication among contributors, publishers, reviewers, and editors.

9.2.3 Life-Long Learning 終生學習

Changes over the 40 years of my career have been dramatic and exciting. Most of you have seen only part of it because you grew up during the last 20 years. Many of these tools that I consider *"new"* have always been there for all of you.

So, what does this all mean? Why am I talking about this Information Technology Age that we are a part of?

Well, what it means to me is that all of you will find, over the next 20 or 30 years, the tools that you will be using, the environment that you will be working in, and the processes that you'll be using will change, perhaps dramatically. Things will be very different from they are today. This will surely affect the way we design and construct structures.

Change is inevitable and more than ever, it is quite rapid. So, in my opinion, you need to hone your ability to change: you need to embed continuous change in your thinking and work processes. You will need to embrace change as your way of life. This change will not only affect your working environment, it will surely affect your personal and family life. Try to make the best out of it.

Focus on the fact that with your education and the experiences that you have been accumulating, each of you has the potential to continuously change, to continuously learn. Simply put, *Life-long learning* must be an integral part of your life.

This is indeed an exciting time to be engineers in the Information Technology Age.

9.2.4 Flexibility in Your Career Choice 彈性選擇

You must be flexible in your career planning and choice. Life is more than just goal-setting as it will have a lot of surprises. This is what happened to me.

I never planned to be a professor but ended up as an academician. I never planned to take an administrative job in a university but ended up as a dean in a major public university. Life is unpredictable but the lessons I've learned – to be a life-long learner and to be flexible – have helped me greatly in achieving success.

Forty years ago my dreamed universities were Cornell and Stanford, but somehow I ended up at Lehigh University in Bethlehem, PA. At the time, Lehigh's structural group was focusing on the development of a new steel design method known as *"Plastic Design"* to replace the century-old *"allowable stress design."* It was a revolutionary concept pioneered at Cambridge University during World War II.

Lehigh's Fritz Laboratory had the world largest testing machine and I participated in the full scale testing of steel structures. The method was later adopted by the American specifications for steel construction and quickly became a world-wide standard. The steel group became world famous and I received my MS degree and became part of this group.

As my studies at Lehigh came to an end, I received an offer from Cornell University to continue my steel research and earn my Ph. D. degree. It was a dream-come-true but for some reason, I was attracted to Brown University in Providence, Rhode Island for its highly theoretical work in solid mechanics in general and mathematical theory of plasticity in particular. As a result, I enrolled at Brown and earned my Ph. D. there.

While doing my studies, I found that highly mathematical theories and their applications frequently complicated matters. Design simplifications were always necessary for practical implementation. I came to the conclusion that nothing can be more practical than a simple theory, which is one of the four lessons that I would like you to remember.

Simple limit analysis methods were developed and widely used in steel structures. I focused on the application of this new method, not on steel, but on soil and concrete materials. This was the first time that this type of application was done. It became instantly popular and made me a celebrity of sorts. During the period, I wrote a series of books on this and related subjects.

9.2.5 Concluding Remarks 總結

I was surprised and honored to be selected as one of the ten giants in engineering science in American during the last forty years in the 2003 book by Dr. Beg in UK. The book was entitled *"Giants of Engineering Science"* which included the well-known figures like Y. C. Fung (馮元

楨), C. L. Tien (田長霖) and T. Y. Wu (吳耀祖).

My specialty and contributions were identified in the area of *Structural and Geotechnical Engineering Science*. This was a special honor as it was the validation of a lifetime of work that I have greatly enjoyed.

Now I realize that everything in a lifetime is a *journey*. So I say to each of you, your professional career that's ahead of you is a journey. There is no end destination. You need to continue to learn, to grow, to be flexible. And with that, I believe that you will find your career exciting, invigorating, challenging, and everything that you want it to be.

Your career and your life will get stale and old and tired only if you allow it to happen, only if you don't enjoy what you are doing. Do what you love to do and continue your journey; keep things simple, balance your career advancement and family life. It will be a fulfilling life.

9.2.6 Summary 摘要

Let's summarize the four lessons we learned today:
1. If you do what you love to do, you will be successful.
2. You must be a life-long learner in this information technology age.
3. You must be flexible in your career choice to be successful.
4. Keep things simple for your engineering work to be useful.

I believe these lessons that I've learned will help you along your life's journey.

I leave you with one quote by Oprah Winfrey, a famous and well-known figure in TV talk show in America:
"You don't become what you want, you become what you believe".
Have a great journey.

Acknowledgment

Mr. Michael Matsumoto, a CEO and a well-known structural engineer, was the keynote speaker of this year's College of Engineering Convocation at the University of Hawaii. I fully agree with his viewpoint and have intertwined what he said in his speech with my thoughts in this discussion about the Information Technology Age.

Visited a highway construction site near Taipei in 2004. From left: Y.L. Huang (黃玉麟), W.F. (惠發) and site engineers.

9.3 A Convocation Address 結業典禮演說

College of Engineering
University of Hawaii, Honolulu, Hawaii
December 22, 2001

Assistant Dean Tep Dobry, Faculty, Graduates, Ladies and Gentleman:

When Assistant Dean Dobry asked me two days ago to address this year's convocation, instead of inviting an external speaker as we usually do, it is an honor but also a great challenge for me. I was not sure what I should talk, where I should focus and what I should avoid. To be honest, I do not remember any of the speeches that I heard over the last forty years. I don't expect that you will remember any of mine. This is your happy day. The focus should be on you, not me. I think what you really want to hear in this address is the following:

1. You want to know the best way to make decisions you will need to live life
2. To build a career, and

3. How a leader can lead in this new landscape that is emerging from the mist. Based on my own life experience, not esoteric theory, this is what I will tell you in my speech today.

But before I start, I want to put you at ease by quoting a famous Chinese scholar, Mr. Lin Yu-Tung. He said I quote *"A good convocation speech is like a girl's skirt, the shorter the better"*. So I will follow this guideline. For your enjoyment, I promise that I will make it short but will cover the essential messages.

In the rapidly changing time of the 21st century:

1. You must be flexible in your career choice.
2. You must be willing to take some risk.
3. You must be a life-long learner. And,
4. Most of all, you must do what you love to do.

First: I must emphasize that you must do what you love to do.

Don't make a choice of any kind, whether in career or in life, just because it pleases others or because it ranks high on someone else's scale of achievement or even it seems to be, perhaps even for you at the time, simply the logical thing to do at that moment on your path.

Make the choice to do something because it engages your heart as well as your mind. Make the choice because it engages all of you. Remember that, as a graduate of this outstanding college, with your degree in Civil, Electrical, or Mechanical engineering, the freedom of choice is now yours. You must do what you love to do.

Second: You must be flexible in your career choice.

The driving force behind the rapid change is computers, communications, and the ability to automate processes in various ways. Boeing 777, for example, was designed virtually. We can use network program to design, develop and manufacture on a computer screen. As a result, the product cycle continues to shrink from years to months.

We can now use Internet to communicate instantly, wirelessly, almost free. This creates a lot of problem for the Mother Bell, AT&T. Remember Lucent Technology, used to be a part of AT&T Bell Lab, a

Wall Street daring only a year ago, it guessed wrong in its strategy, not investing in fiber optics early enough, and is now suffering the consequence, nearly going into bankruptcy.

As we move toward global economy, the market force is swift and the penalty will be severe and merciless. To be successful, you must be flexible in your career choice and always be ready to adjust with the new environment.

Third: You must be a life-long learner.

The half-life of higher engineering education in the United States today is no more than 10 years. In many fields, that is rapidly coming down. In telecommunication, for example, it is five or six years at most. This means that you are going to lose half of your useful knowledge in communications in just five to six years after your graduation.

It will be common for people like you to come back to college and get updated knowledge. It may not be a traditional college campus as you see it today. It may be an electronic campus provided through satellites, fiber optics, Internet or wireless systems. These programs are designed to teach the latest advances to practicing engineers from industry. Here at the College of Engineering, we are also developing and will offer these courses through the distance learning facilities or Internet.

Against this background, for you to be successful, you must be a life-long learner. The good solid education you received here at UH and the engineering fundamentals that you love to hate have already prepared you for this, a life-long self-learner.

Fourth: You must be willing to take some risk

The world is changing and the change is dramatic. Who would imagine that telephone and electricity would be deregulated? You can choose your own service vendors. Who would dream that California's Pacific Gas & Electric would file for bankruptcy? Nothing is safe. When you are young, you must be willing to take some risk. Give your dream a chance. Give yourself a chance. What are you going to lose? Worst of the worst, you can still move back to your folks.

You may want to joint a startup company if you have technology conviction, market conviction, or people conviction. Some of our faculty members are quite successful by every traditional measure. They received prestigious awards for teaching, research and service. Still, there was that sense of coming up short. A decade ago, few university professors envisioned being entrepreneurs. Yet, in these days, pressure to be success in classrooms as well as in boardrooms is mounting in some fields of engineering.

"If you are good, why aren't you a billionaire at one of those start-up companies?" That was the unasked question to our engineering professors sensed at a gathering of friends and family in the California Bay Area a few years ago. The news media have created transition of prestige from academia to money.

For example, in the last three years, UH Electrical Engineering faculty members have been involved in at least nine startups or Initial Public Offering (IPO) high technology companies. Companies range from networking to wireless to optical communication. In the last two years, one former EE professor sold his startup company to AOL for $250M, and another sold his company to Cisco Systems for $130M. These UH EE professors have all moved to the mainland with most going to Silicon Valley.

Well, their success is our success. Life is never fair. Life is not always smooth. Sooner or later, we all will face difficulties or even crises. I view these as challenges. In fact, the word *"crisis"* in Chinese connotes opportunity, opportunities for changes and opportunities for innovation.

Now the dot-com bubble is finally busted, who says this is not an opportunity for further innovation. You must give yourself a chance. You must be willing to take some risk. When you succeed, we will be successful. Someday, you may even want to endow a chair in my name to the College of Engineering if you still remember me.

To summarize, this is my advice to you:
1. You need to be flexible in your career choice,
2. You need be a life-long learner, and
3. You need be willing to take some risk.
4. You should not worry about your job and careers.

You are part of this exciting and rapid changing time. Enjoy the ride and live a rich life. You may even want to tell your grand children about this once-in-a-life time experience.

Good luck and best wishes!

9.4 Changing Jobs - Lessons Learned 換工作的經驗談

When I first searched a teaching job after my doctoral degree, I was not exactly focused. I got the list of university faculty openings in professional magazines and university placement office and sent out 100 resumes. I ended up with the faculty position at Lehigh University simply because I happened to have an interview at Bethlehem Steel and dropped in to visit one of my former professors at Lehigh.

For first time job seekers, my experience seems to be common; unsure what kind of job I can get and what they want me to do. So I used the scattershot approach and it did not get me the result I wanted. It was the connection and network that landed me the right job. It was quite risky to jump at the first opportunity that came along. We all wanted to jump into the right one. I was lucky to jump into the right one by chance, not by plan or design.

In the early stages of our career we may not know exactly what we want; but it will be very helpful to know a sense of what tasks we like and dislike. During the interview we can ask what kind of training we will get. What is the day-to-day work like? What is the natural career path for this position within the company? What is the company's culture like?

I worked at Lehigh for ten years with a good resume and achievement. My salary was adjusted through the usual annual review and modest adjustment. There were no jumps. I never thought of making a lateral move to a new university unless there is a tremendous change in prestige or title or salary adjustment. Purdue fit in this model and I made the move and received a major salary jump. I was not a good bargainer at the time and did not even ask for an instant tenure. Looking back, I should have negotiated for an instant tenure as a full professor. May be, I was quite confident on myself. Even so, I should have done it differently.

In the later stage of my job search, I was really focused. Rather than apply to a large number of universities, I developed a narrow list that truly interested me. I put all my energy into the identified few and made the right contacts. These included Stanford, UC Berkeley, and UC San Diego. I took a sabbatical leave at Stanford as a follow up for my West Coast search of new opportunities.

I was the finalists of all these three universities in the process. I understood well the culture of these universities in general and their civil engineering departments in particular. In this process, I was going toward something for my career, not going away from something where I felt very comfortable.

Once I reached the top of my teaching career at Purdue, it was my accomplishments that could be marketed prominently. The education section of my resume became much less important. It was important to start keeping track of what I had done to make a difference. Moving from Purdue to UH in my new job search was all about my accomplishments, nothing else.

The criteria I used and made sense to me at this later stage of job search were that I felt I was being challenged or some other aspect of the job attractions, such as title, ego, brand name or quality of life that made it appealing. The issues I am now dealing with as a dean are vision, strategic plan, resources, governance, and personnel. It is all about big pictures and ceremony.

One thing I learned is that when you are changing jobs; make sure you only move on if the new opportunity is truly going to be better than your current job. Don't try to move *away* from something, rather, move *toward* something. Interviews are not just for the employer to find out more about you. It is also your opportunity to find out more about your employer. Do not hesitate to ask what about the career paths of your interviewers? What are their backgrounds and day-to-day work life like?

Nowadays, you can research companies by searching the Web for background, talking with professionals you meet on the job, reading trade publications, getting advice from mentors, and making contacts with your network at companies that interest you. You need to understand what the companies need and what are your interests and strengths.

   There is a famous Chinese military strategy, Sun Tze (孫子), saying: "Know yourself and know your enemy; you will win every battle" (知己知彼，百戰百勝)

   In the real world of life, we know that: *"money is not everything in life; but without it, there is no life"* (錢非萬能，沒有錢，是萬萬不能). At the end of the day, we all understand that job seeking and changing job are all about ego and money (為名為利).   This is real life. It is true life.   Money can make the difference.   It can make things happen and perform miracle and cannot without it.

Does this make sense to you!

Gave a special seminar at the National Central University (國立中央大學) in Taipei, Taiwan, 1991.

A list of rules of thumb based on my own experiences may be summarized as follows:

1. Know what are your strengths and interests and the needs of the company.
2. Move toward something, not away from something. Timing is critical.
3. Focus on your search, not a scattershot approach.
4. Do not move when you are angry, but move when you are ready to change.
5. Do your network and share your thoughts with your family members.
6. Do your homework on the company's pension; start up, medical, benefits and salary range.

9.5 How to Make an Effective Presentation 如何有效表達

When I was a graduate student at Lehigh University in the 1960's, Professor Lynn Beedle, a renowned structural engineering researcher and educator and Director of world-known Fritz Engineering Laboratory, had a tremendous influence on us. His dedication resulted in many colleagues working with him and many students studying under his direction, creating one of the country's most active structural engineering research centers, steel structures in particular.

As a result, the University produced advanced and experienced students who later became leaders in education, industry, the armed service, and the government in US as well as abroad. To many of us, perhaps, Beedle's greatest title was that of teacher teaching us how to make effective presentation. Although he received many research and educational awards, the effect he had on his students and colleagues was priceless. He offered a class on effective communications for the graduate students, and his students were pushed to their limits in pursuit excellence in communications.

Beedle's presentation was always concise, to the point, and in good humor. His slides were excellent and easy to understand. Some of the tips on effective presentation are discussed in the following with much input from him and my own experiences.

Deliver your message. Don't expect the audience to come to get your message. If you have only 15 minutes, don't put in 50 slides or dazzled by a PowerPoint presentation with quick speed. It takes time to audience to understand the purpose of that slide before they can get the points.

Be yourself. A good joke is always welcome, it can make audience relaxed, but they will even be more nervous if you are tense. The one-liners will be good to start.

Introduce yourself. Tell the audience who you are and what your company does and why you are here – in 30 seconds or less. The best way to convince the audience is that your expertise or organization is worthy of their attention. Stick to your topic because this is why they come to listen to your talk.

Focus on the message. Talk to the audience, not the moderator, state your message at the very beginning, and then briefly outline the steps you will take to achieve it. So the audience can follow the progress of your presentation.

Do not rush the materials. If time is pressing, skip some slides and points. Go directly to conclusions with some brief elaborations. Rush the materials with either speed or cut into coffee break time is a bad policy.

Know your audience. Know your audience's background so that you can deliver your message effectively. Tailor your level of presentation to the background of the audience. Listening is considered a virtual among Asians while Americans are encouraged to speak out. Europeans tilt toward Asian's reserve. When answer a question, try to repeat the key points of the question since not everyone may understand the question.

Talk more on your own experiences. Share with the audience about your own experiences on the topic. People feel touched when you talk about your own story on success and failures.

Repeat the message at the conclusion. Tell the audience again about your message. Great endings awaken the sleeping ones and dispatch the audience with a final punch, by underscoring your points before the questions and answers session starts.

During the Internet bubble time in Silicon Valley, many start-up companies are seeking for venture capital investment. Since there are so many to seek funding, time is precious and opportunity comes and goes quickly. There is a new term in Silicon Valley called *"elevator talk"*, that is, you need to get your ideas and points across during an elevator ride. We often wonder what we take home after a beautiful PowerPoint presentation with Walt Disney quality of show. The answer is usually close to nothing or something that could have been said in two minutes, or the *"elevator talk"* is sufficient. This is the art of effective presentation.

Taipei Mayor Ma Ying-Jeou (馬英九) gave an effective speech in Honolulu, Hawaii on his vision for Taipei.

Presented a paper at the Joint US-Taiwan Symposium on Disaster Mitigation in Chicago, July 1994. The session was chaired by L.W. Lu (呂立武).

Dean Chen gave a welcome speech for the College of Engineering annual Banquet held at the Hawaiian Hilton Village for fund raising.

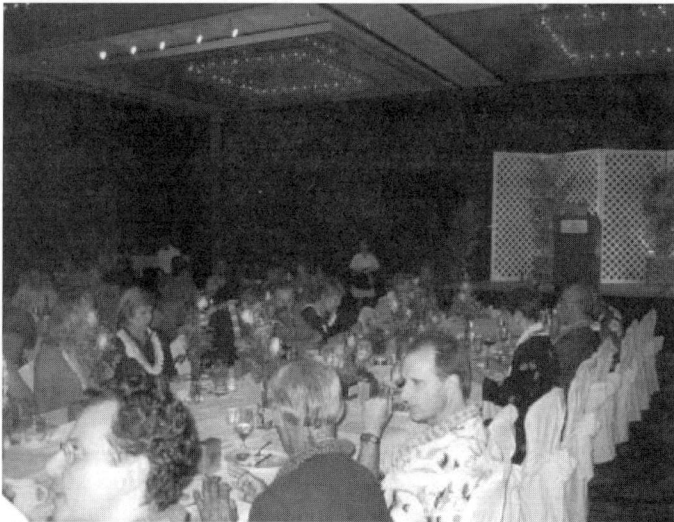

The VIP table at the annual College of Engineering Banquet for friend-building.

10

My Journey in the Rise of China
我見証崛起中的中國

10.1 Background 背景

On November 20, 2003, I attended a Business Outlook Forum sponsored by the First Hawaiian Bank in Honolulu. The speaker was Ken Curtis, Vice Chairman of Goldman Sachs Asia. His talk was about China and the economic opportunity for the people of Hawaii. I was expecting to hear and prepare to accept a significant amount of criticism regarding trade issues with China and the difficulty of making deals with China.

I was really surprised to hear that according to Curtis, the 20 century was the rising of American power replacing European economy, and the 21 century is the rising of China power replacing the American economy. Instead of trying to isolate China, it is best to engage China; so that we can control it and to direct it to the betterment of the world. He concluded by saying that "*I am a simple minded person, I prefer to have rich strong neighbors, than poor and weak ones. It is good for all of us*".

During my forty year career in academic life in America, I had opportunities to visit and make lecture tours in China during the last thirty years, and witnessed the dramatic changes of China during the critical periods: from 1978 Deng Xiaoping's (鄧小平) "*reform and open*" policy, to the crack down of 1989 June 4th student unrest in Tiananmen Square, and to the rapid economic growth of an average of 9.4% annually over the last two decades. Whether in Taiwan or in the U.S., my Chinese roots and traditional education have always made me deeply connected with China and I have constantly China watched with great interest and emotion.

China has been struggling for hundreds of years to get out of poverty, to erase the past humiliations by Major Powers, and to gain respect in the world scène. I feel that this is the first time in the modern history of China that changes have finally started to penetrate Chinese society, especially in recent years. Let me begin by sharing some of my own experiences and thoughts on China in what follows.

I happened to be amongst the first group of scholars to visit China right after President Nixon's historic visit. Also, I happened to be at Tsinghua University in Beijing during President Mikhail Gorbachev visit and eye witnessed the subsequent peaceful students' demonstration in Tiananmen Square in 1989. Those were the most exciting periods in the history of the rise of China.

Looking at China from Honolulu, I really feel that changes are now gathering force in China and taking place in a more persistent and no-return manner starting from economic development, to military build-up, to democratic reform, to more tolerance on the Taiwan issue, and to less control of government and more sensitivity to the pulse of people. It is now more pragmatic and much less ideological.

10.2 The Rise of China 中國崛起

For the last 20 years China's economic growth has made it the world's sixth-largest economy. Over the same period real per capita income has increased more than five times. And the average real wage has doubled during the last decade alone. As a result, China has lifted 400 million of its people out of poverty since 1981. China has managed to accumulate foreign exchange reserves second only to Japan.

In a recent PowerPoint Presentation prepared by the *World Magazine* (天下雜誌) in Taiwan in 2006, I was very impressed by its simple but powerful slide show illustrating the rise of China by a few simple questions and answer pictures as showcased in the following:

o　Do you think you are in Hong Kong in this picture? No, you are wrong. This is Chongquing, China (中國重慶).

o　Do you think you are in London in this picture? No, it is Shanghai, China. (中國上海).

o　China has the highest foreign exchange reserve in the world, is the second largest oil consumption country in the world, the third largest trading partner in the world, and the fourth largest GDP country in the world.

o In the last 15 years, China has built 800,000 kilometers of highways; that can encircle the earth more than 19 times.

o 32.5 kilometers of bridges were built in Shanghai alone to create the world's largest cargo port for Yangtze River (長江) delta.

o Just completed the incredibly difficult 5000 meter altitude 1118 kilometer long Rail connecting Tibet (西藏) to Beijing in just five years.

o Over a ten year period, the number of universities has doubled with a student population of nearly 20 million.

o Through MBA programs with international cooperation, its next generation of leadership in civil service is being trained with management skills, global perspective, and the big system view.

Yes, I have been following these impressive figures from newspapers and magazines, but to my surprise, the leadership in China has not been focusing on boasting their achievements in their talks, but rather constantly worries about the sustainability and economic bubble and possible collapse as happened for the Pacific Rim countries during the period of Asian Economic Crisis in the late 1990's. Now, they are rightfully focusing on the agricultural issues for the peasants, the unequal distribution of wealth, and the development of a democratic government process with election in the town and city level as a start. It is encouraging to see the discussions on further privatization of the state-owned enterprises, restructuring of the banking system, more flexibility on foreign-exchange policy, and the protection of intellectual property rights. Indeed, China is moving towards a market economy with market reforms and is continuing opening up the economy further. It is good to see that China's leadership puts more focus on challenges than on achievements. Complacency can lead to stagnation and future problems.

It is inevitable that I compare China's development with India. China has built impressive infrastructures while India has built impressive high tech businesses, especially software development. India is the largest democratic country with a truly free market economy; while China is the largest central planning economy. Both are successful in developing impressive economic growth and reducing significantly the poverty of its huge population. China was very much influenced by the Soviet Union and its shocking collapse during the Cold-War. Its dismantling afterwards had a profound effect on the Chinese leadership; and as a result, political stability becomes its hallmark in governing and policy setting. It has helped justify a single-party system of governing— something the Western democratic countries do not agree; but which resonates in a country where chaos has always been seen as the greatest social danger. From the Opium War of 1840 to the founding of the People Republic of China in 1950, Chinese people have been plagued by foreign invasions, civil wars, and poverty.

India, on the other hand, was very much influenced by British rule and a free market economy with skillful managerial advantages. At the end, China is portrayed by the Western powers as an unfair competitor in countries that have lost market share to it. India is favored for its openness and fairness. As China builds up its military force, it is portrayed by Western powers as a threat of world peace. This is hard for Chinese people to understand how they can be perceived as a threat when they have been humiliated for hundreds of years of invasions, civil wars and poverty. Realistically, it will take decades before China can claim to be a modernized, medium-level developed country.

China knows that the changes Deng Xiaoping (鄧小平) began in 1978 after the culture revolution are far from complete and a complacency resulting from present achievements will make it real hard, if not impossible, to call for the continuing efforts to triple its per capita GDP between 2000 and 2020, as its goal. They must pursue reforms with an added vigor and political skill, and not to be short-changed by the sensitive issue on Taiwan with America. If so, China's success should be seen as a great economic opportunity for the world as a whole in view of the size of its present and future markets. This growth will in turn press China to go further and faster with more market reforms along with political reforms as well. As Curtis pointed out that "*it will direct China to the betterment of the world.*"

10.3 The NIT Lecture Tour in 1981 講學之旅

Let me begin by sharing some of my own experiences on the reshaping of China. I happened to be in Copenhagen, Denmark, to deliver a keynote lecture on *"Constitutive Equations for Concrete"* for the International Association of Bridge and Structural Engineers (IABSE) Colloquium on Plasticity in Reinforced Concrete on May 21, 1979. That was the beginning of China's open door policy. A group of Chinese delegation headed by Professor Ding Dajun（丁大鈞）of Nanjing Institute of Technology (NIT)（南京工學院）was sent to the Conference by the Chinese Government. They arrived at Copenhagen only a day before the conference and were required to stay in the Chinese Embassy for security reasons. When they attended the conference's reception that evening, they were surprised to find out that their papers were not scheduled for formal presentation. They were thinking about asking their Embassy staff to negotiate with the conference organizer to change the program.

Shu-Park Chen (陳樹柏) (Left of Deng Xiaoping 鄧小平) led a group of Chinese scholars to visit China in 1982. Second row from left: Wai-Kai (惠開) and Professor Pen-Ming Lin (林本民) of Purdue.

As a Chinese-American and the only Asian that appeared at the reception, I explained to Professor Ding（丁大鈞）how a professional or international conference was organized and worked. The paper presentation issue should not be handled as a diplomatic affair. They could be solved easily with a personal touch. Since they had not pre-registered and pre-paid for the conference fee, the organizer had assumed that they would not attend the conference. I helped them contact the organizer and get their papers on the program. This showed how closed the Chinese society was and how nervous people were after the Culture Revolution (文化大革命). They had few opportunities and experiences to attend international conferences organized by Western countries and present their work.

Professor Ding (丁大鈞) was obviously impressed by my presence at the Conference as a keynote speaker and invited me to visit Nanjing Institute of Technology (NIT, 南京工學院) and give lectures for his faculty and students. This type of lecture tour for Chinese-Americans was quite fashionable at that time. Since I was born and raised in Nanjing (南京) during my childhood and remembered living in a "*big*" house near the old Ming Palace (明故宮) in Nanjing I was very much interested in going to China to see my country and the house, a place I had not seen since we left mainland China to Taiwan in the 1950's.

I accepted the invitation and talked to Linlin (玲玲, Lily) about the visit. Linlin had to take care of the three boys at home; and could not go to Nanjing with me and at the end; we decided to let me take Eric (中傑) to accompany me for the trip. My trip to Nanjing Institute of Technology (南京工學院, NIT) was scheduled from July 17 to August 19, 1981 and Eric was about 13 years old at the time. He was a very heart warming kid and tried to take a good care of me during our stay in Nanjing. You can't find anyone more close, sweet and caring for a kid of his age. He acted like a "*little Mom*" and took care of the daily routines for us. To this day, I still feel his presence and see the images vividly during our lunch time pouring the Nanjing beer to my glass; rushing to friendship store after the lunch to play chess with the girls in the front desk; and enjoying so much of his daily chocolate candy bar during his chess play.

It was a very hard work for my daily lectures. Two lectures a day for 10 days: one in the morning and the other in the afternoon each lasted at least two hours. I don't think you could image the students' drive to learn during the period because they were hungered for knowledge after the ten-year wasteful and destructive Cultural Revolution. They really wanted to catch up on their missing years. They lost an entire generation on education. What a tragedy!

There was a communications gap. I prepared my lectures in traditional 35mm Eastman Kodak slides in English and mixed Chinese and English during the presentation. The classroom was large but crowded with many students standing in the back and on the sides. The students might have been intimidated by my presence and seldom asked questions but kept taking notes. It was an impressive experience. To help the students, I provided and they reprinted 20 of my lecture notes in the areas of "*In-Plane and Space Behavior and Design of Beam-Columns*" and "*Concrete Elasticity and Plasticity*" which were quite advanced topics at the time even in U.S.

Top-left: W.F. (惠發) and Eric (中傑) were ready for lunch.
Bottom-left: Eric with the assigned chauffeur.

Below: Eric (left) with his assigned playmates.

We stayed in a Friendship Guest house （友誼賓館） and were completely isolated from the general public. I remembered the 60-Minutes CBS reporter, Mike Wallace, was also stayed in the same guest house at the time. It was a large compound with garden, swimming pool, stores, and kitchen. When we arrived at Nanjing （南京）, we were picked up by a special limo from the airport and were welcomed by the NIT students, faculty and staff at the front gate of the University with flags and long lines. It was a memorable moment for both of us. My distant relatives were happy to know we were coming; and felt proud of the relationships and asked to join us at the guest house. They enjoyed so much the luxury of shower bath, good foods, and special shopping stores reserved for foreigners with special currency. This was the state of the culture and economic environment at the time. Looking back, this kind of sharp differential treatments for overseas visitors can not be sustainable and the "*open and reform*" (改革開放) policy seems to be inevitable for the good of people of China.

During our stay in Nanjing, I was fortunate to know Xila Liu (劉希拉) through some of his contacts. Xila's father was a famous figure in Chinese medicine and medical education in particular. Their home was in Nanjing. After I returned to Purdue, I began to accept and support graduate students from China. Xila Liu and D.J. Han (韓人健) were my first two graduate students from China. They had suffered the painful experience of the Culture Revolution after their college educations. They were a group of very mature students and very eager to learn. It was a pleasure and once-in-a-life-time opportunity to work with such a group of high quality students. Both of them returned to China after their Purdue degrees. D.J. graduated in 1984 and returned to the South China University of Technology in Guangzhou (廣州) and was eventually promoted steadily and later became a Vice President of the University in the process.   Xila graduated a year later and returned to Tsinghua University with his wife, Chen-Chen (陳陳), who also received her doctoral degree in Electrical Engineering from Purdue.   Xila was promoted quickly and steadily and eventually became a Distinguished Professor and Head of the School of Civil Engineering in the process. They were both elected in later years as members of the National People Congress (政協委員) in China, representing the academic profession.

With the help of NIT's foreign office, we visited my old Nanjing (南京) home. It was a big disappointment. The house was small by today's American standards and five families lived there. It had not been maintained and looked somewhat rundown. I was surprised to see that on the "*2 Garden Street*" (公園路二號) where the house was located, not much had been changed over the last 30 years. No new construction, no upgrades and no improvements of the streets, neighbors, and facilities. Most people appeared to be treated equally but everyone was poor, very poor. There was no economic development over the last 30 years, period. What a shame!

After my long and hard work at NIT, they arranged a tour for two of us. We went to the beautiful city of Yangzhou (楊卅) near Nanjing and also visited the famous West Lake in Hangzhou (杭州西湖), the hometown of Linlin's (玲玲) mother. Again, it was a special arrangement and a special treat. We traveled there by a train ride. We were assigned to the special soft seat sections of the train. It was almost empty with air conditioning, tea service and large seats. It was very comfortable but I felt uncomfortable because the hard seat sections were very crowded and some with standing room only. The class and positions made the difference. I was wondered, doubted, and felt sad that a classless society as claimed by the communist party had such a clear class separation. If you were high government or military officials, your treatments could be very different from that of the average population. A day and night difference!

Top: The soft-seat train section was empty and Eric (中傑) was about 13 year old.
Bottom: Eric loved to play Chinese chess on the train.

During our stay in Hangzhou (杭州) guest hotel, we were treated with a small private boat cruise to enjoy the famous West Lake scenery. We were very surprised to hear the singing of the pedal boy. He sang the familiar Taiwan song, *"The Story of a Small Town"* (小城的故事) by the famous and popular Taiwan singer, Teresa Deng (鄧麗君). What a surprise! I was even more surprised to discover that there was a night club dance going on in the hotel with such familiar old songs like "*夜來 香*" and "*何日君再来*". Those were considered to be unhealthy music according to the government guidelines. They were definitely prohibited in public but were readily available in the guest hotel. We were also treated with a Chinese opera on *"Good Official and Bad Official"* （*清官與貪官*）. To my big surprise, Eric （中傑） also enjoyed the show probably because the story was simple and easy to follow, and the dresses were very colorful.

I remember that during our Hangzhou train ride, two Chinese People Liberation Army （中國人民解放軍） generals came on board at Shanghai Station. When they found out that Eric liked to play chess and also learned that we came from American, we started to have some conversations and interactions. I was amazed to hear one of the generals asking Eric the question *"You are from American, right? Do you have comfortable trains like this one in your country?"* This was how they brain-washed their public and military people about American. No wonder, there are wars, because of such a miscalculation or misunderstanding.

During the founding of the People's Republic of China in the 1950's, China was slow to learn lessons from abroad and adjust its thinking and policy, and it paid the price in its inability to compete with Western powers. These days, the tables are turned, and now we need to learn from China. To me, the central challenge for this century will be how to engage and integrate China into the world market for the betterment of the world without repeating the tragic human history of isolation and confrontation.

10.4 The Journey on June 4th 1989 六四經歷

In May 1989, I was invited by the World Bank to serve as a consultant for the Chinese University Development Projects. I arrived at Shanghai first to deliver my keynote lecture on *"Cap Plasticity Model for Embankment: from Theory to Practice"*, for the Second International

Symposium on Environmental Geo-technology, on May 15, and then continued my journey to Beijing to meet Xila Liu（劉希拉）and to present a seminar at Tsinghua University. My next stopover was Changsha University in Changsha（長沙）, and my final stop was Guangzhou（廣卅）to meet with D.J. Han（韓大健）and visited South China University of Technology（華南理工大學）before returning to U.S.

Since I was a guest professor of Tsinghua（清華）and Xila Liu（劉希拉）was Head of School of Civil Engineering at Tsinghua, I stayed at Tsinghua guest house on campus. The guest house was located very close to Peking University（北大）. The next morning Xila met me at the guest house and we walked to the auditorium for my seminar. I realized immediately that something was not right; because there were very few students present. The central administration then decided that I would give my talk at an office setting. Very few questions were asked and everybody seemed to be quite tense. Later, Xila（劉希拉）told me that the students of Peking University had just posted handwritten posters known as "*Big Letter News*" on its campus walls last night. This type of posters was used widely as an effective communication tool among students during the infamous Culture Revolution（文革）. In fact, Chairman Mao（毛主席）wrote the first poster to initiate his Culture Revolution in 1966.

The theme of students' grievances was focusing on corruptions and briberies of high government officials and their kin or sons, daughters and siblings in particular commonly called "*Gang of Princes*"（太子黨）. Quickly, the posters were spread to Tsinghua campus and I had plenty of opportunities and time to read these informative and emotional posters, learned about what was going on at the time, and felt the strong resentments of people against some high government officials, and sensed the strong undercurrent that was going to erupt.

In the next few days, the movement was spread to other campuses around the country. In the meantime, the university students in Beijing under the leadership of Peking University started and continued their daily demonstrations and gradually established their semi-permanent sites in Tiananmen Square in the process. I happened to be at Tsinghua campus during that period of time. I eye witnessed all these historical developments and participated in some of its events before the crack down of the movement by the government on June 4th.

From the summer of 1988 to the end of 1990, it was the most exciting period in the history of the former Soviet Union. Mikhail Gorbachev had successfully shattered the ultra-hard-core Soviet system that had been used to govern the Soviet Empire for nearly 70 years. His perestroika and glasnost policy produced drastic changes at home and abroad. The students were obviously inspired by the development and a few well publicized corruptions and scandals involving government officials and their kin made the situation explosive. People now believed that there had to be political reform and change. On the other hand, the Berlin Wall crumbled and the Cold War ended the Soviet Union overnight and this made the Chinese political leadership extremely nervous and insecure. During this period we all witnessed a revolutionary change in Soviet Union and Gorbachev had to disappear from the scene after he failed to reshape his disintegrated Soviet Empire. With the vivid lessons from the collapse of the Soviet Union, the decision by the communist party to crack down the student movement for political reforms was inevitable. Political stability had to be maintained at any cost in order to achieve economic reforms.

Let me share some of my experiences with you during the period of students' demonstration and their peaceful sitting at the Tiananmen Square. It was amazing to see the sea of changes in daily life during the period. There were no police and no military personnel in sight, no bus services, no public transportation and the broadcasts of CCTV（中央電 視台）changed their Mao's dresses to simple and common dresses of everyday people. There was no chaos and no disorder and people became very courteous to each other. There were rumors going on about the movements of troops and I saw big buses parked by the bus drivers at every critical junctions and bridges with an intention to block the possible movement of these troops. The situation was indeed tense but calm. Any new developments were posted on the campus walls and the news was quite up-to-date and very informative and rather reliable. Obviously, there were disagreements among the high government officials on how to handle this historical event. The secrets were leaked out and passed on to student leaders and they in turn posted on the campus walls.

One of the afternoons, Xila（劉希拉）and I bicycled to Tiananmen Square from Tsinghua and visited the impressive scene of many enclaves set up by students in Tiananmen Square. Since I was a guest professor of Tsinghua, I was allowed to enter the Tsinghua enclave. I saw a few students in hunger-strike lying on the ground in tents. They passed on many grievances materials to me; it reminded me of the re-birth of the famous May 4 Movement（五四運動）by Peking University in the early years of the Republic of China demanding for political change and reform. I felt so proud of them. That movement made a big difference for China at that time. Now, with the rise of China, I would say that the tendency for change has become very solid at the business level, individual level, and even at the political leadership level. It has been a long struggle for Chinese people.

By the end of May, I was told by Xila（劉希拉）that the University central administration had informed him that it was time for me to leave the campus immediately because some actions on student demonstration might take place very soon. So, they took me to the airport and I was on my way to Changsha（長沙）for my next visit. My flight was first delayed and then took off to Changsha. For some reason, the flight returned to Beijing after about 30 minute's flight time. Later, we were told that the central government needed the airplane for important missions. So I returned to Tsinghua campus and tried to figure out a way out of Beijing. Then, more bad news followed, all domestic flights had been cancelled and I was told to go back to the U.S. directly and immediately. After some struggles and consultations with World Bank personnel, I called Linlin in the U.S. and asked her to buy a Japanese Airline ticket for me using an American Express card so I could pick up the ticket at its Beijing downtown office. This was the way I got out of Beijing on June 1 and returned to Purdue campus on the same day.

The June 4th event on Tiananmen Square（天安門廣場）made the headline news around the world. I happened to be in Beijing at the time with first-hand materials and information about the development. I was asked immediately by the Chinese Student Associations to give special seminars on the event; the newspapers interviewed me on my experiences; and the World Bank was felt a sigh of relieve for my safe return. It was understandable that Purdue's mainland Chinese students in particular were very angry at the time. Some even burned their communist party cards openly and declared to withdraw their memberships from the party. Another group of students on campus started to form a Chinese Solidarity Union in support of the students in China. The June 4th event leads to a large group of Chinese students in the U.S. seeking political asylum. The U.S government passed a special law granting their requests.

10.5 The Culture Tour in 2004 文化之旅

Visitors to China are always astonished by the new highways, new transportation systems, and new skyscrapers and by the endless construction projects that cover the landscape of many cities. These are "*hardware*" development, but its investment in modernization for "*software*" development such as human capital improvement, high tech industry creation, and new business opportunity for world market may be even more impressive. With brains, labors, money, market and backing from the top, China is not only creating a domestic high tech industry, it is getting ready to be a global major player.

For example, my second son, Arnold (中毅), joined a start-up optical-chip company called Genoa in the San Francisco Bay Area during the Internet bubble time. The company produced optical amplifiers that could help light move efficiently around the Internet; that could reduce the existing suitcase size to a sugar cube size; and that could sell for as little as $1500 instead of current cost of $30,000. They had a very good product but the timing was terrible. There was no demand in the communication gears at the time of the burst of Internet bubble. So, Arnold got laid off and was looking for new opportunities.

He found an opportunity to join a new start-up company in China producing critical parts of wireless communication gears. Since he was just married a year ago, they were debating whether they should go to China to take the Chief Technology Officer (CTO) job in Dailein (大連),

China. If they went, they had to learn Chinese quickly. All of my three boys: Eric, Arnold and Brian had attended Chinese Sunday School when they were in primary schools, but their ability to speak and write were not adequate for serious career development, although Arnold's wife, Lin Ng（黃慧琳）was much better in her Chinese education from Malaysia. At the end of the day, they decided not to go. This was the motivation for Linlin and me to plan a private culture tour for the family together to tour China in the summer of 2004.

The first stop was Beijing. In addition to the traditional sightseeing of the Great Wall, the Old Palace（故宮）, the shopping malls, the Peking ducks and Chinese operas, the main attraction was the visit of the assembly lines at TPV Technology founded by Linlin's brother, Jason Hsuan and his friends in Taiwan over 30 years ago. The TPV Technology is currently the largest monitor maker in the world after a recent merge with Phillips. According to Jason, the TPV is a leading display solution provider in the world, specializing in design and production of a wide spectrum of CRTs, LCDs, and CKD/SKD. Listed in both Hong Kong and Singapore Stock Exchanges in October 1999, they have now commanded a 40 percent of the market share in the PRC.

Jason provided us a five star guest apartments and food services at his TPV headquarter facility in Beijing site. He also assigned a SUV driver for us to go around and tour the city. His son, Michael and daughter-in-law, Elaine Lin, hosted a welcome banquet for us and held an engagement celebration dinner for Brian（中宇）and his wife to be, Christine De Asis at the Banquet room since Jason was traveling at the time. We were served daily with excellent home-made Chinese meals with special selections tailored to our needs. It was a very happy family gathering and occasion.

From a distance the TPV assembly lines looked like just row after row of Chinese workers sitting at the benches diligently carrying out their tasks while waiting for their long day to end. Upon closer inspection, however, they were not just doing simple tasks, rather, crafting precision components with measurements and reliability check or quality control. They were producing state-of-the-art monitors for major customers such as IBM, Hewlett-Packard, Dell, and Mitsubishi Electric.

In a recent news article commenting on the new partnership with Royal Philips Electronics, Dr. Jason Hsuan (宣建生), Chairman of TPV said, "*The alliance signifies the attainment of TPV's long-term corporate objective of becoming the largest PC monitor manufacturer in the world, as well as a big leap forward in our strategy of diversifying quickly into the growing flat screen TV manufacturing and development market. The integration will allow both TPV and Philips to concentrate on their respective core competence whilst providing consumers with cost effective multi-media solutions and entertainment.*" It was an impressive visit and the TPV R&D skills were also improving and getting better, and much better. They were not just copycatting the multinationals or just buying communication parts to assemble. They are also building up their own brand names with innovative design and quality services for the world market.

I read in a recent article about the Shenzhen's Huawei Technologies (虎威) who produced the cutting-edge optical switching equipment that competed directly against the communication giants Cisco Systems or Nortel Networks in international markets. The gap was closing so fast that made Cisco trying to slow its inroad with lawsuits. It was amazing to see a nation that has not produced a noteworthy high tech product since the abacus has emerged as the cradle of power house of hardware and software companies. To challenge the Information Technology (IT) might of the U.S. and Japan by China was unthinkable just a few years ago, it is now probable that China may be the next tech super power.

Our next destination was Xian, the old imperial city of China. We visited the famous Emperor Ching's Tomb (清陵), watched a colorful ancient palace dance of Tang Dynasty (唐朝), toured some old city walls, and climbed some old Pagodas and Buddha temples. There were long histories behind each of these historic sites but very few were taught in American schools for our *American Born Chinese* (*ABC*) sons. They are 100 percent Americans with the nickname ABC's in the Chinese community. This sometimes made our tours boring and less appreciative for them because of their lack of culture understanding of the events. Yes, it sounds like a different world for them!

Our final stop was Shanghai and Hangzhou（杭州）. Shanghai's skylines and skyscrapers along with the endless construction projects made us even forget that we were in China, not in San Francisco. We did the usual tours like many other tourists including shopping, boat cruise, and Buddha temple and drank coffee at Starbucks, just like we would do in San Francisco. This change of landscape was amazing and incredible! We went to Hangzhou by car and stay at a five-star hotel near West Lake（西湖）. We visited temples, ate at several famous restaurants, cruise around the Lake, and bought some fresh but expensive teas from a local producer. Our tourist guide spoke fluent English with some British accent. This was rare and we were very impressed. He told us that he was a college graduate majoring in English but had never been going abroad. If this was true and widely spread, it would be a very important development for the future of China.

In the context of Internet age and globalization of business and trade, it is a very good strategy for China to dramatically strengthening its English education including the possibility of making English China's second official language in the future. The increasing globalization will lead to increasing mobility of people, which, in turn, will increase global contact and interaction with other cultures. This will help China integrate smoothly into the world and make Chinese nationals responsible world citizens.

To sum up on how effective was the culture tour to educate our next generation and get them connected with our roots in China; I came across these Chinese Proverbs that may help explain the thoughts behind the tour:

o　Tell me and I will forget.
o　Show me and I may remember
o　Involve me and I will understand
o　口到　眼到　手到　心到
o　百聞不如一見，百見不如一試
o　聽不如看，　看不如做
o　耳聞是虛，眼見是實
o　不經一事，不長一智

10.6 Quotation from the President of US NAE in 2006 引述美國國家工程院院長對中國長期科學與工程發展計劃談話的驚愕

Quotation from a newsletter I received from the President of the US National Academy of Engineering on China's science and technology, July 2006.

....

Now to relate a stunning experience I had about a month ago when I was in Beijing to attend the Annual Meeting of the Chinese Academy of Engineering. The Chinese Academies of Science and Engineering hold annual meetings at the same time, so they share a common opening session in the Great Hall of the People. Of the 3,000 people in attendance, perhaps a dozen were non-Chinese.

On the dais were the president of China, the premier of China, and every member of the Politburo — the entire top leadership of the country. President Hu (胡) gave a 40-minute keynote address, the subject of which was making China an innovation-driven nation. He was not talking about a nation of low-wage garment workers. He was talking about an innovation-driven nation! The items on his agenda included education, research, intellectual property rights, and making China an attractive place for people from inside and outside China to pursue science and engineering. These are the same things we talk about in our recent report *"Rising Above the Gathering Storm"*.

Like Tom Friedman, I believe the world will be safer as developing countries become more prosperous. But we can not be complacent and assume that we are entitled to be the world's leading innovative nation and to reap the material benefits of innovation. Sitting on the floor of the Great Hall surrounded by 3,000 of the best scientists and engineers in China and listening to the simultaneous translation of President Hu's speech made an enormous impression on me. The Chinese are serious about education and research, the very things we are mostly still talking about. It's time we started taking action!

Wm. A. Wulf
President
The U.S. National Academy of Engineering

11

My Students at a Glance
學生略述

In the course of our lives there are three modes of being. In the *"education"* mode we learn the basic skill to become *"competent"*. Then, we go like crazy to further establish our competency and to apply our core competency during the career path. Finally, there is the *"reflective"* mode, which is where I am now; and also for some of my students too. I try to cover all three ways of being while I am writing this book. I also invite my former students to reflect on their lives by writing a brief biography of them; so that we can share about the wealth of our experiences on career, life and family.

Each of us has a different definition on success. What was your expectation? What brought you here from there? What was your life goal towards which you were driving to achieve? Did you achieve that goal? Let me give you my definition of success, for what it is worth. Here, like Economics 101, there are two sides to our souls: supply and demand. Everyone has some built-in talents and gifts, or what we are especially good at doing. For example, you may be a good observer and listener. You may have a quick calculating mind with photographic memory. You may be good in learning different languages. You may be born with an optimistic attitude. These are your *"core competencies"* in addition to those you learned through your education – this is the supply side of your personal equation.

The other side is the demand side – your goal or your ambition or your desire. But there is also an intrinsic motivation to your soul. What really excites you when you do something just for the joy of doing it? If your goal coincided with your intrinsic motivation, you will sure be more inspired and more creative, and will make you more successful. The real success of your life is of course finding a way to match these two souls –

241

supply and demand – and achieve your good life. This is how I define success.

For me, my graduate students furnished me the highlights of my professional career; expanded the breadth of my technical interests; contributed to my professional achievements; and connected me to the real world of engineering. They came from different parts of the world and characterized by their desire to learn and to hard work. Through them, steel structures and constitutive modeling became my professional niche. I have produced a total of 56 doctoral students with nine from Lehigh University (1966 to 76), forty six from Purdue University (1976 to 99), and one from the University of Hawaii (1999-06). All of them coauthored papers with me; while ten of them coauthored or co-edited books with me. I kept good contact with most of them, some very closely. But I lost contacts with some of them after their graduation; and most time when they changed jobs and moved to new locations or countries.

Based on the information I received; or downloaded from Google search, I prepared a brief description for some of them when appropriate. Some even provided me with one or two digital photos that could be easily edited into my writing. Some described their career paths; some talked about their life struggles; while some traced their technical journey in a nutshell. I tried to keep each of their descriptions in its original form as much as possible to reflect the diversity of this group of students I had the privilege to supervise; each with his or her own destiny. The balance between the "*supply and demand*" sides of each student's life and career was defined by the individual student. This chapter describes briefly some of my doctoral students' life and career in a nutshell.

Dean Y.B. Yang (楊永斌) of National Taiwan University and Dean Chen at Yang's office in Taipei during an Engineering College evaluation visit.

A full description of three individuals: two are my former doctoral students (T. Atsuta of Japan from Lehigh and Xila Liu（劉希拉）of China from Purdue) and the other is almost like my student since his graduate student time (Y. B. Yang of Taiwan from Cornell). These three individuals were most senior and quite representative of that particular cluster of student group at the time. Their life stories were given in three separate chapters (Chapters 15, 16 and 17 respectively).

11.1 Students from North America 北美學生

Dr. CHARLES SCAWTHORN (1968, MS, Lehigh)

After graduating from the Brooklyn Technical High School in 1962, Charles majored in structural engineering at The Cooper Union for the Advancement of Science and Art (New York City), receiving his BS degree in 1966. His senior thesis was the design of an orthotropic plate girder steel highway bridge and received first prize in the student category of the US Steel *International Steel Highway Bridge Design Contest* as a joint project with another student (C. Hofmayer).

In 1966, he entered the MSCE program at Lehigh University, and performed research on analytical models of foundation and slope stability. He was my first graduate student, when I had just returned to Lehigh University after receiving my Ph.D. from Brown. Charles received his MS in 1968 and his thesis "*Limit Analysis and Limit Equilibrium Solutions in Soil Mechanics*", was published in *Soils and Foundations* (Tokyo) in 1970 with me.

From 1968, he worked for the Consolidated Edison Company of New York on a number of projects, including the tornado effects on a cable crossing, the effects of towers impacting a nuclear containment structure, and design of upgrades for major transmission tower structures. He also oversaw development of a new transmission structure, and then the design of several hundred of these structures serving New York City.

In 1974 he joined Bechtel Corporation (San Francisco), Refinery and Chemical division, where he was involved in several projects, including a buried radioactive waste tank, a nuclear fuel reprocessing plant, and a water treatment plant for Saudi Arabia.

While working in San Francisco he became interested in how to reduce losses due to earthquakes, and decided to focus on this topic in graduate study at Kyoto University (Japan), where he received his doctorate in 1981. His doctoral dissertation was entitled *"Urban Seismic Risk: Analysis and Mitigation"*.

In 1981, he joined the consulting firm, Dames & Moore in San Francisco, with major research and engineering projects, including development of a fire following earthquake model for US cities, research into the seismic vulnerability and strengthening of low strength masonry and seismic analysis of numerous industrial facilities.

In 1987 he joined the consulting engineering firm, EQE International, where he originated and developed the *EQEHAZARD* family of software, which developed into a major business unit for EQE (*EQECAT*) serving the global insurance industry. Developed the *rapid visual screening methodology* for the US Federal Emergency Management Agency (FEMA), performed the *"Seismic Vulnerability and Impact of Disruption of Lifelines in the Conterminous United States"* for FEMA, and had increasing responsibility with projects such as the development of seismic retrofit measures for San Francisco fire department facilities, and for the City of Seattle as well.

During this period, he became increasingly interested in systems and network reliability issues for infrastructure such as water, power and telecommunications systems, and performed major reliability analyses for such large facilities for several major cities in US and Canada. In these studies, the analyses were typically for earthquake and/or other stress events, with the goal of improved reliability via the development of a mitigation program combining strengthening and design of robust and redundant system features.

He also continued field investigations of major disasters, including the 1988 Armenia, 1989 Loma Prieta, 1993 Hokkaido Nansei, 1994 Northridge, and 1999 Marmara and Duzce (Turkey) earthquakes, several major hurricanes, and numerous wild land and structural fires.

On the evening of 16 January 2005, he arrived in Osaka, Japan for a US-Japan meeting, and was shaken early the next morning by the

Hanshin (Kobe) earthquake, which was a close analog to the scenario event that was the focus of his Ph.D. research at Kyoto University 15 years earlier. He over flew Kobe that afternoon, observing the fires and damage, and then led EQE's reconnaissance team during the next two weeks. As a result of the Hanshin earthquake, while he'd come to Japan for a few day visit, he founded and ran EQE's Tokyo office for the next two years.

On returning to the US in 1997, he assumed technical direction of a major project for the National Institute of Building Sciences involving the development of a national flood loss estimation model. This project was completed in 2003, required over 100,000 man-hours of effort, and resulted in a state-of-the-art software package which permits detailed examination of flood hazards, loss potential, and mitigation benefit-cost returns for buildings, infrastructure and people, anywhere in the US. Concurrent with this project, Charles managed the Infrastructure practice for EQE, and engaged in various structural and other analyses. A notable project was the seismic analysis of the Arecibo Radio Telescope – the cable suspended structure is the world's largest radio observatory and has been featured in several motion pictures (see photos).

In 2003, Charles was appointed to the Chair of Lifelines Earthquake Engineering, Kyoto University, Japan, which position he currently holds. He is Professor in the Department of Urban Management and heads the Earthquake Disaster Prevention Systems Laboratory. Research projects focus on infrastructure risk and reliability, supported by the World Bank and corporate sponsors. Since coming to Kyoto University, he has actively investigated three major disasters, leading the US team investigating the 2004 Niigata Chuetsu (Japan) earthquake, which included the first ever derailment of a *Shinkansen* 'bullet' train; leading the lifelines portion of a JSCE team investigating the 2004 Indian Ocean earthquake and tsunami, and serving as the seismic advisor to the joint World Bank – Asian Development Bank team, which assessed recovery needs following the 2005 South Asian Earthquake (which killed 80,000 in Pakistan).

In 2002 he co-edited with me the *"Earthquake Engineering Handbook"*, and in 2005 co-edited the ASCE TCLEE Monograph *"Fire*

Following Earthquake".

Charles has been married to Nini Jensen since 1981, and they have one son.

Suspended Gregorian truss work, 2001 (Inset: aerial view showing truss suspended 500 ft above the dish)

Charles Scawthorn, 2006.

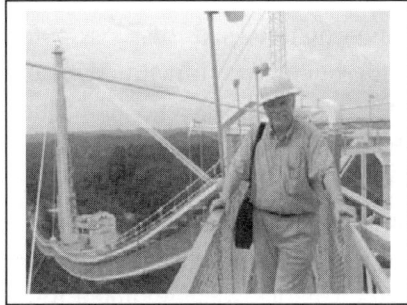

Dr. GLENN P. RENTSCHLER (1979, Lehigh)

Glenn received his BSCE from Pennsylvania State University in 1969, MSCE and Ph.D. from Lehigh University, in 1972 and 1979 respectively. Since joining WJE in 1985, he had investigated numerous deteriorated and distressed structures. He performed investigations and provided litigation support on the subjects of structural fire damage, parking garage collapses, structural roof collapses, concrete masonry walls, and steel impact damage. He is currently Unit Manager and Principal of the WJE consulting firm headquartered in Chicago. His Ph.D. thesis was on *"Tests and Analysis of Steel Beam-to-Columns Moment Connections"* under the joint guidance of myself and Professor George Driscoll. Glenn was my last Ph.D. student at Lehigh when I moved to Purdue in 1976.

Dr. JESUS LARRALDE (1984, Purdue)

Larralde came to Purdue in 1980, received his MSCE in 1982, and Ph.D. in 1984. He remembered me as a diligent, dedicated, and enthusiastic support for him to excel at the work in his thesis. He remembered his years at Purdue with great respect for the University and with joy for being able to receive part of his education as an engineer there. The lasting impression of the school of engineering in general and the civil engineering program in particular was their academic excellence. Jesus is currently the coordinator of the civil engineering program at California State University, Fresno.

Dr. WILLIAM MCCARRON (1985, Purdue)

Bill arrived at Purdue University in August of 1982, having driven through the corn fields of the Midwest. In 1983, he began working with me on his dissertation "*Soil Plasticity and Finite Element Applications*", and received his Ph.D. in August of 1985. He worked along side with many fine students.

In September of 1985, he started work at Amoco Production Research in Tulsa, Oklahoma. Early on Bill spent time on the Alaskan Beaufort Sea coast west of Prudhoe Bay. Those were the days when oil companies were heavily involved in exploring the Arctic offshore for oil reserves. His travel resulted in a great appreciation of the Alaskan wilderness. The exploration did not lead to any significant discoveries and low oil prices in the late 1980s resulted in a broad drawback from the Beaufort Sea. Over the next twenty years, the industry witnessed several episodes of increased offshore Arctic interest. However, poor economics and limited discoveries did not lead to any sustained basin developments in U.S. waters.

While working at Amoco, Bill had the opportunity to travel much and work on several interesting projects. These ranged from the instrumentation of offshore platforms to acquiring geotechnical and environmental data in Russian Siberia. One year he spent three or four months traveling in Russia. He met his wife Eileen, a geophysicist, while working at Amoco in Tulsa. They were married in 1990.

Oil exploration in the deepwater environment of the Gulf of Mexico resulted in the opportunity for Bill to work on two tension leg platforms in project engineering capacities beginning in 1994. He was responsible for the design all the subsea foundation components for the

Shell/Amoco/Exxon Ram-Powell TLP installed in 1000 meters of water. Following that, he was a project engineer on Amoco's Marlin TLP responsible for its foundation, tendon and production riser components. These were rewarding projects with many engineering and fabrication problems to overcome. He was privileged to work with many talented individuals.

Having lived in Houston for five years, Eileen and Bill were interested in moving on to a new location with more interesting scenery and fewer mosquitoes. The merger of Amoco and BP in 1999 gave them the option of leaving the company with two severance packages. They moved to Denver, Colorado, where Bill worked briefly in a small office of Schnabel Engineering. Eileen worked for a while with a couple non-profit organizations. In late 2000 the Schnabel office closed and Bill started his own one-man consultancy, ASGM Engineering. Eileen moved on to teaching high school mathematics. She works about three times harder for a fraction of the pay of a geophysicist.

Working for himself provides freedom. Bill can choose projects and schedule *'free'* time in the summer to renovate the 75 year-old house they bought in Denver.

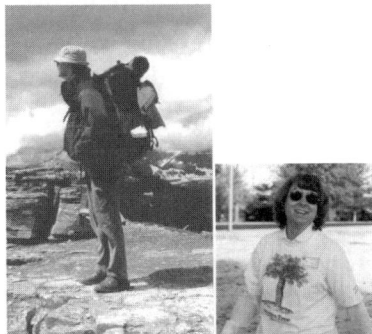

Bill backpacking in Colorado.   Eileen at a tree planting in Tulsa.

Part of the motivation for leaving Amoco was that project engineering work led him away from the technical side of the industry. While working for himself, he has been lucky to work on a variety of projects involving concrete fracture mechanics, supervision of laboratory testing programs, buckling analysis of oil and gas subsea flowlines, and numerous finite element studies. Recently, Bill has had the opportunity to become involved with a challenging geomechanical modeling project.

We collaborated on a few works in the early years following his graduation, and continue to exchange ideas in areas where our experiences intersect. Bill says his three years at Purdue were quite enjoyable, especially after the classroom work was completed. Sharing Indiana sweet corn on summer afternoons with fellow graduate students left him with good memories.

Dr. ERIC M. LUI (MSCE, 1982; Ph.D., 1985, Purdue) 呂汶

Time flies. More than twenty years have passed since Eric first stepped foot on the campus of Purdue University in 1981 to begin his graduate studies! American culture wasn't foreign to him then. He grew up in Hong Kong watching American movies, game shows and TV programs as well as eating American fast food. And since his undergraduate degree was from the University of Wisconsin-Madison, he felt pretty much at home when he arrived at the West Lafayette campus. Nevertheless, he did have that feeling of trepidation, which is human nature when one arrives at a new place in a new environment. This apprehensiveness quickly dissipated during his first meeting with me. According to Eric on his impression on me, he stated that *"Professor Chen is a very intelligent and erudite individual, yet there isn't a trace of pomposity in his demeanor. He always greeted me with a warm and welcoming smile and he constantly showed interest in my opinions, even though I was just an unfledged young man in my early twenties."*

For the following five years, I served as his academic advisor and thesis supervisor for both his M.S. and Ph.D. degrees. Needless to say, Eric felt that *"he is indebted to me and a number of other professors at Wisconsin and Purdue for our indefatigable nurturing and guidance. They have all imparted to me not only their tremendous technical know-how, but a piece of their wisdom, which has helped shape my thinking and prepare me for a challenging career in academics."*

Eric is from a family of teachers – his mother and aunt were teachers, so are his brother, sister and brother-in-law. It was just natural for him to follow their footsteps and aspired to be a teacher as well. Upon graduation from Purdue in 1985, he interviewed for a few faculty positions, received offers from three universities, and chose to begin his academic career at Syracuse University.   One reason that he chose Syracuse was the warmth, affability and collegiality that he felt about the department faculty during the entire interview process.   Little did he know that despite his loath of cold weather and his distaste for snowy winters, Syracuse has become his home since!

When he was at Purdue, his research was on semi-rigid connection effect on column and frame behavior.   He has since branched out into other areas such as limit states and performance based designs, parallel computing, structural dynamics, damage identification, and bridge and earthquake engineering, etc. His academic pursuits have been rewarding, but his proudest achievement is to see that all his students successfully graduated and received good job offers.   Like what I have done, Eric always treats his students with respect, listen to their ideas, provide them with the necessary guidance, and give them encomiums when deserved. Many of his past students are now his friends, and He takes on their achievements.

One of Eric's favorite indoor activities is to play the piano.   His most beloved classical music composers are Beethoven, Chopin, and Rachmaninoff.

Although Eric (呂汶) is hardly an outdoorsman, he does enjoy the beauty that nature accords.   The mesmerizing scenery in the background is Yellowstone Park.

At present, in addition to being a faculty member, he is also serving as Chair of the Department of Civil and Environmental Engineering. According to Eric, doing administrative work is as challenging as (if not more challenging than) scholarly pursuits since many problems can not be quantified using numbers or solved using equations, and the number of variables that are in play always seems to exceed what one can handle comfortably. Nevertheless, engineers are problem solvers, and he sure will endeavor to do the best job he can during his tenure so as not to disappoint his colleagues who have entrusted him with this leadership position.

11.2 Students from Taiwan 台灣學生

Dr. ANDRIE CHEN (Lehigh, 1973) 陳啓宗

After spending two years at the University of Kansas for his MS degree and one year into his Ph.D. study, Andrie Chen decided to transfer to a more research oriented university. He entered Lehigh in September 1969 as a graduate teaching assistant. With his gift for mathematics and theoretical work and interest in engineering applications; we decided his thesis topic on *"Constitutive Relations of Concrete and Punch Indentation Problem"*. At Lehigh at the time, almost everyone was doing steel research on beam-columns, connections or plastic design; he took this challenging topic on concrete mechanics, not a popular one at Lehigh's Fritz Lab, the Mecca of steel research in the nation at the time.

It took more than three years of concentrated efforts; Andrie formulated a beautiful kinematical hardening rule to simulate the elastic-plastic behavior of concrete. He also developed his own linear elastic-strain hardening-fracture finite element program. His research resulted in three technical journal publications. Andrie was awarded his Ph.D. in 1973. Andrie and Julia were married at Lawrence, Kansas in October 1968 and their first child, Natasha, was born at Bethlehem, PA in 1971.

Andrie moved to Harrisburg, PA immediately to start his work at Gannet, Fleming, Coddry, and Carpenter consulting firm. His work was designing highway bridges. Their son, Amos, was born in the same year. The following year he joined Gilbert Associate in Reading, PA and worked on nuclear power plant design. In 1977, Andrie joined the oil giant, the Exxon Production Research Company (EPR) in Houston; and worked in its Arctic Operations Section. Andrie was pleasantly surprised to find out that the Arctic group had been using my plasticity models to predict the load carrying capacity of moving ice. Andrie was even more surprised to find out that a training course was offered by the Platform Design Section to describe the constitutive relations of concrete materials that were developed by us as part of Andrie's thesis.

With his strong structural mechanics background, he was able to solve several analytical problems that had not been solved at the Arctic Section for a long time. He loved what he did in theoretical work but to his surprise; he was asked to lead a team to the Arctic to conduct some field tests on ice mechanics. The same test was failed once in the previous year. Since he had no training in running experiments; he had to learn these procedures quickly. At the end of the day, everything moved smoothly and he completed the mission. The manager praised him as one of their best field engineers and project managers. Consequently, he had to say goodbye to his beloved analytical work and changed course. From thereon, his work in the Arctic Section was exclusively experiments on ice.

In 1979-1980, Andrie led a team to conduct a large scale field ice strength tests in Prudhoe Bay, Alaska. The test results were reported in a paper; that won the best paper award in the 1986 Offshore Mechanics and Arctic Engineering Conference held in Tokyo, Japan. In 1984, he led a team to conduct spray ice strength test in Harrison Bay, Alaska. The strength data were used by the Bureau of Land Management as standard for spray ice island design. Since 1982, he had served as the group leader for the Arctic System Group in the development of various arctic and sub-arctic structure conceptual designs. He had two U.S. patents on Arctic structures related inventions.

The Exxon Valdez oil spill incident took place in March 1989. Shortly after that he assumed the leadership position in oil spill research in EPR, supporting Exxon's clean up efforts in Prince William Sound. His group prepared a handbook on "*Exxon Oil Spill Response Manual*". This manual was widely circulated in the industry and hailed as the industry standard at that time.

After Andrie retired from Exxon-Mobil in 2004, he wanted to do some charity work and began to help teach English at Yu-Shan

Theological College and Seminary (YTS) as a missionary. About 95 percent of the YTS students were aboriginal Taiwanese. It was a rewarding and rich experience for Andrie. Despite all the hardship, Andrie enjoyed the warm and loving relationship with these people greatly. This is what he loves to do in his retirement years.

Andrie Chen (陳啓宗) with his YTS students. Andrie at the Arctic.

Dr. STEVE S. HSIEH (1981, Purdue) 謝錫興

Steve Hsieh was a graduate of National Cheng-Kung University in Tainan, Taiwan with a MS degree in 1976; and received a David Ross Fellowship from Purdue in 1978; and was awarded a Ph.D. degree in 1981. His thesis topic was on "*An Elastic-Plastic-Fracture Analysis for Concrete*" under my joint mentoring with Professor E. C. Ting.

He joined Gilbert/Commonwealth in Reading, Pennsylvania in 1981 and performed nuclear power related structural analyses and engineering consulting services. At Gilbert, he developed and used for the first time a nonlinear finite element model for a safety analysis of a nuclear power plant in Philadelphia; and saved the client millions of dollars. In one instance, his finite element model consisted of 400,000 degrees of freedom and required the use of the powerful software: MSC.Nastran. This was his first encounter with MSC.Nastran and led to a life long work with the software company in the later years.

In 1984, Hsieh moved to Los Angeles to work for Karagozian & Case Structural Engineers which was a subcontractor of the U.S. Air Forces. This job provided him an opportunity to apply the Hsieh-Ting-Chen failure criterion and the elastic-plastic fracture model to the analysis and design of MX missile launch silos against nuclear attacks. Karagozian &

Case also initiated a joint research project with Purdue and acquired a copy of NFAP computer program from us in 1986.

In 1987, he joined MacNeal-Schwendler Corporation (MSC) which later changed name to MSC.Software Corporation in 1999. The MSC.Nastran was the most widely used structural engineering software around the world at the time. He helped the company to expand its nonlinear analysis capability. He was promoted steadily and became general manager of MSC Hong Kong in 1992. In 2001, he was promoted to general manager of Greater China and Southeast Asia. He retired from MSC in October, 2004.

In 2004, he joined the Center for Aerospace and Systems Technology (CAST) of the Industrial Technology Research Institute (ITRI) in Taiwan to help identify potential technologies developed at the Center ready for spin-off. In 2006, the ITRI CAST restructured and renamed as ITRI RFID Technology Center (RTC). He was named Deputy General Director of the new Center. He is expecting to lead one of the Center's spin-off opportunities in the future utilizing his long time experiences in international business development and operation.

His research work at Purdue and his gift in large scale programming and concrete constitutive modeling undoubtedly influenced his field selection and career choice. In fact, during the summer break in 1979, I sent Steve to Professor Paul Chang (張智勇) of the University of Akron to learn his powerful NFEAP program; and asked him to bring back a copy of the program consisting of a big box of punched cards with pillow-size printout of the program listing. One of Professor Chang's Ph.D. students helped him to store, pack and collect the huge program. In 2001, 22 years later, Steve was surprised to meet this student again at a business meeting with Goodyear Tire company at the MSC.Software headquarter in Los Angeles. In 1993, when Steve stationed in Hong Kong, he met again for the first time with Professor Chang, who moved to Hong Kong Science and Technology University. Through Professor Chang introduction, Steve met my former student, Xila Liu (劉希拉), who was at Tsinghua University; and worked with Professor Chang on a joint research project at the time. What a small world!

When Steve returned to Purdue from Akron in 1979, he was asked by the Purdue Chinese Student Association to help pick up a new girl student just arrived from Taiwan for a welcome dinner party at that day. Six years later, the girl, Jennifer Hou, became his wife. She received her M.S. degree in nutrition in 1981. They married in 1985. Steve still felt regrettable that he did not proceed to write a joint book with me on "*Introduction of Nonlinear Finite Element Analysis*" while he stationed in Hong Kong. I was surprised to learn that he still remembered that proposed book. It was long time ago.

W.F. and Steve Hsieh (謝錫與) on March 24, 2000 during W.F. (惠發) visit to National Center for High Performance Computing in Hsin-Zhu (新竹), Taiwan.

Dr. MING-FANG CHANG (1981, Purdue) 張明芳

Ming-Fang came to Purdue University in August 1978. He participated in a group research on the application of the theory of plasticity to geotechnical problems involving seismic forces. His assignment was to develop a method for analyzing seismic lateral earth pressures on retaining walls using the upper bound limit theorem. He was awarded a Ph.D. degree in 1981 for his thesis on "*Static and Seismic Lateral Earth Pressures on Retaining Structures*"

Ming-Fang received his B.S. degree from the National Taiwan University in Taipei; and his M. Eng. from the Asian Institute of Technology (AIT) in Bangkok. Before coming to Purdue, he worked firstly at AIT as a research associate for one year; and subsequently as a soil engineer for the China Engineering Consultants, Inc. (CECI) in Taipei for over four years.

After graduating from Purdue, he worked for Ardaman and Associates, a geotechnical consulting firm in Orlando, Florida, as a project engineer for nearly two years before moving to Singapore to join the then Nanyang Technological Institute as Senior Lecturer in the

School of Civil and Structural Engineering. He was promoted to Associate Professor in Geotechnical Engineering in 1991 and continued his teaching career at the present Nanyang Technological University until December 2006. Throughout his teaching career, he had maintained a close contact with engineering practice through consultancy services. Major projects he had worked on included the Singapore-Malaysia Second Link Bridge at Tuas, and the Land Reclamation at Changi East in Singapore.

Ming-Fang left his academic career after more than 23 years of service; and joined Shannon and Wilson, Inc. a geotechnical and environmental consulting firm in Seattle, Washington in 2006. He plans to continue his professional practice, especially in promoting the-state-of-the-art design of geotechnical engineering. His main specialties include in-situ testing, pile foundations, deep excavations, soil improvement, field instrumentation, and lateral earth pressure problems.

Dr. FU-HSIANG WU (1988, Purdue) 吳福祥

福祥於 1988 年 11 月得博士學位，論文題目是"Semi-Rigid Connections in Steel Frame". 學業完成後，立刻回到母校中正理工學院 (Chung Cheng Institute of Technology) 服務，擔任土木系主任，至 1990 年 8 月 15 日退伍，轉到交通部台北市區地下鐵路工程處擔任總工程司，不久再轉任交通部高速鐵路工程籌備處總工程司及副處長，至 1997 年 1 月底交通部高速鐵路工程局正式成立，即升任副局長及局長 (2005 至 2007 年)。在交通部服務期間全程參與負責台灣高速鐵路建設計畫的推動執行，歷經綜合規劃、基本設計、環境影響評估、用地取得與拆遷補償、細部設計、徵求民間投資案甄審、議約與簽約、合約執行與管理、工程興建監督、竣工監查、營運準備查驗、系統整合測試及履勘等工作，終於在 2006 年 12 月讓全世界矚目的台灣高速鐵路正式通車營運。

台灣的高速鐵路建設計畫，是以民間投資興建、營運後移轉政府 (Built-Operate-Transfer, BOT)之方式辦理，交通部於 1998 年 7 月 23 日與民間的台灣高鐵公司簽訂高鐵「興建營運合約」，總建設經費約新台幣 NT5133 億元。高速鐵路路線全長 345 公里，行車時間只需 90 分鐘，其工程最主要的特色有:(1)是目前世界上由民間投資興建營運最鉅大的鐵路工程；(2)是日本新幹線高速鐵路首次外銷到國

外成功的案例；(3)興建完成世界上最長的預力混凝土箱型梁高架橋，由台灣中部的彰化縣至南部高雄縣，總長度為 157 公里；(4)在最短時間 2 年 7 個月完成淨斷面積 90 平方公尺，全長 7.4 公里的八卦山隧道；(5)在高鐵土木工程接近完工時，福祥率工程人員以徒步 26 天踏勘高鐵主線 344.2 公里及支線 5.98 公里。

　　台灣高速鐵路於 2006 年 12 月 7 日通車營運，是世界上高速鐵路家族(日、法、德、西班牙、義大利、韓、美)的一件大事，由綜合規劃至通車營運，歷經 16 年 5 個月，
有幸全程參與，深感榮耀，也是思考退休交棒的時刻，因此已簽准提前自 2007 年 1 月 2 日退休。往後的日子，希望能在私立大學找到一份教職，把工程經驗及中醫養生傳授給學生，並擔任服務社區的義工，照顧弱勢族群，讓社會更祥和更溫暖。

W.F. (惠發) and Director F.H. Wu (吳福祥) at his office of the Bureau of High Speed Rail in Taipei, 2006 before his lecture.

Dr. WON-SUN KING (1990, Purdue) 金文森

Born in 1954 and served in the Taiwan Army during the period 1977 to 97, Colonel King is best known as a teacher in the Taiwan Military College - Chung-Cheng Institute of Technology (C.C.I.T.) (中正理工) – in the department of military engineering for over 20 years. He was professor of military engineering and head of its graduate school. In 1990, he was awarded a Loyalty and Diligence Medal from Taiwan President. In 2000, he received an achievement medal from Taiwan Prime Minister.

　　With a full scholarship from the government, Won-Sun entered Purdue in 1987 and received his Ph.D. degree in structural engineering in 1990. His thesis was on "*Simplified Second-order Inelastic Analysis for Frame Design*". In 1997, he retired from the Army; accepted a faculty position as professor of construction engineering; and also an administrative position as dean of School of Science and Engineering at the Chaoyang University of Technology (CYUT) (朝陽大學), Taiwan. In 2002, he was promoted to dean of Academic Affairs. He has served as a consultant for Taichung City Government on Committee on Public Safety Supervision.

Won-Sun King (金文生) at his office as Dean of School of Science and Engineering at Chaoyang University (朝陽大學) in Taiwan.

Dr. TIEN-KUEN HUANG (1990, Purdue) 黃添坤

Born and educated in Taiwan, T.K. Huang received his undergraduate degree in civil engineering from the National Chung-Hsing University (NCHU) in central Taiwan in 1986. In the same year he continued his graduate study at Purdue in the area of soil plasticity. He developed practical numerical procedures to implement the highly mathematical theory of plasticity to solve realistic geotechnical engineering problems. His research covered theoretical exploration, numerical simulation, and practical application of soil plasticity.

In 1990, he received his Ph.D. and returned to Taiwan as an associate professor of civil engineering at NCHU (中興大學). He continued his research in soil plasticity. Later, he expanded his study to include an emerging field of Discontinuous Deformation Analysis known as the DDA. He used the DDA method to study the mechanical behavior of retaining wall made of H-type blocks. He also carried out a full-scale field test to verify his analysis. He published a sequence of papers on the subject. One of the papers won the first prize in the Journal of Chinese Institute of Civil and Hydraulic Engineering in 1997. T.K. was promoted to full professor in the same year.

As a professor at the NCHU, he put a lot effort on teaching and research. In addition, he worked closely with practitioners; and consulted extensively in the real world geotechnical engineering. T.K. has been active in professional societies and served as committee members on city planning, construction examination, and developing test problems for the engineer license examinations.

Dr. JUI-LIN PENG (1994, Purdue) 彭瑞麟

瑞麟住在台灣, 大學畢業於國立中興大學, 碩士, 國立台灣大學。於 1991 年赴美國普渡, 於 1994 年畢業獲得博士學位, 博士論文題目為「高挑空鋼管鷹架系統之分析模式及設計指引」"Analysis Models and Design Guidelines for High-Clearance Scaffold Systems"。

瑞麟於 1991 年進入普渡大學時, 我 指導兩年後, 第三年由於申請計畫經費補助關係, 瑞麟與我的研究群中年輕之潘德恩教授, 共同撰寫鷹架支撐相關計畫書, 該計畫書通過審查後, 因為獎學金補助關係, 於第三年由潘教授掛名為指導教授, 我為共同指導教授。不過潘教授主修混凝土耐震, 而我為鋼結構穩定方面, 由於二者研究領域差異太大, 故由我為主繼續指導他完成博士學位。

碩士畢業後, 瑞麟於台灣的航空工業發展中心-起落架小組擔任飛機結構工程師工作四年, 之後才來美國普渡大學攻讀博士學位。博士畢業後回台灣在一家營造公司工作二年, 後來在私立朝陽科技大學營建工程系服務四年, 任職副教授職位。目前則是任教於國立雲林科技大學營建工程系任職教授職位。

瑞麟已結婚, 妻子在一家貿易公司工作, 有兩個可愛的女兒, 分別為 13 歲及 10 歲, 目前就讀小學 6 年級及 4 年級。

瑞麟在赴普渡大學求學前, 或許曾經在航發中心工作四年的關係, 我覺得他對於研究常較其他研究生有較實務的考量。我原先規劃他延續之前的「支撐、再撐」(Shore & Reshore)等相關研究議題, 瑞麟則提議改作「鋼管鷹架」(Scaffold)議題。他認為「支撐、再撐」用在內部挑空高度較低的結構物, 如辦公大樓、集合住宅等;「鋼管鷹架」則用在內部挑空高度較高且面積較廣的結構物, 如禮堂、倉庫、音樂廳等。若有支撐倒塌災害發生, 後者造成生命與財產的損失要較前者大的多, 後者的研究價值顯然要較前者高。我經過考量後, 同意瑞麟選擇成本效益較高的研究議題; 而且由於他還規劃畢業後要回台灣, 建議鋼管鷹架選擇台灣使用的各類型尺寸,

以方便未來回台灣的發展。迄今他一直是以這種態度，先檢討工程實務價值後才進行研究，令我印象深刻。

現今的鋼結構設計概念過於落伍，一切以構件安全為依據建立構件設計規範，事實上對於靜不定度高的結構物，單根構件的破壞並不會引發整體結構物的失敗。由於電腦發展迅速，可以考慮利用電腦強大的計算能力來直接進行分析及設計的工作，這給瑞麟非常大的啓示。

瑞麟在航發中心設計飛機時，必須要作「設計驗核」的工作，完成的原型飛機上要裝設各類量測計，在試飛階段需量測相關數據，藉以驗核原始設計是否有誤，這與汽車設計作的設計驗核是一樣的。奇怪的是我們土木工程結構物的設計，絕大部分沒有作設計驗核工作，所有的驗核只等地震來時結構物倒塌後再討論，這樣的設計是否合理呢？

瑞麟的支撐研究是屬於營造施工中的假設構造物(Falsework)，由於整體系統載重試驗的費用尚可容忍，因此可以作全尺寸結構系統分析及設計的驗核測試。其研究除了進行單組支撐結構破壞的試驗外，亦進行室外全尺寸整體支撐結構系統破壞之載重試驗，以驗核分析之支撐結構系統的破壞模式及承載力。目前他的研究，即是採用非線性分析方式直接驗核大量的試驗數據，想從系統破壞觀點處理支撐結構設計問題，避免再去討論構件設計時單根構件之有效長度係數（K）的問題。他大概是我所知道，第一個想將飛機設計之「設計驗核」概念，直接引用在營造工程支撐結構設計上的人。

我與瑞麟在普渡大學農曆春節除夕聚餐時照片 1993. 瑞麟近照

由於支撐結構用於房屋建築工程以及橋樑工程不太相同，因此，配合工程型態的支撐組搭就必須隨之變化，這幾年來瑞麟的研究極具多樣性，各類型的支撐結構承載力及營造載重特質均有涉獵。迄今研究成果已能直接用於營造工程界，瑞麟也配合台灣相關營造工程主辦機關如行政院勞工委員會等，協助他們處理營造工程模板支撐倒塌防治的工作。台灣當地工程的支撐若有發生倒塌，瑞麟多半會在第一時間陪同勞委會人員親赴意外現場收集倒塌資料，這對於他在支撐破壞機制驗核的研究上極具意義。

Dr. I-HONG CHEN (Purdue, 1999) 陳奕宏

I-Hong Chen comes from a very nice and warming family in Taiwan. All his family members have high academic achievements (his parents- a Ph.D. and a Master; his elder brother-a Ph.D.; and his wife- a Pharm. D.) and he is no exception.

I-Hong attended Purdue in 1995 and received his Ph.D. in 1999. His thesis was on "*Practical Advanced Analysis in Seismic Design of Steel Building Frames*". To gain practical engineering experience for his specialty, he started his structural engineering career in the U.S. West Coast. Over the years, I-Hong has become expert in different types of construction materials – concrete, steel, and wood. I-Hong had worked on over 10 building-towers of 30-story and higher. Among those included the "*42-story St. Regis Museum Tower in San Francisco*" which currently is the tallest concrete building in the west US continent; the "*58-story Millennium Tower (301 Mission)*" which is currently under construction and would be the new tallest concrete building in the west US when finished. His project experiences ranged from the 73-story high-rise building in New York, modern art museum, base isolation, roadway, to single wood residential house addition. I-Hong had been affiliated with SOM (Skidmore, Owings & Merrill), DESIMONE, etc.

As a practicing engineer, I-Hong has not lost touch with academy; and I have continued to encourage him to do what he loves to do. I-Hong had published papers on the AISC seismic code revisions over the years in different parts of the World including India, U.K., Belgium, and Hong Kong. He is currently at a point where he is considering being a professor to share the experience, or to start his company, or to continue as practicing engineer.

Dr. YIMING LAN (1998, Purdue) 藍一鳴

Yi-Ming, born and grew up in Taiwan, had an interesting career path. He always enjoys building things. However, his dream profession is not in civil engineering, but rather in aerospace engineering. As a step toward his dream, he studied advanced composites (known as FRP or Fiber-Reinforced Plastics) in his master program, since these materials have been extensively used in aerospace industry.

Yi-Ming was not so sure whether he should pursue a Ph.D. degree in civil engineering when he first arrived at Purdue University. But it turned out to be a good choice. The FRP applications to civil engineering structures were just taken off, when he chose this as his research topic. He felt this is the best ever happened to him.

Since the potential of FRP market in civil engineering applications is enormous, several organizations approached Purdue team to promote such a technology transfer. As my first graduate student studying this subject, Yi-Ming learned quickly, became an expert, and participated in conferences, laboratory and field demonstrations, and pilot programs. During this period, he helped me gain research funds, presented seminars to professional societies, and published articles in trade journals. Yi-Ming's other specialties are advanced composites, concrete plasticity, constitutive models, computational mechanics and finite element analysis. He received scholarships from the Civil Engineering Department, and Purdue Research Foundation during his undergraduate and graduate studies at Purdue.

After graduation, Yi-Ming served as the Transportation Department Manager and engaged in all phases of civil engineering projects from planning, design to construction inspection. In addition to structural engineering, he has worked on a variety of other areas including roadways, drainage, inspections and survey. Yiming is currently employed by Louis Bowman, the "*ASCE Civil Engineer of the Year*" in his downtown Chicago firm.

As for his personal life, Yiming has an unforgettable love story. He was engaged with a Caucasian American. How could an Irish princess be fascinated by an "*ancient Chinese warrior*" (as she described him)? To her, he is a very traditional person; whereas to him, she is unusually sophisticated. This might be their mutual attraction. She even likes how Yi-Ming eats pizzas with chopsticks. Everything was going so well. She even flew to Taiwan to meet his parents. There was one problem, however. She enjoyed everything in Taiwan, but cannot imagine living overseas; and leaving her parents alone in the US (being the only child in her family). On the other hand, Yi-Ming cannot give up his option of returning to Taiwan someday in the future either. They almost break up. In the end, they chose their affection over devotion. They married on July 12, 2007 in Taiwan.

Reunion dinner with former students at the Howard Plaza Hotel in Taipei, 2006. From right: First row: F.H. Wu. (吳福祥), W.F. (惠發), Linlin (玲玲), and J.L. Peng. (彭瑞麟).

Second row (from right): Y.L. Huang (黃玉麟), T.K. Huang (黃添坤), T.F. Lee (李騰芳), Y.K. Wang (王永康), Ken Hwa. (華根).

Dr. HSIAO-LIN CHENG (2000, Purdue) 成曉琳

曉琳來自臺灣，1987 年畢業於中正理工學院 (Chung-Cheng Institute of Technology) 獲學士學位；1991 年畢業於臺灣科技大學 (National Taiwan University of Science and Technology) 獲碩士學位；1997 年春季進入普渡大學，2000 年底獲博士學位，博士論文題目為「*A Study of FRP Wrapped Reinforced Column Columns*」。

　　曉琳碩士畢業後返回母校(中正理工學院)擔任講師，負責相關的教學與行政工作，並於 1996 年獲選為學校優良教師。1997 年曉琳獲國防公費獎助來普渡進修，由本人擔任其指導教授，1999 年因本人生涯規劃，轉任夏威夷大學擔任工學院院長，曉琳則由普渡大學土木系教授 E. D. Sotelino 接續指導，而我本人則擔任其共同指導教授繼續指導他完成博士學位，2000 年曉琳博士畢業後，返回母校軍事工程系 (Civil Engineering) 服務，任職副教授職位。

　　曉琳已婚，妻子負責家管，並育有 3 位可愛的子女，兒子目前 12 歲，女兒則分別為 7 歲及 2 歲。

　　由於曉琳獲臺灣政府公費獎助來普渡進修，其修習年限僅為 4 年，因此，在指導研究期間，充分感受到他在研究上的主動與積極性。同時，複合材料(FRP)在土木結構上的應用在當時相當流行，不僅其可應用於新建結構上，亦可應用於既有結構物之補強及維修，以提昇及改善結構效能，曉琳對此研究方向相當感興趣，並選定以「*複合材料包覆鋼筋混凝土柱*」進行相關研究。1999 年臺灣發生芮氏規模(ML) 7.3 的集集大地震，造成重大的人員及建築的損傷，其中結構物的補強及維修，複合材料(FRP)的應用就佔了重要的一部份。

　　曉琳目前除了教學及研究工作外，並擔任臺灣國防部重大工程查核的內聘委員。在研究方向，除了持續進行「*複合材料在土木結構上的應用*」研究外，由於單位任務的特性，目前也進行「*結構防護技術*」等相關研究，並積極參與國際相關的研討會。近年已完成的研究計畫：複合材料包覆混凝土在不同應變率下之材料組合率研究；複合材料貼附於鋼筋混凝土版在不同載重速率作用下之研究(I)；軍事陣地掩體結構安全性設計標準程序建立之研究。執行中的研究計畫：複合材料貼附於鋼筋混凝土版在不同載重速率作用下之研究(II)。未來的研究方向：大應變、高應變率及高壓下混凝土材料模型之數值研究；投射體衝擊及貫穿混凝土之數值模擬研究等。

曉琳近照, 2006

參加日本國防大學舉辦之國際研討會（1st International Conference on Protective Design and Analysis against Impact, Impulsive, and Shock Loads）.

Dr. KEN HWA (MS, 1995, Purdue, Ph.D., 2003, UH) 華根

1995 美國普渡大學 (Purdue University) 土木工程碩士; 2003 美國夏威夷大學 (University of Hawaii at Manoa) 土木工程博士. Ken 是我擔任

夏威夷大學工學院院長七年之中所指導的唯一的研究生，也是我所指導的博士生中最近與最年輕的一位，因此，對於他的印象也較為深刻。

Ken 的碩士學位也是由我指導完成的。他於 1994 年 8 月進入美國普渡大學碩士班，那時，大部份的碩士班學生都選擇以修課而不寫論文的方式修習，Ken 在三個學期中跟隨我修了三門課，我想他是屬於刻苦認真，而非天才類型的學生，他的成績相當不錯。1995 年 12 月在我主持的口試委員會中，他通過了口試而取得碩士學位。

1999 年 9 月我剛到夏威夷大學擔任工學院院長不久後，就接到 Ken 的 e-mail，表示希望到夏大來請我指導他的博士進修。雖然院長的工作相當忙碌，但基於過去對他的認識，我欣然答應了他的請求。由於 Ken 在獲得碩士學位後曾在台灣中壢的南亞工專建築系擔任講師，教授工程力學及結構學等科目，所以夏大土木系給予他獎學金，並擔任助教的工作，後來分別在 2001 年及 2003 年的春季班還請他擔任「應用力學 I (靜力學),的授課老師，學生對於他的教學反應相當良好，他並於 2003 年 1 月獲得夏大土木系推薦申請傑出助教的 Frances Davis 獎。

除教學之外，Ken 在夏威夷的第一年閱讀了大量有關 Advanced Analysis 的論文及書籍。我自 1990 年代開始就與我的學生們致力於 Advanced Analysis 應用於鋼結構分析設計之研究，此一設計法係以功能日漸強大的電腦軟硬體設備執行結構的二階非彈性分析, (second-order inelastic analysis); 因而可以免除繁瑣的有效長度因數 K 的計算及個別構件強度的檢核，從而大幅簡化結構分析的步驟。其時，Advanced Analysis 的理論已發展的相當完善且已成為鋼結構設計的新趨勢，我希望 Ken 能將之商用化而易於實務工程師的使用或將之應用於特定的工程實務議題中。

2001 年 9 月 11 日美國紐約世貿大樓攻擊事件之後，Ken 即開始思考將 Advanced Analysis 應用於鋼構架的耐火性能設計 (performance-based fire resistance design)。在我的指導與協助下，Ken 於 2002 年底發展出以試誤過程估算在增溫情形下，固定負載鋼構架能保持穩定的極限溫度的方法及步驟，結構自受火害至失穩的時間與破壞模式也能一併預測。之所以要採用試誤過程乃是因為原本的 Advanced Analysis 並非是針對結構物火害來發展的。Ken 的博士論文: *"Toward Advanced Analysis in Steel Frame Design"* 於 2003 年 4 月完成並通過論文口試。

    Ken 於 2003 年 7 月返回台灣中壢的南亞技術學院 (Nanya Institute of Technology) 建築系並晉升為副教授。由於南亞技術學院是一所小型的私立學校，研究人員及經費均亟缺乏，因此我鼓勵他能有機會往行政職位上發展。Ken 曾分別於 1997 年與 1999 年與另二位普渡大學土木系的校友李台光博士及田乾隆先生合作翻譯了二本工程教科書「結構分析」(編者為 Aslam Kassimali)及「鋼結構設計」(編者為 J. C. Smith)，另外，他也分別在 1998 及 2006 年以中文編寫二本大學部使用的教科書「*結構學*」及「*工程力學-靜力學於材料力學*」，對於土木工程教育也有相當程度的貢獻。
Ken 尚未結婚，我現在最關心的是他的終身大事。

Left: 2003 年 5 月與 Ken 攝於夏威夷大學校園.

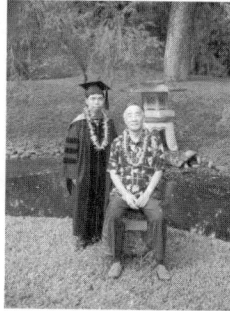

Right: 2003 年 5 月與 Ken 攝於夏威夷大學中之日本花園.

11.3 Students from China 中國學生

Dr. DA-JIAN HAN (MS, 1982, and Ph.D., 1984, Purdue) 韓大健

After passed a very competitive national examination in 1978, Da-Jian was selected by the Chinese government and sent to US for graduate study. On February 14, 1980, at the age of 39, she arrived at Purdue University to start her graduate program under my supervision. In four and a half years at Purdue, she received her M.S. in 1982 and Ph.D. in 1984, respectively. Then she returned to China and continued her teaching career in Civil Engineering at the South China University of Technology（華南理工大學）in Guangzhou（廣州），Guangdong Province（廣東省）.

    "Professor Chen is a leading researcher as well as an outstanding teacher", stated Da-Jian in her memory, "He always gives the students opportunity to extend their capability". Da-Jian continued by saying, "It is only under his direction and encouragement she could finish her M.S. on "Cyclic Inelastic Behavior of Steel Tubular Beam-Columns" and Ph.D. on "Constitutive Modeling in Analysis of Concrete Structures", and later publish two books coauthored with Professor Chen, entitled "Tubular Members in Offshore Structures" and "Plasticity for Structural Engineers" as well as two book chapters.

Da-Jian further explained that "The experience of studying and working with Professor Chen at Purdue has widened her academic view and plucked up her academic courage and thus created favorable conditions for her successful career".

Several years after her graduation from Purdue, Da-Jian was promoted to full Professorship; and started to supervise her own Ph.D. students. With economic booming in China there were lots of civil engineering projects such as high-rise buildings, bridges, highways and tunnels, etc. And there was a great demand for research related to those projects. Since the late 1980's, she has been leading a group of talented engineers and students working on a variety of research topics, such as construction monitoring and control for large-span cable-stayed bridges, wind field simulation and analysis of wind induced vibration of large-span bridges and large-span space structures, analysis and seismic design for a submerged tunnel, damage identification and bridge health monitoring, bridge condition evaluation and bridge management system, etc.

Da-Jian also involved in the constructions of four bridges with spans of 338m to 480m; and the first submerged tunnel in Pearl River (珠江) in China used their research results. She has been well recognized for her research in the field of structural engineering and structural mechanics. She has published 115 journal papers; produced 20 Ph.D. and 43 M.S. students. In 2005, her research team won a top award in Science and Technology by the Guangdong Provincial Government.

From 1993-2000, Da-Jian served as the Vice President of the South China University of Technology (SCUT). The SCUT is one of the top 20 best universities in China. Also, since 1998, she has served as a standing committee member of the prestigious Chinese People Political Consulting Committee (CPPCC) （中國人民政治協商會）, the equivalent of a member in the US Congress.

Da-Jian's family is a quite traditional old-fashion Chinese one: with four generations living together. Her father is 98 years old, while her grand daughter is only 5. Her husband was also a faculty in the same university and now retired. They have two children. Both are engineers. Their son is working in Columbus, Ohio in US; while their daughter is working in an architectural design institute and frequently coming home to stay with them.

Left: Da-Jian with her team on the construction side of the Guangzhou （廣州） Panyu Cable-Stayed Bridge (1997).
Right: Da-Jian at the opening ceremony of the Guangdong Cultural Week (Marseilles, France 1994). （廣東省政協副主席韓大建在馬賽・中国廣東文化周開幕式上講話 04-6-19).

Dr. LIAN DUAN (1990, Purdue) 段鍊

Lian Duan is a Senior Bridge Engineer and Structural Steel Committee Chair with the California Department of Transportation. He received his B.S. in Civil Engineering in 1975; M.S. in Structural Engineering in 1981 from Taiyuan University of Technology; and Ph.D. in Structural Engineering from Purdue University in 1990. He had worked at the North China Power Design Institute from 1975 to 1978 and taught at Taiyuan University of Technology from 1981 to 1985. His Ph.D. thesis was on *"Stability Analysis and Design of Steel Structures"*.

His research interests covered a wide area including inelastic behavior of reinforced concrete and steel structures, structural stability, seismic bridge analysis and design. With more than 60 authored and co-authored papers, chapters, and reports, his research focused on the development of unified interaction equations for steel beam-columns, flexural stiffness of reinforced concrete members, effective length factors of compression members, and design of bridge structures.

Duan is an esteemed practicing engineer. He had designed numerous building and bridge structures. He was lead engineer for the development of Caltrans *Guide Specifications for Seismic Design of Steel Bridges*. He co-edited the *Bridge Engineering Handbook* with me in 2000. This Handbook won a Choice magazine's Outstanding Academic Title award for 2000. He received the prestigious 2001 Arthur M. Wellington Prize from the American Society of Civil Engineers (ASCE).

Dr. WEI-HONG YANG (1997, Purdue) 楊卫红

Born in China and entered a three-year college at the age of 16, Yang get lost completely in his first two years of study, but managed to graduate with an average GPA in 1985. After graduation, he worked as a design engineer at the Industrial Design Institute, Changsha（長沙）, China. He had a very close girl friend. Unfortunately, the girl married to someone else. For Yang, this waked him up. He studied extremely hard in the following three years to prove to his girl friend that she made a big mistake. With his hard work, he was awarded a research assistantship from Harbin Architectural & Civil Engineering Institute in Harbin（哈爾濱）, China from 1988 to 1992.

In 1993, Wei-Hong won a scholarship from Chinese government to study in UK; and entered the Manchester University, Manchester, UK. His officemate was Osmen Anwar Beg, who ten years later turned out to be the author of the book on *"Giants of Engineering Science"* in which he featured me as one of the ten giants in engineering science in American over the last 40 years. Osmen was at least 5 years younger than Wei-Hong. He was working on his M.S. degree on some sort of fluid mechanics subject at the time. Wei-Hong believed that Osmen knew my name by reading one of my books on soil mechanic. Wei-Hong knew me by reading my book entitled: *"Plasticity in Reinforced Concrete"* around 1991-1992. Those days, according to him, finite element analysis on reinforced concrete was new and quite hot in China. Since the university had only few original books from overseas, his school (then Harbin Architectural and Civil Engineering Institute) set aside a room for reading those *"original foreign books"* in the library through a loan form the World Bank. Wei-Hong came cross the book there, checked it out and studied it from beginning to the end; all were brand new information to him. From that moment on, he said, he was quite determined to be my student some day in the future. He wrote me a letter afterwards.

As an officemate and close friend, Osmen talked a lot about me. With that constant talk and further influence, Wei-Hong decided to transfer quickly to Purdue University to become my doctoral student in 1994. His thesis topic was on *"The Behavior and Design of Unstiffened Seated-Beam Connections"* under the guidance of myself and Professor Mark Bowman. He was awarded his Ph.D. in 1997 with a perfect GPA of 4.0; and received the Maple Point Foundation Award in the same year.

Wei-Hong's dreamed profession used to be either in medicine or in financial business, but opportunity led him to structural engineering, which was quite close to his childhood dream to become a world famous architect. He worked one year at a San Francisco structural engineering firm; and then moved to Weidlinger Associates, Inc., Mountain View, CA in 1999 as a senior research engineer. Through his self-study, Wei-Hong has become a specialist in blast engineering. He loves his work as a structural engineer; and plans to become an independent consultant in building design in the future.

11.4 Students from Korea 韓國學生

Dr. SEUNG EOCK KIM (1993, Purdue) 김승억

I was pleasantly surprised to learn from Seung-Eock that when he enrolled to our graduate program, he was *"anticipating and looking forward to the stability of structure lecture, the first course he would take at Purdue University"*. The reason for the excitement was, according to him, that *"the lecture was given by the world renowned Professor Chen. Among many lectures he took in his life, Professor Chen's teaching was the best by far. Professor Chen began the very first lecture by asking the difference between elastic and linear".* Seung-Eock promptly raised his hand and gave an answer, which left a lasting impression on me. By receiving the highest grade in the course, I offered him a research assistantship; and assigned him to work on the advanced analysis for steel frame design.

Before starting his Ph.D. program in 1994 at an age of 34, Seung-Eock worked in Korea for 11 years as a structure engineer. Since his age was rather old for a Korean to start a Ph.D. program, it was necessary for him to finish his degree as fast as possible and return to Korea. Considering this situation, I advised and worked with him closely to achieve his goal. Seung-Eock said that *"I was a hard and fast worker. If he gave me a draft of paper on Friday afternoon, I always returned the corrected version of the draft filled with my red remarks on the following Monday morning".* The repetition of this working process allowed him to finish his degree in 30 months and to publish 6 papers in national journals.

During his Ph.D. years, he said he always wanted to publish a book. So when I asked him to consider writing a textbook based on his thesis, he jumped on the opportunity and published a textbook on *"LRFD Steel Design Using Advanced Analysis"* with me as a coauthor in 1997. These accomplishments greatly helped him find a faculty position in Korea. Looking back, Seung-Eock felt that it was a great fortunate meeting me and became my graduate student.

Seung-Eock spent the past 10 years at Sejong University in Korea as a professor. By following my teaching of *"fast and hard work"* he said, he was able to publish several dozens of paper and to be involved in numerous research projects. Among many accomplishments in Korea, the most prestigious one was the receiving of the NRL (National Research Laboratory) grant, which was financially supported by Korea Science and Technology Department from year 2000 to 2005 in the amount of one million US dollars. The NRL project is coveted by all of scientists and researchers as well as professors in Korea. Therefore, by receiving the grant, his laboratory received a national recognition and is respected by his peers in Korea. As a matching fund to his NRL research project, Sejong University constructed a structure testing laboratory for his department. This laboratory gave his department faculties and students easy access to structural testing. The main topic of the NRL project was *"Inelastic, nonlinear analysis of steel frame structure"*; and the representative output includes a development of practical analysis and design program software using inelastic, nonlinear analysis entitled *"3D-PAAP: three dimensional practical advanced analysis program"*.

The photo shows Seung-Eock's wife and his two sons. The eldest is currently in US as a high school student and second son is a junior high school student in Korea who is especially interested in Math and Science.

Seung-Eock would like to especially thank his wife for her continuous prayers, loving supports, and unending sacrifices for the better of their family.

Dr. CHANGBIN JOH (1998, Purdue) 曺彰彬

Changbin received his B.S. and M.S. degrees in civil engineering from the Seoul National University (SNU) in Korea. He was awarded his Ph.D. degree from Purdue in May 1998. His thesis topic was *"Application of Fracture Mechanics to Steel Moment Connections"*

He started his career as an adjunct professor at the SNU. His research covered two major topics: the seismic behavior of a concrete-filled steel pier; and the design of void slab bridges. In 2000, he joined me at the University of Hawaii as my research associate and continued his study on the seismic behavior of steel moment connections. In 2002, he joined the Korea Institute of Construction Technology (KICT) as a senior researcher. His research was focused on the fatigue behavior of prestressed concrete deck slab.

Changbin can think of at least four milestones in his life after his MS degree. The first was to propose to his wife, Hwajeong Park. She has been his best friend and advisor since then. The second was going abroad for his doctoral study. The third was to become my graduate student. The fourth, which is currently underway, is to join a diet program.

Before and after the diet program.

Changbin is very excited about his new project on the development of a hybrid cable stayed bridge with more than 600 meter main span using ultra high performance concrete (UHPC). This is a 6-year project for him to accomplish. His dream is to become a well-respected expert on bridges, long-span bridges in particular.

11.5 Students from Japan 日本學生

Dr. SHOUJI TOMA (1980, Purdue) 当麻庄司

Toma was a graduate of Kobe University, Kobe, Japan in 1967; and joined the Kawasaki Heavy Industry Corporation as a structural engineer for ten years. In 1977, he came to Purdue University with a research assistantship; and worked on the offshore structural engineering program. He received his MSCE in 1978 on "*Analysis of Axially Loaded Fabricated Tubular Columns*"; and his Ph.D. degree in 1980 on "*Analysis and Design of Fabricated Tubular Beam-Columns*".

In January 1981, he returned to the Welding Research Laboratory of the Kawasaki Heavy Industry. A year later, he accepted an associate professor offer from the Hokkai-Gakuen University, Sapporo. He was promoted to full professor a year later. He took one year sabbatical leave at Purdue in 1990 to 1991 as a visiting professor in the School of Civil Engineering.

Toma visited Dean Chen at the University of Hawaii in March 2002. The photo was taken with Toma in Dean Office in engineering.

The photo above was taken at the time in Sapporo with (from left) Prof. N. Kishi, Prof. Y. Takahashi, Prof. K. Matsuoka (now President of Muroran Institute of Technology), Prof. Chen and Prof. Y. Honda, who all (except Prof. Y. Takahashi) were visiting scholars of Prof. Chen at Purdue University.

In the summer of 1990, with the support of a Fellowship for Research by the Japan Society for the Promotion of Science, I and my second son, Arnold, came to Japan for a one-month lecture tour. In 1995, I made a second trip with my youngest son, Brian, to Japan. Since I had many former students and researchers all over Japan, who worked with me; Toma arranged itineraries for me for each trip to meet with all of them and give lectures at most national universities in Japan. The tour started from south to north including Fukuoka, Hiroshima, Kobe, Kyoto, Nagoya, Tokyo and Sapporo.

Professor Chen visited the site of the Kobe Earthquake on July 21, 1995, six months after the major earthquake to see the damages. The photo was taken with my former student, Prof. Y. Ohtani (right), and my son, Brian (中宇, left).

Dr. EIKI YAMAGUCHI (Ph.D., 1987, Purdue)

Even though it is more than twenty years ago, Eiki still remembers vividly his first impression on Indiana: white, white, white.

As he started his graduate study for Ph.D. at Purdue University in January, 1984, he left Japan just before Christmas in 1983. He had a good friend, Makoto Obata, studying at Northwestern University at that time, so he landed in Chicago and stayed in Makoto's place in Evanston, Illinois, for one night. Makoto drove him to West Lafayette, Indian, next day.

The scenery during the trip in his memory is just white: everything was covered with snow all the way from Chicago to West Lafayette. Except in some northern part, it doesn't snow so much and is not so cold in Japan. The weather seemed completely different from what he got used to and seemed depressing. He was wondering where he was heading for.

After several hours of driving, a small town finally appeared along the interstate and that was West Lafayette where Purdue University is located. He was rather disappointed and sad, thinking he was going to stay in this seemingly remote town for some years to come.

Since I had hired him as a research assistant, we had a meeting next day. The details are blurring now, but he remembers well that the meeting changed the entire mood he was in: he was very much encouraged to pursue the degree.

Hard work under my guidance got him to earn the degree successfully. His subject was mechanics of concrete materials. In 1986, there was the US-Japan Seminar on this issue in Tokyo. I was invited to deliver a lecture there and presented his work. He was very happy about it. After the graduate study at Purdue, he returned to Japan and has been working in academia in Japan ever since. He dealt with concrete materials for his Ph.D., but like many of my former students, it may be better stated that his major is applied mechanics. He can handle any materials. I intentionally educated my students that way. In fact, his recent research is more related to steel than concrete.

In addition to the academic experience and knowledge, he treasures the people he got acquainted with at Purdue. Eric Lui (呂汶), Xila Liu (劉希拉), Lian Duan (段煉), and Hong Zhang (張宏), to name a few, are those who got Ph.D. under my supervision about the same time, and have been his good friends.

Although they live in different parts of the world now, he keeps in touch with them: he has visited Lian and Hong in California and Eric in New York. He saw Xila in several occasions such as an international conference. He will probably see Xila in Shanghai early next year.

Friends are important and helpful for professional activities as well, he feels. For instance, whenever he needs information regarding steel structures in US, he can e-mail Lian and/or Eric without hesitation and they e-mail him back with what he needs right away. It is just a couple of weeks ago when he asked Lian for the information on design practice of steel bridges in US, which was needed in the activity towards the revision of Japanese design codes for highway bridges.

He also met many Japanese during his stay at Purdue. Among them, he must mention Professor Yosiaki Goto and Professor Norimitsu Kishi, who came to Purdue to work with me for about a year. He says they have been his mentors and are also role models after he started his academic career in Japan. Last March, he visited me in Hawaii with Professor Kishi (Photo below).

He is now a professor of civil engineering at Kyushu Institute of Technology, Japan. The school is in the hometown of his wife, Masami, who he met at Purdue! Purdue means really a lot to him in many ways.

Life in Japan is busy, requiring a hard work. Academia is not the exception. Teaching, research and administration keep one busy at school. Social services such as the involvement in technical committees and attendance in conferences make one even busier. Especially so these days, as the university system in Japan is changing drastically. Nevertheless, he likes the life in academia and doesn't mind working hard. He particularly enjoys interacting with students, hoping that the college life will be one of their good memories in the future, just like Purdue life is one of his fondest memories now.

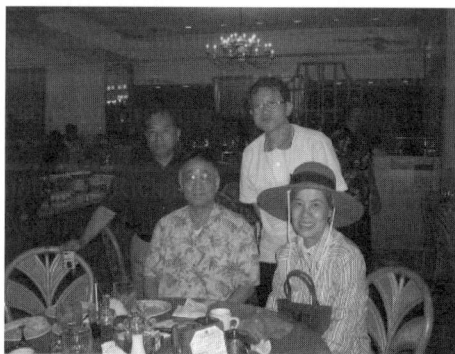

Dr. Kishi (left, standing) and Eiki with W.F. (惠發) and Linlin (玲玲, seat with hat) in Honolulu in 2006.

11.6 Student from Singapore 新加坡學生

Dr. J Y RICHARD LIEW (Purdue, 1992) 劉德源

Richard Liew received his B.Eng degree from the National University of

Singapore (NUS) in 1986, and M.Eng degree from the NUS in 1988. He received a scholarship from NUS to pursue a Ph.D. degree. The scholarship included an obligation to teach in NUS for five years after returning from overseas studying. The scholarship awarded by NUS included full salary, college tuition fees and all overseas expenses. It was an incentive to encourage Singapore citizens to become academics in the government-funded university and to train teaching staff in strategic disciplines. Richard was chosen to specialize in structural steel. During that time, Singapore's economy went into its first recession after independence because of oil crisis (1985 to 1987), and many of his classmates could not even find jobs in the construction and building service industry.

Richard decided to study at Purdue University because its School of Civil Engineering ranked top five in the United States. He acquainted my reputation through his research into my beam-column work, and read many of my publications and books during his Master's program in NUS. Richard believed that I was the world's renowned authority on steel structures and he was glad that I accepted his application. He realized later that his trip to the United States had changed his life when he began to pursue an academic career in NUS.

Richard married Liang-See in 1988. She resigned from her work as a high school teacher to accompany Richard to America. All this was done within 3 months and they arrived at Purdue in August 1989. His wife did not feel well after arriving at Purdue. Initially they thought this was due to jet lag (the time difference between Singapore and State of Indiana is about 13 hours). But it turned out later from her medical examination that she was pregnant at that time. Richard was overjoyed but realized the burden soon after. He was motivated to complete his Ph.D. degree as soon as possible to return home to pursue his teaching career. Meanwhile, his wife did a Master's degree specializing in gifted education at Purdue. She gave birth to their daughter, Jane, at St. Elizabeth Hospital, West Lafayette, in April 20, 1990, by that time Richard had completed his Ph.D. qualifying exam.

Richard remembered meeting me for the first time in my office. He said that we chatted for about half-an-hour, and I wrote on a piece of paper about his research scope and my idea. I gave him many reference materials (many of them were the research reports from my former Ph.D. students) and told him to study them carefully and to prepare for his research plan. Richard was anxious at the time to tell me about his ambition to complete his Ph.D. work in as short the time as possible. I referred him to the scope of work and said, "*as long as you can finish your work, you are free to go.*" Richard said he took my words very seriously, and immediate worked on the research project since the first day we met.

He completed his task in two years time and handed me the draft of his thesis and we discussed the possibility of his Ph.D. exam. I also advised him to publish his work and I agreed that he could take his exam. Richard passed his Ph.D. exam and stayed for another three months to complete a series of journal papers. Richard was a very task-oriented person; and learned how to organize work in a very efficient manner. Richard believed that his approach to organize work was influenced by me; because I was one of the most prolific researchers during that time, and an effective use of time was the key for my success. Richard returned to Singapore in March 1992 as a lecturer in NUS, and his wife obtained her Master's degree and a teaching position in a secondary school and their daughter was about 22 months old.

When he was in NUS, we continued in close touch; and worked on several book chapters and books. We published a book on "*Stability Design of Semi-Rigid Frames*" together with Professor Y. Goto from the Nagoya Institute of Technology in 1995. In this book, Richard included a computer program with FORTRAN source code that he developed for second-order plastic hinge analysis of two dimensional frames. This helped those research students and practicing engineers; who wrote to us indicating their intention to further improve and extend the program to other applications. The main contribution to this book was to provide a design guide and computer source codes with connection database for designing semi-rigid steel frames considering direct second-order inelastic effects. I

We continued to develop close bond in our subsequent work on the second editions of the Civil Engineering Handbook, and a number of book chapters in the areas related to plastic and stability design of frames. From this, Richard learned to mentor and nurture his own Ph.D. students and maintain close contact with them after they graduated. Many of his Ph.D. students are currently holding senior positions in the steel construction and offshore oil and gas industries. He continued to work with his post graduate students on many challenging projects and solved complex problems that were outside the scope of the design codes and standard textbooks.

Singapore economy was doing well from 1992-1996 with annual economy growth averaged about 8 percent. The good economy prospects provided greater opportunities to obtain research funds and working in major industrial and building projects. He continued to involve himself in technical committees and in professional works. After 1996, Singapore entered into a long period of slower economic growth and the construction industry was hurt. He became the President of Singapore Structural Steel Society (SSSS) from 2000-2002. He gathered a group of very capable individuals to form a strong council. With generous funding from the industry, SSSS began to gain wider acceptance. The steel industry gradually gained strength and more projects adopted structural steel and composite designs. The Steel Society continues to grow even after he stepped down as the president in 2002 till now. He began to network more with the professionals and contractors in the steel industry. He established a reputable structural steel research group in NUS capable of drawing supports from the industry.

In 1990, many of my students including Richard worked on nonlinear inelastic analysis of frames. This was a logical extension from my early work on theory of beam columns. Richard's task was to focus on the overall frame behavior rather than on individual member's behavior, which was the subject of research for many decades. The Structural Steel Research Council (SSRC) decided to add a new task group on advanced analysis of steel frames chaired by D W White, who served as Richard's thesis co-supervisor. Richard's first paper from Purdue was on *"Methods of Effective Length Considering Story Buckling Effect"*, but then devoted most of his work to develop a technique for direct second-order inelastic analysis that would eventually remove the limitations and use of effective length method for frame design. Later, he met his good friend Prof. S.L. Chan （陳樹禮） from the Hong Kong Polytechnic University; who spent 6 months at Purdue for his sabbatical leave and he told Richard that he could gather a group of people to be the pioneer members of the "CAMEL" Club. CAMEL means "*Campaign Against the Method of Effective Length*". They decided that Don White deserved the honor. It took more than 10 years for AISC LRFD to include second-order analysis methods that do not require the use of column effective length. S.L. Chan and Richard worked together and successfully got the techniques to be included in the Hong Kong's steel code in 2005.

According to Richard, he took a number of post-graduate modules used in my teaching, including plasticity for structural engineers, plastic design of structures and structural stability. The subjects were taught in such a concise manner that they provided sound fundamental background and applications that can be done in a simple manner using the available computer resources. In early 1990, the speed of computer was double every 1.5 years. Richard's research then was to integrate the theory of plasticity and stability and to develop a computational method for designing semi-rigid steel frames using personal computers. The emphasis was on design implementation in order to make impact to the design professions. Richard said that his teaching in NUS was largely influenced by the way he learned from my teaching. Richard won several teaching awards in NUS. He included practical examples that he learned from his professional practices and he always tried to connect theory to practice in his teaching.

Richard Liew with his wife LiangSee, daughter Jane and sons, John and Jovin.

Richard has engaged in an interesting journey which requires him to learn continuously. He grew wiser both intellectually and emotionally. He learned to adapt fast and remained flexible. Richard firmly believes that his training at Purdue allowed him to work on blast and fire engineering, composite systems, forensic investigations, complex fabrication and construction problems, etc. He still loves what he is doing. He learned to keep things simple just like what I have preached. He became directors of several research initiatives including a public listed company in Singapore. His involvement in new business and technology venture allowed him to learn financial management, risk assessment and human resource skills. These are invaluable experience for him to continue his journey.

11.7 Students from Egypt 埃及學生

Dr. ATEF SALEEB (1981, Purdue)

Saleeb is a professor of civil engineering at the University of Akron, Akron, Ohio since 1992. He came to Purdue with a graduate research assistantship in 1977; received his M.S. in 1979; and Ph.D. in structural engineering in 1981. He did one-year post doctoral research work at Purdue; and returned to Egypt in 1983. He taught one year at Cairo University, and then returned to US as an assistant professor at the University of Akron in 1983. He was promoted to full professor in 1992.

  Born and educated in Egypt, Atef received his undergraduate degree in civil engineering from Cairo University with Distinction with Honors Degree (ranked first) in 1974, and received the Egyptian Ministry of education Special Award for Nations Top Ten Students entering the University (total of about 117,000 students). During his doctoral study at Purdue, he received the 1981 Nellie Munson Award for Best Teaching Assistant in the School of Civil Engineering.   During his academic career at the University of Akron, he was awarded both the Outstanding Research Award; and the College of Engineering Louis A. Hill Award for Outstanding Achievement by the University in 1995. He authored or coauthored 67 technical papers and 20 conferences proceeding articles, and produced more than 22 Ph.D. students over the last 22 years. Saleeb and I coauthored the volume one of the book "*Constitutive Equations for Engineering Materials*", John Wiley & Sons, New York, 1982. Atef was nominated as a candidate for the distinguished professorship at the University in 2005.

Dr. SALAH El-DIN E. EL-METWALLY (1986, Purdue)

Born and educated in Egypt, Salah received his undergraduate degree in civil engineering at Mansoura University, Mansoura, Egypt with the highest distinction among 320 students class in 1977. He received a Peace Fellowship for graduate study in USA in1981-1982 and was awarded a MS degree in structural engineering from the George Washington University, Washington, D.C. in 1982. In 1986, he was awarded a Ph.D. in structural engineering from Purdue with an Egyptian Fellowship for graduate study in USA, 1983-1985. His Ph.D. thesis was on "*Nonlinear Analysis of Reinforced Concrete Frames*".

Salah's work has spanned several civil engineering areas over the last 20 years as a faculty at Mansoura University; he had been honored with several awards in recognition of his accomplishments. In 1991, Salah won the Alexander von Humboldt fellowship for two years. In 1995, he received an Award of Distinction of First Class by the President of Egypt. In 1996, he was granted a State Prize in Structural Engineering by the Egyptian Academy of Science and Technology. During his Ph.D. study, he received a Nellie Munson Award for the best teaching assistant in the School of Civil Engineering, Purdue University in 1985.

He has been active in the real world of structural engineering through his extensive consulting work; and has produced 18 M.S. students and four doctoral students during his teaching career. He authored and coauthored 54 technical papers and one textbook on reinforced concrete design.

11.8 Students from Thailand 泰國學生

Dr. NIMITCHAI SNITBHAN (Lehigh, 1975)

Born in July, 1945 and educated in Thailand, Nimitchai entered a graduate program at Lehigh University in 1970, and received MS in civil engineering in 1972; and Ph.D. in 1975. His thesis was on *"Plasticity Solutions for Slopes in Anisotropic Inhomogeneous Soil"*.

Nimitchai spent his first 20 years as engineer and directors of the State Railway of Thailand (1975-1995). From 1996 to 2005, he worked independently as a transportation and logistics consultant for the International Institutions, i.e.: IBRD, JBIC, United Nations, and the Thailand Development And Research Institute.

As Project's Chief Bridge Engineer, Nimitchai carried out stress analysis and design for the rehabilitation of the King Rama I Memorial Bridge, the first multi-span steel trussed bridge across the Chao Phaya River in Bangkok. He designed railway bridges and mono-block ties of prestessed concrete construction; which have become a *"Standard Design for the Thai Railway"* since 1978. He took part in developing the Trans Asian Railway Network linking ASEAN member countries with Europe and the Middle East.

Nimitchai married to Chutatip Phornphiboon from Thailand who received her MBA from Lehigh University in 1975. They have a daughter who received her BA in psychology from Lehigh in 2001 and MA in applied psychology from New York University. They have a son who received his BS in biomedical engineering from Johns Hopkins University in 2002, and MS from University of Pennsylvania in 2003.

Nimitchai and his wife Chutatip Phornphiboon.

Nimitchai has recently been appointed as a member of the Board of Directors of the two state-owned enterprises: State Railway of Thailand, and Mass Rapid Transit Authority of Thailand; and there would be work to do for the future years. Other than that, he has traveled extensively with his wife to places in Thailand and abroad. This will be their main activities for the years to come.

Dr. KAMAITON WONGKAEW (2000, Purdue)

Born in Thailand and educated mostly in the US, Wongkaew received all his three degrees from Purdue: BSCE in 1996, MSCE in 1997, and Ph.D. in 2000. He joined AECON, Inc. in Bloomington, Indiana as a project engineer in 2000; and then moved to Parsons Brinckerhoff Quade & Douglas, San Francisco, California, in 2002. During his consulting work, he performed engineering analysis and design of structures according to applicable design standards. When explicit design codes were not available, he conducted research, applied engineering theory and developed practical analysis and design procedures. His practical experience included the design of segmental steel fiber reinforced concrete lining for sewer tunnel in Portland, Oregon, the design of un-reinforced concrete lining for the light rail twin tunnels in Sydney, Australia, and the design of the Radar Building in Taiwan.

He has gift in mathematics and his thesis on *"Practical Advanced Analysis for Design of Laterally Unrestrained Steel Planar Frames under In-plane Loads"* was highly theoretical and most challenging involving inelastic out-of-plane flexural-torsion buckling in the design of planar frames. He coauthored with me *"Chapter 1: Introduction"* in *Practical Analysis for Semi-Rigid Frame Design*, World Scientific, Singapore, 2001, and a paper *"Consideration of Out-of-Plane Buckling in Advanced Analysis for Planar Steel Frame Design"*, Journal of Constructional Steel Research, 2002. Wongkaew also published 2 additional papers on the design of underground structures.

11.9 Student from Norway 挪威學生

EINAR DAHL-JORGENSEN (M.S., Lehigh, 1975)

Born and raised in Norway, and went to a commercial high school there before was drafted to a compulsory service in the army. This was in the early sixties where the Soviets were testing hydrogen bombs up the Arctic not too far away from the northernmost part of Norway. Einar was up there watching the Soviets in binoculars across the river dividing the two countries.

After dismissal from the Army he decided that economics was not for him. It would have to be engineering. He didn't have the right high school degree to be accepted to an engineering program at a university. So he went to Sweden with the purpose to enroll at a school that would give him an engineering degree after two years at what could be compared with a community college level.

While waiting to be admitted, Einar started to work as a draftsman in an engineering firm. Although the salary not much to write home about, he could support himself and even spend an occasional evening out with friends. Instead of attending the two-year engineering curriculum, he enrolled at a night school where the same program took four years; but he could work full time and not pile up any debts. It made sense to him at the time. The consulting company he worked for specialized in design of steel structures. So design of steel structures was what he learned to do.

Einar returned to Norway in 1968 and worked first for a steel fabricator and later for an engineering consultant as their steel design specialist. He felt he did a better job within his field than his colleagues with a university degree from Norway, and for less pay. At the same time Einar went back to night school and took courses in math, physics, and mechanics at the university level.

Einar read an article by a Swedish steel designer who had done post doctoral work at an American university. At this university they had many steels specialists on the faculty. The name is Lehigh University. It was almost next door neighbor to Bethlehem Steel in Pennsylvania. He actually met with the Swede. His name is Göran Alpsten; and he recommended Einar to apply at Lehigh.

So Einar did and was accepted with credits covering the freshman and sophomore years. Still two more years to get a B.S.; and Einar felt a bit too old to be an undergraduate student. Well, what the heck. He accepted, sold his car, sent his earthly belongings in a big trunk by ship to New York, and got on a student flight to the US.

After the junior year he was looking for work over the summer, preferably on a steel project in Fritz Engineering Laboratory at Lehigh. Nothing was available. However, I was a young professor at the time and could help him. I was embarking on a polymer concrete project that would be starting out in the fall. The project aimed at repairing concrete bridge decks and involved two departments at Lehigh and a group at Penn State University. I needed a student to start on the project over the summer. Einar was only pleased to get something to work on. That was the beginning of our long time relationship. Little did know that that would be the end of his career in steel design.

Einar was allowed to stay on the project during his senior year; and got credit for the work as a special project. The polymer concrete the team made was four times stronger than the same concrete without the polymer. With his friend and co-worker, Harshavardhan C. Mehta, they were making polymer concrete in the basement of Fritz lab. The problem was that it smelled so bad so professors on the seventh floor evacuated the building for fresh air. He was not so sure they all that popular. The polymer concrete was strong, but also brittle. Einar tried to modify this by adding a more ductile polymer to the mix. It worked very nicely. They got a strong, but also flexible concrete that had some interesting structural properties. A paper was written as part of the senior year special project.

Einar was chosen to represent the Civil Engineering Department of Lehigh in a contest arranged by the American Society of Civil Engineers' Student Chapter with the paper. A great honor of course. The paper was submitted to a committee in advance followed by an oral presentation. He was coached by Professor Lynn Beadle, Director of Fritz Lab; and was escorted by Professor David van Horn, the department Chairman, to the presentation that took place at Princeton University. After the presentation by students, the committee withdrew for their evaluation. In the evening they announced their decision. It was just like the Oscar award nomination. *"And the winner is...Einar Dahl-Jorgensen from Lehigh University"*. Till this day Einar wondered if that might have been the peak of his career. Professor van Horn was elated. Lehigh had won.

Einar spent another two years at Lehigh as a research assistant, most of the time working on the polymer concrete project for me. This resulted in several papers where he was also an author/co-author in addition to the topic of his master's thesis.

In 1975 he accepted a position as an assistant professor at the Norwegian University of Science and Technology (NTNU). At NTNU, Einar was working with cement, concrete and polymers. He also got his Ph.D. there; and also joined the research organization SINTEF in their Cement and Concrete Institute. One of the items of the research there was the development of a tool to test concrete strength in the building structure. This could be used to predict when the concrete had developed sufficient strength to remove the formwork. It was called the Break Off tester and eventually became an ASTM standard. I was by then appointed Head of Structural Engineering at Purdue University. I was happy to say we still kept in touch. Einar brought a Break Off tester with him on a visit to Purdue; and I included Einar's tester in my research program.

In 1985, Einar was asked by the Division of Petroleum Engineering at NTNU to participate in a research program with the purpose to make a better material than cement to be used in oil wells. Again this was another shift in his career. This time it was goodbye to cement and concrete and the building industry.

Development, including two patents using a polymer material, followed. At the end of 1991 Einar left the R&D group and the university. A new company was founded to bring the result of research to the market in the oil industry. Einar started out as the president. Since this included too much administration, he later switched to become the vice president in charge of technology for the company WellCem AS. WellCem's ThermaSet® products have been used with success in oil wells on the Norwegian and British continental shelf, in France, the Middle East, and the Gulf of Mexico. Einar has recently initiated another project with the purpose to develop a new polymer material to shut off the production of water from oil wells. A patent for this has also been filed. More than 500 million barrels of water and about 80 million barrels of oil are being produced every day from oil around the world. The staggering cost for the oil companies is about 190 billion dollars annually to treat and handle the water. His system can be pumped into the well and selectively shut off the water in the well, but the production of oil and gas will not be obstructed.

Einar believes it is fair to admit that this all started in the basement of Fritz lab at Lehigh University under the auspices of me. He was proud to have had me as his mentor; and now as a friend; and we have kept in touch over the years. What has been important for Einar was that the initial work for him in Fritz lab pointed him in the direction that he has since followed.

Einar Dahl-Jorgensen

11.10 Student from India 印度學生

HARSH MEHTA (1972, MS, Lehigh)

Attended the civil engineering studies at Jadavpur University, Calcutta for a five-year undergraduate program in 1966; and received his BE in civil engineering with honors in 1972. He was awarded two gold medals for standing first, with merit scholarship awarded in 4th & final year of his studies.

Harsh joined MS in civil engineering program at Lehigh University in January 1972 and worked as my research assistant. I put him in charge of a new project on polymer impregnation of concrete to seal pores; and improve its in-situ strength and durability. The new project was specially aimed to solve the problem of corrosion and durability facing the highways and bridges in winter due to severe damage by deicing salts and freeze thaw cycles. Harsh with his colleagues worked day and night for 3 years establishing pioneering studies in this field with immense benefits to other problem areas of concrete structures.

At the end of the research cycle, we successfully demonstrated the use of this new technology on a bridge near Lehigh campus by in-situ polymer impregnation and steam curing.   Harsh was awarded several citations and a *"Second Award"* for design of welded aluminum impregnation chamber for highway applications by James F. Lincoln Arc-Welding Foundation.

There were several benefits of using polymers with concrete in our research to other concrete problem areas. I gave the team full freedom to investigate such applications in various areas such as: Applications of polymer concretes to improve strength and durability of concrete facing severe corrosion in chemical factories, storage and handling areas, structures in marine and saline environments, waste disposal and treatment plants, and development of high-voltage concrete insulators, etc.

Harsh received his M.S. degree in January 1974 with major in structures; and MBA in 1976, both from Lehigh.   His MBA was on finance and management model development. He had published 14 research papers as author/coauthor in the fields of polymer concretes, wire reinforced concretes, and use of waste materials in construction and earthquake resistant foundations.

He had worked on several other research projects such as testing of multi-story composite steel structure, Use of steel fibers in concrete, developments of better electrical concrete insulation materials, use of waste materials like sulfur for concrete impregnation for strength and durability, improving bamboo properties for cheap tensile reinforcement in concrete' and earth mats for soil stabilization, utilizing other waste materials like rice husk for lightweight concrete and soil stabilization, and developing two stage non-linear portfolio optimization model under conditions of uncertainty.

After completion of his studies at Lehigh, he decided to return to India. For the last 30 years, he has been doing project consultancy in various fields which has culminated in many rewarding projects for his clients.

He married to Varsha and celebrated their 30 years together on August 1, 2006! Harsh has two children: daughter Falguni married to Aashit Shah and teaches Economics at Raheja College in Bombay; and son Jayvardhan, who will soon get married to Bhavika, works in a diamond firm in Antwerp, Belgium.

Left to right:
Varsha, Mehta's wife; Bhavika, his son's fiancée; his son, Jayvardhan; and Mehta.

11.11 Concluding Remarks 評論

If there is any great secret of success in life, it lies in the ability to learn your past experiences and to put yourself in the other's place and to see things from his point of view – as well as your own. All life is an experiment. The more experiments you make and you know the better. The purpose of this chapter is to share our experiences, career, and life together; so that good judgments will come from these shared experiences. As we all know that experiences and wisdom come from past poor judgments.

11.12 Wai-Fah Chen at a Glance 陳惠發略述

Most Recent Positions
○ Professor and Dean of the College of Engineering at University of Hawaii, 1999-2006.
○ George E. Goodwin Distinguished Professor of Civil Engineering and Head of the Department of Structural Engineering at Purdue University, 1976 to 1999.
○ Taught at Lehigh University, 1966 to 1975.

Education

- Ph.D. in solid mechanics, Brown University, RI, 1966.
- M.S. in structural engineering, Lehigh University, PA, 1963.
- B.S. in civil engineering, National Cheng-Kung University, Taiwan, 1959.

Research Interests

- Constitutive modeling of engineering materials, soil and concrete plasticity, structural connections, and structural stability and design.

Awards and Honors

- 1984 US Senior Scientist Award, Alexander von Humboldt Foundation, Germany.
- 1985 T.R. Higgins Lectureship Award, American Institute of Steel Construction.
- 1985 Raymond C. Reese Research Prize of the ASCE.
- 1988 Distinguished Alumnus Award, National Cheng-Kung University.
- 1990 Shortridge Hardesty Award of the ASCE.
- 1991 Honorary Fellow, Singapore Structural Steel Society.
- 1995 Elected to the U.S. National Academy of Engineering.
- 1997 Awarded Honorary Membership of the ASCE.
- 1998 Elected to the Academia Sinica (Taiwan's National Academy of Science).
- 1999 Distinguished Engineering Alumnus Medal, Brown University.
- 2003 Lifetime Achievement Award, America Institute of Steel Construction.
- 2003 Featured in the biographical monograph "Giants of Engineering Science" as one of ten of the world's leading engineering scientists, UK.

Distinctions

- Author or Co-author of more than 20 engineering books.
- Author of more than 345 peer-reviewed publications.
- Serve on editorial boards of 15 technical journals
- List on 35 Who's Who publications.
- Editor of four engineering Handbooks.
- Consulting editor of Encyclopedia of Science and Technology.

12

The Teacher D.C. Drucker
師生情誼

12.1 Background 背景

A former professor and chairman of the Division of Engineering and of the Physical Sciences Council at Brown University in Providence, Rhode Island, and a renowned engineering educator and former dean of engineering at the University of Illinois at Urbana-Champaign, Daniel C. Drucker's accomplishments range from the unification of the classical theory of plasticity to quiet, unsung moments with his students.

Throughout his career, Drucker's developments have benefited the entire engineering community. His methods—simple, yet powerful—analyzed the complexity of structural materials and components under various load conditions. He also focused on the development of the fundamental theory of soil plasticity and contributed greatly to the critical state of soil mechanics in particular. He understood the behavior of structural materials and their idealizations and simplifications according to their environment and applications, including fatigue and brittle fracture.

Even today, engineers in reinforced concrete design apply Drucker's knowledge concerning the behavior and design of *"shear strength"* in deep beams and the simple equilibrium concept of *"truss model"* in designing beam to column joints. Although Drucker's ideas were profound, his analysis and design techniques were practical, using an engineer's intuition and physical visualization.

Among his life achievements, Drucker was honored with almost every major award given by major engineering societies including the Von Karman Medal from ASCE, Timoshenko Medal from ASME and Lamme Medal from ASEE. He was a member of the National Academy of Engineering, received honorary doctorates from five different

293

universities, and served as president of five U.S. and international societies. One of Drucker's most prestigious honors was given by a U.S. President – the National Medal of Science.

As a former graduate student and long time friend, I was deeply impressed by his wide range of interests in properties of materials, structures, structural mechanics, photo-elasticity, material science, and soil mechanics. His fundamental theory on inelastic material behavior with deep physical insight provided the basis for an inspirational education. It was no accident that many young engineers were attracted to study under him.

Looking back, I was indeed fortunate to have had the opportunity to study under him and have a life long association with him as a friend and mentor. Daniel Drucker was known throughout the world as a brilliant scholar, a leader in education and spokesman for engineering. Those who knew Professor Drucker well knew him as a thoughtful, kind, generous, wonderful human being who will be sorely missed.

12.2 The Teacher 恩師

Drucker was born in New York City in 1918, attended Columbia University, where he earned three degrees, including a Ph.D. at the age of 21. He went on to teach at Cornell University and worked at the Armour Research Foundation before spending a year in the US Army Air Corps during WWII.

In 1947, Drucker went to Brown University where for two decades; he helped build one of the best programs in the country in materials engineering and solid mechanics. He is best known for his pioneering work in the theory of plasticity and its applications to analysis and design of metal structures.

In 1968, Drucker became Dean of Engineering at the University of Illinois, where he is credited with improving the quality of the faculty. In 1984, Drucker left Illinois to become a graduate research professor at the University of Florida where he was a kind of senior statesman on campus and was always available to help, especially the younger faculty members.

To many, perhaps Drucker's greatest title was that of teacher.

Among many of his former students and colleagues, our comments on Professor Drucker may be summarized as follows: alert, perceptive, insightful, and persistent but always in good humor and graciously courteous to even the most difficult of us.

After receiving my M.S. degree from Lehigh University in 1966, I moved to Providence and went to Brown University to register. The simplicity of the procedure astounded me. The lady who helped me to register was a senior secretary. She looked at my registration form with only Drucker's initial "*DCD*" at the bottom of the form and wondered why I did not get the signature from the graduate chair of the Division of Engineering. She then told her associate sitting next to her and said to her that I think it is okay, since his adviser is a "*big shot*". As for my English requirement for foreign students, she looked at my Lehigh transcript and thought that I had no problem with the language. Since I had a research assistantship from engineering, I had no financial problems. She therefore decided that I had completed all paper work, and that was that.

The whole process would be inconceivable in Taiwan or China, where every such decision had to be checked and substantiated by documents and approved by higher administrators who may not be available. It would have taken days in Taiwan or China to arrive at the decisions that took her ten minutes at Brown to make without consultation with chairs or my major adviser. This left a profound impression on me about how Americans do things and delegate authority. They work efficiently and deal directly with the problem, with a minimum of bureaucratic encumbrances.

At the opening session of Drucker's beginning class on "*Plasticity*" which I took during my first semester in solid mechanics, he pointed out that the materials in technical reports were at least one year old; materials in technical papers were two to three years old, textbook materials were at least five years old. Accordingly, there was no point for him to present these old materials in class since we could all read these for ourselves. Instead, he expected that we would spend our time on reading the textbook and listing all the mistakes and incorrect concepts we found in our class notes and submitting the notebooks by the end of the semester as part of our grade. I was shocked since I fully expected to learn the theories from his class presentation.

Instead, Drucker went on to mention that if you knew what theorem should be proven, it would be done sooner or later. The key questions were: What were the motivations for such a theorem? , How you came up with the idea? , What was the physical basis for the existence of such a theorem? And how it could be applied for engineering practice? He then spent a lot time on the illustration of the historical development of the famous "*limit theorems*" he established and proved with his colleagues Prager and Greenberg in the early 1950's. It was an eye opening session on the teaching and a high bar that he set for us.

In the mid-year examination, he collected our course notebooks before the start of the examination. He gave three problems with Problem 4 stating that "*if you have run out of useful things to say about Problems 1-3, discuss uniqueness and stability for a nonlinear irreversible time-dependent system.*" In the final examination, he again collected the course notebooks from us and asked us to answer in Questions 1 and three of the remaining four questions. However, "*you may, if you wish, substitute the discussion of a topic on which you are well prepared for any one of Problems 2-5.*"

When I went to him concerned over what I believed to be an extraordinarily poor performance by me on one of his exams, he said "*If anyone gets 100% on my exam, I have not done my job because I have not been able to help you do bette*r". I was relieved to find that I had, in fact, done well.

As a doctoral thesis adviser, he was sensitive to guiding, but not directing my thinking. His view on an acceptable thesis is that a student can independently identify a problem, reduce it to a simpler model, work out its solution, and then interpret the results as they apply to the original problem. When we started to write a paper, he insisted that I drafted an abstract first and then the conclusions from which the table of contents was prepared to support the conclusions. His English writing was very elegant and concise and he seldom wrote a long reference letter, mostly just a few paragraphs on his observations about your strengths and achievements.

My memory of my Brown weekly meetings was that of extensive discussions in Professor Drucker's office. After an involved discussion of a subject, he would make a well-phrased and accurate account of our discussion with a simple model to deal with the complexities of the problem. I was always amazed to find that he could use that simple thin-walled tube to address to any complex issues about engineering materials problems.

Photo with Dan Drucker at Eric (中傑) and Arnold (中毅)'s graduation ceremony when Professor Drucker served as Dean of Engineering at University of Illinois.

Professor Drucker's office was originally in an old applied mathematics building and later in a new engineering building. In the new building, the space was large, somewhat like a small classroom with a conference table. His office was quite orderly and neat despite the fact that he was chair of the Division of Engineering with significant administrative duties. His office was definitely not like most of the administrators with whom I was normally associated.

12.3 The Researcher 研究學者

Some 50 years ago, Brown University was a Mecca for a number of young renowned professors who taught in the Division of Applied Mathematics and Department of Engineering. After completing my first graduate degree from Lehigh University, I was drawn to Brown University, as I desired to study under Professor Drucker. From our first meeting in 1963 and throughout the many years of working under and along side him and trying to follow his footsteps, he played such a major role in my life.

12.3.1 Steel Structures 鋼結構

Professor Drucker's contributions to the theory of plasticity, limit analysis, geo-mechanics, engineering science, and structural engineering are incredible. In the early 1960's, the computer was in its infancy while the theory of plasticity was in its golden time. We had a rigorous theory but few practical solutions. For engineering practice, we must develop simple theory that is realistic and practical. There is nothing more practical than the simple limit analysis theorems and techniques developed by Drucker and his co-workers at Brown University in the 1950's.

As a result, the limit analysis methods or the plastic analysis methods for steel construction were widely applied to steel structures almost 50 years ago. As a student of plastic design in the early 60s, I read plastic design books from Cambridge University and participated in full scale testing of structural members and frames at Lehigh University for practical implementation of plastic design methods to steel. The American Institute of Steel Construction adopted the plastic design method officially in 1963 as the new design code.

12.3.2 Concrete Structures 混凝土結構

With the extension of the limit design theorems to continuous media by Drucker, Greenberg and Prager in 1952, applications of the powerful limit analysis techniques were expanded to plates and shells for both metal and reinforced concrete materials as well as soil and concrete mechanics. Only recently has the lower bound stress field method been adopted by the America Concrete Institute specifications for the design of structural members and joints in reinforced concrete, known as the *truss model*.

The upper bound techniques of limit analysis, known as the *yield-line theory* for slab design, have long been used in engineering practice, but the application of stress fields to reinforced concrete design, based on the concept of lower bound theorem of limit analysis, was first applied by Drucker in 1961 to a simple reinforced concrete beam. This new conceptual approach is now being extended and expanded to the design of concrete structures in a very practical manner.

The significance of this new conceptual change in the design of reinforced concrete structures can best be described by an observation made in 1984 by Professor Macgregor of Canada:

"*One of the most important advances in reinforced concrete design in the next decade will be the extension of plasticity based design procedures to shear, torsion, bearing stresses, and the design of structural discontinuities such as joints and corners. These will have the advantage of allowing a designer to follow the forces through a structure*".

12.3.3 Soil Mechanics 土壤力學

The development of the modern theory of soil plasticity, as a new field, was strongly influenced by the publications of the two classical papers by Drucker and his co-workers: "*Soil Mechanics and Plastic Analysis or Limit Design*" in 1952 and "*Soil Mechanics and Work-Hardening Theories of Plasticity*" in 1957. In the former, they extended the Coulomb criterion to three dimensional soil mechanics problems. In the later, the concept of work-hardening plasticity was introduced into soil mechanics.

The innovative ideas in these two papers have led in turn to the generation of many soil models, leading to the development of the *critical state soil mechanics* at Cambridge University. In recent years, these new soil models have grown increasingly complex as additional experimental data are collected, interpreted, and matched. This extension marks the beginning of the modern development of a modern consistent theory of soil plasticity.

12.3.4 Limit Analysis 極限分析

Between 1950 and 1965, the concept of perfect plasticity (i.e., no work-hardening) and the theorems of limit analysis form the central and most extensively developed part of the theory of metal plasticity. However the corresponding extension to problems in soil mechanics occurs much later. Perhaps the most striking feature of the limit analysis method is that no matter how complex the geometry of a problem or loading condition, it is always possible to obtain a realistic value of the collapse load without the use of computers.

When I returned to Lehigh University in 1966 to start my teaching career, I started to apply these limit analysis techniques to soil mechanics and developed simple methods for obtaining practical solutions. The upper bound method of limit analysis deals with work equation and failure mechanism. The development of work equation from an assumed mechanism is always clear to an engineer. However, many engineers find the construction of a plastic equilibrium stress field for a lower bound solution to be quite unrelated to physical intuition. Intuition and innovation are discouraged by unfamiliarity and apparent complexity.

To this end, and Professor Drucker and I wrote a paper in 1968 "*on the use of discontinuous fields to bound limit loads*". This concept of constructing stress fields is familiar to the civil engineer and can be utilized by the practitioner as a working tool.

Motivated by the success of this method, I wrote a book on "*Limit Analysis and Soil Plasticity*" in 1975. It soon became a very popular one, because geotechnical engineers can now reproduce almost all the existing solutions in the well-known textbook by Terzaghi in a simple and original manner. Professor Drucker was very pleased to see this publication and wrote a Foreword for the book in his usual concise and right to the point style:

"*An engineering practitioner concerned directly with soil and foundation problems will find many tables of useful numbers and the rationale for their validity in the standard terminology of soil mechanics. The author has brought together in a unified manner so much of what until now was known only to a few in the field, who like Professor W. F.*"

Chen contributed fruitful ideas and techniques".

12.3.5 Constitutive Modeling 本構模型

In the 1970s, our computing power changed drastically with mainframe computing. The finite element methods were well developed and widely used in structural and geotechnical engineering. In subsequent years, our computing environment changed even more drastically. The cost of computing became almost insignificant as we enter into the PC and workstation era. This was also the period that we were able to solve almost any kind of engineering problems with computer simulation.

Computer simulation has now joined theory and experimentation as a third path to scientific knowledge. But now, for the first time, the physical theory is lagging behind the computing power. We need to develop a more refined theory of constitutive equations of engineering materials for finite element types of application and simulation.

The mechanical behavior of all materials is complex and must be drastically idealized in order to make mathematical analysis tractable. I learned significantly from Professor Drucker's approach to mathematical modeling of engineering materials and his teaching on the necessity of proper simplifications and idealizations for a particular application in order to find a reasonable approximate solution. The geometry or compatibility, the stress-strain relations, and the equations of equilibrium must all be properly idealized to accomplish a solution.

I have benefited much from this endeavor by always trying to bring the theory/research to engineering practice. The two-volume book on "*Constitutive Equations for Engineering Materials*" is my effort to introduce this type of materials into a standard civil engineering teaching,

12.4 The Passing of a Giant 巨人仙去

Professor Drucker died August 25, 2001of leukemia in Gainesville, Fla., shortly after the passing of his beloved wife Ann. Ann was his life long companion since their college years. In later years, Ann always accompanied Professor Drucker to conferences or official events. I prepared the following article for Professor Desai's journal for news release when I heard the sad news and learned the details of his last few days at home with his daughter Mady.

A mere ten months away from the departure of his beloved wife, Ann, it is truly unfortunate and it causes a great pain to me that my graduate thesis advisor and life-long mentor, Professor Daniel C. Drucker, one of the greatest leaders in engineering science and a giant in the world of theory of plasticity, departs us. I first learned this sad news from an email from Professor Chandra Desai of University of Arizona who forwarded me an email from Professor Patrick Selvadurai of McGill University reporting this news.

Patrick happened to send Professor Drucker a copy of his book to be "autographed", and found out that Professor Drucker was very seriously ill, and his daughter, Mady, wrote to him on September 1, 2001 that Professor Drucker passed away peacefully a week ago. I was very much moved by Mady's letter to Patrick together with Professor Drucker's note describing the situation,

"It took a huge effort for him to sign and write a note which he asked me to type up properly. I thought you'd prefer it if I sent the handwritten note along with the book since it's the last thing he'll ever write because even sitting up is an effort these days. I can't thank you enough for your kind request. Dad's note reads,

Dear Patrick,

It is a pleasure and honor to return to you an autographed copy of your well-used copy of my text. The request was a great moral booster at a low point. Ann died in December and I am about to follow so you are due a special thank you.

Sincerely, Dan.
Thanks again, Magy".

12.5 My Tribute 追悼

For almost half a century, the advances in the theory of plasticity and the work of Dan Drucker were intimately linked. The familiar terms such as stability postulate, stable materials, limit theorems, limit analysis, plastic design, Drucker-Prager model, and soil plasticity come to my mind immediately, among others. He is best known for his pioneering work on the theory of plasticity and its application to analysis and design in metal structures. He introduced the concept of material stability, now known as "*Drucker's Stability Postulate*." The simple fact that his name is attached to it and that it has survived for half century is indicative of the significance of the person.

Drucker' family members: Son and daughters (Magy, right).

It has been a wonderful experience and rewarding career for me to have the privilege to work with him. As Professor Drucker liked to emphasize, a true fulfillment of engineering research and education is "*a place in practice*". Professor Drucker's impact on engineering practice and education has been tremendous. He remains a role model for us all.

The Diamond Head view of the lower campus from Dean's office at Holmes Hall of the University of Hawaii at Manoa campus in Honolulu.

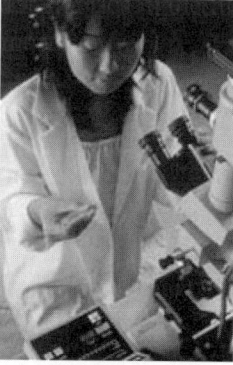

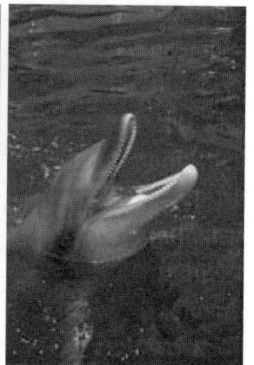

The University of Hawaii is well known for its unique specialties in tropical agriculture, astronomy, and oceanography.

13

Transforming the College
改變中的工學院
Some Parting Thoughts
離別感言

13.1 Background 背景

I was appointed as Dean of Engineering at the University of Hawaii at Manoa effective September 1, 1999. I was with Purdue University for more than 23 years before joining UH. I held the holder of the first Distinguished Professor Chair of the School of Civil Engineering at Purdue University. Since the deans are part of the university's top management team, we have business guys, market guys, and academic guys serving on the team. I was a hard-core academic guy managing the College of Engineering.

The university had several rough years before 1999 and budget cuts were the major problem. Every college had felt the sting of the budget cuts that had been plaguing UH throughout the 1990's. These cuts had been distributed roughly equally over all sections of the university. Across-the-board or "*horizontal*" cuts were the fairest and, as a result, constituted the least political resistance. Unfortunately, this approach ultimately caused the most damage at a research university.

In 1999, the university decided to hire new leadership in the form of three new deans and one center director to revive the optimism. President Kenneth Mortimer was putting hope in the appointments of Ed Cadman, dean of the medical school; Carl Vogel, head of Cancer Research Center, David McClain, dean of business school, and myself as dean of engineering.

President Mortimer introduced me to Governor Benjamin Cayetano as soon as I arrived on campus. Ben was very fond of and proud of Guy Kawasaki, a local grown Iolani high school graduate. Guy was one of the pioneers at Apple Computer and a nationally known hi-tech consultant and evangelist. Guy once told Ben that if Hawaii wanted to develop a hi-tech industry, it must have a world-class engineering school.

I was fortunate to have the opportunity to tell Ben that UH already had a fine College of Engineering. We had all-star faculty including almost 25 percent of college's professors earning the National Science Foundation's Young Investigator award. The award goes to the nation's best and brightest faculty. As a start, I proposed to build the Hawaii Wireless Communication Center because of our current strength and Hawaii's geographical location. Ben was impressed with my academic credentials and my timely proposal, and he decided to augment our base budget by $1 million to enable us to build a center. To be fair, he also gave the other two deans one million dollars each for the purpose of revitalization of Hawaii's economy after the real estate bubble broke and economic slowdown of the 1990's.

Time flies and seven years later, David McClain built his business school and moved upward to central administration and became President of the UH system in 2005. Ed Cadman succeeded in building his new medical school in Kakaako and then stepped down as dean because of his health. I focused on building the college to a top tier engineering school. We all did our best to fulfill this goal. My efforts reflected truly my own interest and expertise in building academic excellence for the University, the College of Engineering in particular (see Also Appendix 13.1, A Conversation with Dean Chen, *Quadrangle*, Fall 2005).

13.2 The Vision and Strategy 憧憬與戰略

My goal for the college was simple and clear--*to elevate the College to the level of the nation's top-tier engineering schools by building academic excellence*. We needed first-rate faculty, outstanding students, and excellent facilities. That was my blueprint for success at the UH College of Engineering. When I arrived at the College, I identified the college's problems immediately. Our operating budget was down 30 percent, faculty ranks were depleted by 20 percent and enrollment had dropped 30 percent during the previous five years. Reversing these trends would be a formidable task, but I did see enough potential to make me confident that the engineering college had a bright future ahead.

I told my faculty and staff at the first College of Engineering meeting that *"our top priority must be to strengthen the faculty, since it is the key to attract students and bringing in research dollars"*. To reach this conclusion, we only needed to ask the following question *"Do we want to build the College as a research university or do we want to maintain it as an instructional college?"* Then the next step is logical. To this end, I said that *"our strategy is to recruit faculty members with national or international reputations. To keep and recruit quality faculty members, we need attractive salaries, strong colleagues, access to equipment and resources, and opportunities to work with top-notch students. Nothing can be taken for granted. We must continue to pay attention to our current faculty while working to attract new people. Recognition is a critical part of the process. It is very important that we demonstrate how much we value our colleagues' work, achievement and success. We can highlight our faculty's work in five areas: Teaching, Research, Scholarship, Mentoring, and Service, and recognize their success with support and rewards."*

Dean Chen in front of UH Engineering Building.

13.3 The Progress 進展

Faculty

One of my most visible accomplishments was the launching of a merit-based salary adjustment for the deserving faculty and staff, something virtually unknown at the UH. This was the critical problem I faced with my turnaround strategy for the College at the time. With the one million dollar added to my budget from the Governor, I reallocated resources and adjusted my internal budget and provided market raises for some faculty. I had no choice but to squarely address this issue and improve faculty retention and recruiting.

The very best engineering schools recognize the fact that recruiting a few superstars will result in a tremendous enhancement of the entire program. This occurs not only because of the research funding these individuals bring with them, but because the top candidates for new faculty positions are attracted to colleges with a prestigious faculty.

During my tenure, the College hired more than 50 percent of our current faculty members at the market price, and increased the number of women faculty. At the same time, we recognized the College's most productive faculty by pursuing an aggressive salary adjustment through a merit raise process that was approved by the Faculty Senate and union. This stood as testament to our commitment to building and retaining a first-rate faculty.

Research
Research is the source of new ideas that create new opportunities. It is also the centerpiece of our graduate education programs. Maintaining the faculty and developing centers of excellence in our acknowledged areas of strength are centerpieces for enhancing the College's reputation. This in turn attracts quality students which leads to preferential recruitment of UH graduates by industry. Providing incentive for our faculty members to work as a team in the focused areas was the second part of my strategic plan to increase our reputation.

We established the Hawaii Center for Advanced Communications (HCAC) in 2000, the Corrosion Research Laboratory in 2001, upgraded the Environmental Engineering Laboratory in 2002, and established optical, microwave, and wireless laboratories in 2003-05, among others. Our goal was to build these centers and facilities to attract more creative researchers and offer them a variety of supportive resources.

Students
In the area of student enrollment, the College was experiencing a steady decline that we had reversed only with a lot of hard work.   We

implemented an outreach, recruiting, and retention program. We introduced a comprehensive enrollment plan. Our alumni and community worked with our administration and the UH Foundation to increase scholarships. Thanks to these efforts, our freshman undergraduate enrollment has nearly doubled and total enrollment was up significantly. In particular, I was very pleased to see the huge success of one of our newest programs, the Native Hawaiian Science & Engineering Mentoring Program. Native Hawaiian students now comprised 10 percent of our enrollment with an impressive retention rate of 70 percent.

Dean Chen conducted a convocation ceremony for graduates of the College of Engineering.

Summary

I definitely had big ideas for the College. My vision for the College was simple and ambitious. It began with scholarly faculty, which conducted better research, which attracted quality students, which leaded to better graduates for the marketplace, which raised the College's reputation. The resulting increase in stakeholders' support improved facilities and infrastructure of the College. This was the logic I had for "*How to Build a First-Rate Engineering College*" in my vision statement which I prepared when I was appointed Dean in September 1999 (see Appendix 14.1 Chapter 14).

13.4 The Milestones 哩程碑

The College has made tremendous progress since 1999. Some of the memorable events, achievements, and activities are listed below: (see also Appendix 13.2, Engineering His Exit, *Hawaii Business*, December 2005).

o   The undergraduate enrollment has nearly doubled.
o   The research grants have increased significantly. They provided more than half of the College's total operating budget.
o   The number of student scholarships has nearly doubled.
o   The number of Native Hawaiian students enrolling in the engineering program has reached a record level of ten percent of its total student body with an incredible retention rate of over 70 percent, a record too.
o   The College has added and hired more than 50% of its current faculty.
o   The College has implemented a merit raise process for its faculty and staff. (See Appendix 13.4: UH Engineering School Rewards Faculty, January 28, 2002, and Appendix 13.5: Editorial: Tread Carefully with University Merit Raises, January 29, 2002, *The Honolulu Advertiser*).
o   The College was planning the construction a state-of-the-art student activity center.
o   The College has built several multi-million dollar research laboratories.
o   The College established the Hawaiian Center for Advanced Communications to do cutting-edge research on wireless communication systems.
o   The College has developed a business plan for the construction of the Civil Engineering Research Park to meet the huge demand of civil engineers in Hawaii.
o   The College was developing new programs in biomedical engineering and computer engineering.
o   The College was providing distance learning courses for its students on neighbor islands.

o   The number of participants of the annual banquet has steadily increased and reached an all time high of nearly 1000.

Dean Chen, UH President David McClain, and US Navy Pacific Commander Admiral Fargo at the 2005 College of Engineering Banquet.

13.5 Management Style 管理風格

As a more than 40 year veteran of academic life, I have been a faculty member of civil engineering for more than 43 years, served nearly 20 years as head of structural engineering and seven years as dean of engineering at private and major public schools. My observations on higher education in America are that people still believe that the most successful institution is the highest cost institution. The most successful president is the one who can raise the institution's prestige by spending money and raise large sum of donations. Thus, most universities are run by business people or politically well-connected people. Very few hard-core academic scholars have the opportunity to run the university with their visions.

UH was no exception and most past presidents and Board of Regents were business people or politicians. They run the university by bean counting and do not delegate to the professionals or deans and departments to run their business. To run an academic institution, one must let the individual units and their deans know what are the priorities and expectations and then let people do their job. This was what I had done at the College. To identify each one's interest and expertise, define clearly the goal and priority, and let each of them do their jobs. Motivation and creativity are the key elements for success. As a dean, we need to ask chairs not *"How are we doing?"*, rather *"Where are we going?"* This is critical in achieving our goals.

President Mortimer recruited me in 1999 but he stepped down the following year. The University recruited a new president, Evan Dobelle in 2001 through a widely publicized national search. President Dobelle acted as a winner of a political campaign, issued letters on his first day of business, terminating the appointments of 230 to 240 deans, assistant deans, associate deans and some upper administrators. Dobelle then said that *"all these jobs will be evaluated through two fronts. Deane Neubauer, UH interim academic vice president will review instructional posts, while he will use different criteria for non-instructional jobs."* But Dobelle seemed to have no understanding that more than half of the current UH deans were already in *"interim"* status because of difficulty to fill these positions permanently through national searches as required by the Board of Regents' policy.

In the meantime, he immediately re-organized his upper administration and separated himself from the deans and chancellors with several vice presidents and chief-of-staff and personal secretary, most of them were his old friends, such as Paul B. Costello as UH vice president for external affairs, J.R.W. "Wick" Sloan as UH chief financial officer, and his wife Elizabeth Sloan as UH Foundation president, among others. He also made promises to faculty members that his top priority would be to adjust their salary to the 80 percentile of the national average over the next few years. In addition, he promised the legislators that he would raise $150 million for the new medical school buildings to be built in Kakaako. He also hired his new Chancellor, Peter Englert, for the UH Manoa campus in 2002. It was music to all. He was in the newspaper headlines almost every day and attended endless welcome receptions and banquets and gave more promises in speeches on various issues. (See also Appendix 13.6: UH Executive Salaries Increase, Ka Leo, May 7, 2003).

The sad story was that nothing was really achieved and he was fired two years later along with his vice presidents. Chancellor Englert was also fired by new president McClain in 2005. During the seven years of my tenure as dean of engineering, I had to work with three presidents (Kenneth Mortimer, Evan Dobelle, and David McClain), and four Chancellors or Chancellor-equivalent (Dean Smith, Deane Neubauer, Peter Englert, and Denise Konan). During the period, I had to deal with one major faculty strike, one major flood, student demonstrations against military research on campus known as the proposed University Affiliated Research Center (UARC), along with budget cuts and faculty unrest. It was a long way path to reach the present state of the college. It was a miracle. This difficult time had demonstrated the true character of our engineering faculty and staff, both their resilience and their compassion for the College. (See also Appendix 13.3: A Letter to Faculty and Staff of the College, February 14, 2005).

Mayor Ma Ying-Jeou （馬英九）of Taipei, W.F. (惠發) and Linlin (玲玲) Chen in Honolulu, Hawaii.

Appendix 13.1
A Conversation with Dean Chen
與陳院長的一席對話

Fall 2005 Quadrangle
College of Engineering
University of Hawaii at Manoa

A Conversation with Dean Wai-Fah Chen
By Marvin Nitta

Interview at Dean Chen Office.

Q & A
In September 2005, Dean Chen will complete his sixth year as the College's leader. As he enters his "*senior*" year at UH this fall and as he steps down as dean upon the recruitment of a new dean, Quadrangle editor, Marvin Nitta, sat down with him in his office to reflect on the College's accomplishments during his tenure to date and to look at what's ahead.

Q: What is your assessment on the progress thus far in achieving your goals since your arrival in 1999?
A: To understand more fully what was accomplished, we need to first know what we were trying to accomplish. The unwavering goal throughout my deanship was, and still is, for the College to follow a strategic plan built around the clear vision of becoming a top-ranked engineering school in the country.

As described in my vision paper in 1999, the strategic plan to become one of the top 50 engineering schools included efforts to strengthen our faculty, improve our facilities, and increase our enrollment. We have made a lot of progress in all three areas. We have added faculty, we have grown our research program, we have improved our facilities, and we have increased our enrollment.

The number of faculty has increased from 45 to nearly 65, and almost half are newly recruited at more competitive salaries. We also implemented a merit raise process in 2001. This has significantly improved the salaries for some of our existing faculty and staff. With the upcoming significant salary adjustment negotiated by the faculty Union over the next five years, our salaries will be much more competitive.

We have built several million-dollar laboratories, including the state-of-the-art wireless test bed and corrosion research facility. We have completely renovated our environmental laboratory and also built two multi-media laboratories, among others.

Last year, we raised $1.5 million in private funds, 50 percent more than the previous year. Our research funds have grown to nearly $8 million, a 60 percent increase from the previous year. Our goal is to increase research funding to $10 million in a year or so. Dr. Vassilis Syrmos, Associate Dean, has made significant contributions to our success and will continue to help the College to achieve this goal while serving in the Office of the Vice Chancellor for Research and Graduate Studies during the coming year.

Our enrollment has steadily increased over the last five years and has reached more than 900 students with 750 undergraduates and 160 graduate students. Our goal is to cap the undergraduate enrollment to around 1,000, and to increase our graduate enrollment to 220 in the next few years with particular emphasis on enrolling more doctoral students. Dr. Song Choi, Assistant Dean, has expanded our current enrollment management plan with particular emphasis on improving the retention rate of our undergraduate students.

Q: What do you believe are the biggest challenges ahead in achieving your vision?
A: Our major challenge is to improve our graduate program by enrolling

more doctoral students; and to sustain the current effort to improve the retention rate of our undergraduate students. Our performance goal is for each of our faculty to produce on the average one doctoral student every three years and one MS student per year. This goal will require a lot of commitment and hard work from our faculty members. With the current high level of research funding, I am very optimistic that we will meet this goal.

As a comparison, at Stanford University, a top-ranked research university, each of their faculty members produces about one doctoral and four MS students per year. Both at Stanford and at UH, the number of undergraduate graduating students per faculty per year is about the same. However, our resources per faculty from the state and tuition revenues are only about half of Stanford's.

To be ranked in the top 50 engineering schools in the country, the size of the College also matters. As you well know, we have only three engineering programs under the College, but the actual size of our engineering programs at Manoa is much larger than the three departments. For example, Ocean Engineering and the Hawaii Natural Energy Institute in SOEST are also engineering programs; and the Ocean Engineering Department in particular has produced many graduate engineering students.

Similarly, the Information and Computer Science Department is under the College of Natural Science; while the Molecular Biosciences and Bioengineering Department is under CTAHR. Water Resources and Research Center is under the Office of the Vice Chancellor for Research. If we can report all these engineering programs and activities at Manoa together, the size of our engineering programs can be doubled overnight in faculty numbers, in research funding, as well as in the number of graduate students. This unrecognized strength in engineering at Manoa can be a great help, if recognized, in promoting our national ranking and reputation. This is the second major challenge with which we have to deal.

Q: Construction on the Manoa Campus is happening at a record volume and pace. What is the College's plan in regards to the future space need in your vision?

A: The Holmes Hall facilities are insufficient to accommodate the growth of research in the College. In particular, the facilities in the Department of Civil and Environmental Engineering are too small to accommodate large-scale testing. The Manoa Campus does not have space for such facilities. Unlike other urban research universities such as UC Berkeley, Purdue and Texas, we do not have off-campus sites to accommodate large-scale civil engineering research.

Since we have enormous need for additional space for growth, we have already developed a business plan for the construction of an off-campus Civil Engineering Research Park consisting of several clusters, each of which will be developed in stages as the need arises.

The first cluster we have identified is the development of a large-scale testing laboratory in structural, geotechnical and transportation engineering. It will accommodate research in construction technologies relevant to both private and military housing, retrofit of deficient bridges, and new pavement.

The plan calls for us to raise the funding required building the CERP from both public and private sources. However, we will be financially responsible for the operation of this laboratory. Professor Ron Riggs, chair of our CEE department, has been leading this development since last Fall. A detailed business plan has been developed, and more actions will be followed in due course.

Q: The strategic plan promotes the College's research and graduate program with similar emphasis to its undergraduate education mission. Why, in your view, should research and graduate education be seen as parallel in importance to undergraduate education?
A: Manoa is a research campus and that is distinctive compared, say, to the Hilo campus or the community colleges. We have made it clear in our vision statement that taking the College to the next level must involve strengthening its research mission. We cannot become preeminent if we don't have a preeminent research and graduate education program – they go together. A quality graduate program will help significantly raise the quality of our undergraduate education. It will help enormously in attracting and retaining top-notch students in Hawaii.

There is a growing realization that the engineering workforce plays a central role in shaping the future of Hawaii's economy. There is a greater awareness by our State's leaders, public leaders, and private sector leaders, that the research mission at UH, and the College in particular, is really pivotal for this development.

As I mentioned previously, our research funding has reached $8 million this year. This funding essentially supports research and graduate education, and has not diverted our current resources from our undergraduate education.

First-rate faculty attracts quality students and quality students attract quality faculty. Each is dependent on the other and together they bring excellence and prestige to the University. This is how we define "the next level" - what we mean by "preeminence". This is something that the College can do, should do, wants to do, and must do.

Q: Because the size of the College is relatively small and resources are limited, the College must concentrate on the effort to strengthen the focused areas of research in the College. One of the focused areas of development in your strategic plan is the establishment of the Hawaii Center for Advanced Communications. What can you tell us now about the progress and achievements of this Center?
A: It took us two years to search nationwide for the right person to lead the Center. We made a spectacular choice and recruited Professor Magdy Iskander from the University of Utah to lead the Center. Magdy and I have worked together closely and in consortia. He has taken the Center to new levels of excellence that promotes multi-disciplinary research, international collaboration, and partnership with industry.

As a research unit, the Center places a significant focus of its activities and efforts on attracting research funding, recruiting doctoral students, and building research facilities. This year, the Center's funding reached a new high of $1.2 million, or $600,000 per faculty, with eight doctoral students, or four per faculty, an incredible achievement by any measure.

Last year, with a major research instrumentation grant from NSF, the Center established three leading-edge laboratories in wireless test bed, microwave network analysis, and indoor antenna range. The Center also joined "Connection One", a NSF sponsored multi- university program located at the University of Arizona, to conduct research in communications in partnership with other universities and industrial partners.

To create a more entrepreneurial environment in Hawaii, and particularly in our College, the Center is working with colleagues from the business school to develop integrated graduate programs that will merge communications technology with entrepreneurship, management and policy. In addition, the Center is developing a distance learning graduate program, primarily targeting students on the neighbor islands.

To bring the research to our undergraduate students through the project-based learning process, the Center launched the Engineering Clinic Program two years ago. The program uses industrial support to enable teams of undergraduate students to do real world research projects. This program has provided better learning experiences for our students, who are the big winners.

Q: Throughout the UH system, all eyes are focused on the remarkable development of the medical school and biomedical research in Hawaii. What is your vision for the College's involvement in connection with this biomedical research development?

A: As we were thinking about future growth areas for the College, a consensus emerged that biomedical engineering research and education held great promise in Hawaii and the College in particular. There is significant interest and desire among existing and newly appointed faculty in the Department of Mechanical Engineering to build a research and teaching program focused on biomedical engineering.

A two-year business plan was developed to guide the department in its effort to develop the Biomedical Engineering Program: first, in the form of a biomedical graduate certificate program; second, in the form of an accredited undergraduate biomedical engineering program in the future. Professor Bruce Liebert, chair of our ME department, has been leading this development since last Fall. The ME department will start to offer new courses and a seminar series for upper classmen and graduate students this Fall.

Q: Is there any other department interested in biomedical engineering?

A: Yes, our Electrical Engineering Department has traditionally offered biomedical engineering courses. In fact, in EE, like ME, there is significant interest among existing and newly appointed faculty on biomedical research. This is why our proposed biomedical engineering program has a bright future in the College.

The current priority for our EE department is to develop an accredited computer engineering program within the department. The department has proposed to change its name to the Department of Electrical and Computer Engineering to truly reflect the real interest and expertise of its faculty members. Professor Galen Sasaki, chair of our EE department, has been leading this development since last spring

Q: After six years as dean, how did you decide to step down at this time and what do you plan to do?

A: This is a good time for me to prepare to step down when the College is in very good condition. This would allow the Chancellor to have adequate time in searching for a new dean. I will not be going far! I have decided to take a six-month sabbatical leave after the appointment and start of a new dean, an option for an administrator returning to academic duty or near retirement.

I plan to stay at UH for a few more years as a research professor, and be actively involved in research activities, mentoring doctoral students, and working with some of our young faculty members in the Department of Civil and Environmental Engineering.

Thus, I am stepping down from the deanship with the satisfaction of having achieved the challenges and goals set up for me by the administration more than five years ago. In addition, I will leave behind a solid roadmap or blueprint for the College to continue reaching its goal of being ranked as one of the top 50 engineering schools in the country. And finally, I have the great fortune of having the opportunity to continue my life's work in the work that I have loved and will always love as a graduate research faculty here at Manoa's research campus.

Dean Chen discussed research with Ken Hwa (華根) at his UH office.

Appendix 13.2
Engineering His Exit
離職机制

Dean Wai-Fah Chen's dream of turning UH's college of engineering into one of the best schools in the nation is finally within arm's reach. So why is he suddenly leaving?

By Jacy L. Youn,
HawaiiBusiness, December, 2005
Copyright ©2005 Pacific Basin Communications, LLC
Permission to reprint, 2006.

He may not be a superstar athlete or a top-rated comedian, but Wai-Fah Chen does have something in common with a couple of celebrities who are. Like Michael Jordan and Jerry Seinfeld before him, Chen, dean of the University of Hawaii's college of engineering, wants to go out at the top of his game. That's why, after having increased the number of faculty from 45 to 65, doubled the school's research funding and paved the path for a $25 million engineering research park, Chen is stepping down from the position he's held for just over seven years.

"Since I arrived, we've made a lot of progress. We've basically created the road map for the next dean to turn the school into one of the Top 50 engineering schools in the country. That has been my unwavering goal throughout my time here, and it will continue to be," says Chen, 69, who will stay on board as a research professor. "But surely now that we've got momentum, the time is right to bring in someone new to continue what we've started."

HB: When you arrived in 1999, you put together a very thorough vision paper. Can you briefly outline your goals then, and provide an update on your progress thus far?
A: I had a simple vision. I wanted the college to be ranked in the Top 50 engineering schools in the nation. I outlined the three things we had to do: No. 1 is hire quality faculty, No. 2 is student retention, and No. 3 is better facilities. So what I did was make the college the first in the university to give merit raises based on performance to retain faculty. Then I recruited over 20 new faculties at market price. Then we raised enrollment from 500 undergraduates to 760. Our goal is 800. Also, we increased external funding to boost our graduate programs. At Stanford University, every faculty produces four masters and one Ph.D. per year. Our faculty only produces a half a master per year and very few Ph.D.'s. When I came in, we had a $4 or $5 million budget, we've since reached $8 million, and this year we may reach $10 million. Finally, in terms of facilities, we have built several multi-million-dollar laboratories.

HB: In what areas do you think Hawaii is excelling compared to other research laboratories in the nation, and in what areas can we improve?

A: One area we can really compete in is corrosion research. The military's worried about rusting, and Hawaii is the only state that can simulate any climate weather condition on one island, so we've received good funding to do corrosion testing. Wireless communication is another one. All this optical fiber from the Mainland to the Far East goes through Hawaii. A lot of satellites are stationed on Hawaii. And then biomedical research.

HB: What plans does the engineering school have to work with the medical school in developing biomedical research?
A: On the research side, we have some overlapping projects with the National Institute of Health. On the academic side, the first step is to produce a biomedical certificate program – a graduate program. Then we could grow to have an accredited undergraduate biomedical engineering program. Then eventually spin it off as a department.

HB: Given Hawaii's massive infrastructure needs, will we be, or are we currently experiencing an increased demand for an engineering work force?
A: We are increasing enrollment, but not fast enough to meet the local need. Construction is a very important part of the Hawaii economy, and yet we're the only accredited engineering school in the state of Hawaii. So we need more civil engineers. The only limitation is facilities. Civil engineering structures are big, but we don't have the space in this building. Also, from the military's point of view, we need more engineers. According to the former Commander of the U.S. Pacific Command, Admiral [Thomas B.] Fargo, the military should be managed by engineers, not by MBAs, because the work is very high-tech.

HB: Once students join the work force, how are we doing at retaining them locally?
A: Right now, mechanical engineers mostly work with Pearl Harbor shipyard. There's going to be a lot of expansion of military in Hawaii, so for those high-tech military jobs, there's good pay and job security. The second part, civil engineer construction, is very booming. They have no trouble finding people to stay here. In fact, 75 percent of all our

construction company engineers are UH graduates. The trouble is, we want to produce more. But we don't have the facilities.

HB: What will be the new dean's biggest challenges as he attempts to fill your shoes?
A: The goal is very simple – try to be ranked in the Top 50. He has to improve our graduate program by hiring more doctoral students. That's one of my goals upon leaving. I want to stay onboard and help mentor the junior faculty to grow and be successful, and then I want to help recruit doctoral students. This is a very good time for the school. That's why I decided to step down. I came in at the worst time, and I'm leaving at the best time.

Appendix 13.3
A Letter to Faculty and Staff of the College
告別書

From: Dean W. F. Chen
To: Faculty and Staff
         College of Engineering
Date: February 14, 2005

I recall an award-winning-author used the words: "*Real success is finding your lifework in the work that you love*".

These words have a real meaning to me as I reflect on the past five years with the College of Engineering. I have found both success and satisfaction in serving the College as Dean.

As Dean of the College, I have accomplished the main challenges as stated in my July 22, 1999 offering letter: "*Your main challenges will be to strengthen the College's undergraduate offerings and to increase enrollment. Moreover, the College's research activity should be enhanced. This should be another major goal*".

To this end, we have recruited a total of more than 20 new faculty members. We have increased our undergraduate enrollment by nearly 50%. We have increased our research, gifts and donations to almost equal our operating budget provided by the State and tuition revenue. We have established five new laboratories and improved five existing laboratory facilities, including the development of centers of excellence in advanced communication and corrosion. We have enhanced the diversity of our College and made our faculty salaries more competitive.

I am proud of our achievements to date and excited about the new opportunities ahead. Over the next two years, our plan calls for the implementation of an enrollment management system for the College, a proposed biomedical engineering program in the Department of Mechanical Engineering, the establishment of an accredited computer engineering program in the Department of Electrical Engineering, the development of distance learning courses in the Hawaii Center for Advanced Communications, and the development of a Civil Engineering Research Park for the State and the Department of Civil and Environmental Engineering. Good progress has already been made and more will soon be implemented.

This is a good time for me to prepare to step down as Dean and let the Chancellor to have an adequate time to search for a new dean. I will not be going far! I have decided to take a six month sabbatical leave, an option for an administrator returning to academic duty, followed by an active involvement in research-focused activities and mentoring doctoral students in the Department of Civil and Environmental Engineering.

The administration has been exceptional in understanding and supporting the College. The College of Engineering is growing and I have been enjoying with our faculty and staff in the College and throughout the engineering community at large. All of my colleagues, alumni and friends have helped to make my tenure here successful and my vision for the College to be steadily realized.

Thus, I am stepping away from the deanship with the satisfaction of having achieved the challenges and goals set up for me by the administration more than five years ago, and having also found the opportunity to continue my life's work in the work I have loved to be as a graduate research faculty here at the Manoa's Research Campus.

Appendix 13.4
UH Engineering School Rewards Faculty
報酬

By Beverly Creamer
Advertiser Education Writer
Posted on: Monday, January 28, 2002
Permission to reprint, 2006

It's the oldest truism in the world: If you pay them, they will come.

But the School of Engineering at the University of Hawai'i is taking it several steps further: If you pay them better, they will stay.

More than a year ago, Engineering Dean Wai-Fah Chen did something virtually unknown at the University of Hawai'i: he launched a bonafide criteria-based system of faculty merit raises in his college.

He found the money by adjusting his internal budget and using some of the money freed up by faculty pulling in research money.

The new system called for bumping up professors by as much as 5 percent, or about $4,000 a year, for outstanding teaching, mentoring, research, service or scholarship. Those who provided exceptional contributions to fund-raising campaigns, founded a research center or major instructional facility, received prestigious awards, or stimulated an increase in enrollment were also evaluated for the raises.

At the end of last year 10 percent of his teaching staff, six people, received merit raises approved by UH President Evan Dobelle, and the union.

A pat on the back may be nice, Chen said, but a pay increase means more.

"The whole idea is to recognize the faculty with good performance and show appreciation for that," he says. "To motivate them, to recognize them, and to keep them here, this is a must."

While the Engineering School's merit system is unique, it's also the beginning of a long-overdue catch-up program to bring salaries more in line with Mainland institutions, and prevent the loss of good faculty to better-paying universities.

But it also serves as an example for much of the rest of the campus as each of the colleges begins an evaluation of teaching staff under stipulations in the new the University of Hawai'i Professional Assembly contract.

In April, the UHPA signed a contract that provides for 1 percent of the salary appropriation for adjustments that could include merit-based raises.

Throughout February, faculty members will be filing applications with their deans for salary adjustments under the new contract, and will be evaluated in the type of process Chen has already put into effect.

According to UHPA executive director J.N. Musto, there have only been a "handful" of merit increases throughout the 3,000-faculty system in the past three years. In the same time period, 140 faculty members have been nominated for increases in pay — primarily to keep them from going elsewhere.

Only now, said Musto, has there been actual money budgeted for salary adjustments, many of them based on merit.

"In the past, we've lost very good faculty because they thought there was no way to get ahead, only the union raises," said Bruce Liebert, chairman of the Mechanical Engineering Department in the School of Engineering, who applauds Chen's initiative.

"If you'd like to keep the better faculty who have other opportunities, then having them get salary increases based on their contribution is definitely going to help keep them," said Galen Sasaki, associate professor in the Department of Electrical Engineering.

"Just like in a business, the people who get more done should get paid more."

Although the merit raise system in the Engineering College went in before Dobelle's arrival, the new president is providing strong support for deans and faculty on this issue.

In the past few months, Dean Barry C. Raleigh of the School of Ocean and Earth Science & Technology has seen about a dozen of his top faculty receive merit pay increases he has requested, even though he has been making such requests for a decade with little success.

"It's been very difficult," said Raleigh. "Now we're getting them through reasonably fast, and I'm very happy to say the faculties are much happier."

It's a far cry from the past.

A May 31, 2000, memo to deans and directors at Manoa from former UH President Kenneth P. Mortimer reads in part:

"Although requests have been received for merit adjustments, very few have been approved even when a strong case was made for outstanding achievement in the profession."

In the past, retention raises to hold onto important faculty were often the only way to reward the best. The system was so faulty, said Chuck Hayes, interim dean of the College of Natural Sciences, that two years ago when he nominated two phenomenal professors for merit increases in his faculty of 125, both were turned down.

"I then had to go to Bachman Hall (the administration building), and ask for retention (raises) for these two people," said Hayes. "One of them I was able to hold, and the other I lost. That would not have happened if I was able to give merit. And it cost a lot more to retain (the one) than it would have if I had got them merit in the first place."

Merit-based salary increases are standard at universities across the country, and Dobelle has pledged support. "If you don't validate your employees tangibly with money then I don't know how else you reward them," said Dobelle. "To me, it's fundamental."

Dobelle said that in comparing salaries with Mainland colleges, Hawai'i ranks in the 20th percentile, and he wants to work together with the union to move that up to the 50th and then 80th percentile.

Part of his commitment to the faculty includes his personal efforts to raise outside money for merit raises, faculty endowments and endowed chairs. That will also be part of the expectation he has for the new chief operating officer and chief financial officer of the UH Foundation, whom he expects will be named in the next week or two. He has high hopes the foundation will raise $250 million over the next five years.

"You are the future of this institution and, in many ways, the future of Hawai'i," he told a meeting of outstanding faculty researchers this week. "If we don't have the appropriate incentives for you, or raise money for you, then we're denying you the ability to make your research happen."

Mehrdad G. Nejhad was one of those at the meeting and one of the six engineering faculty who won a merit raise last year based both on his teaching and his research.

As he moves forward with $2 million in financing from the Center for Space Technology of the Naval Research Laboratory, to build "intelligent" structures that could be used to repair systems problems in space, the raise meant a lot, he says.

"It creates a far-reaching effect in the whole university," said Nejhad. "We lose people to the Mainland because of the differences in salary, but this will fill that gap and will also attract new faculty if they know this reward system is in place."

Appendix 13.5 - EDITORIAL
Tread Carefully with University Merit Raises
功績報酬

Posted on: Tuesday, January 29, 2002
The Honolulu Advertiser
Permission to reprint, 2006

All too often, the best and brightest teachers and researchers at the University of Hawai'i don't receive a financial incentive to stick around until it's too late. By the time those with the hottest prospects are offered retention raises beyond their standard union rates, they're already headed for more generous Mainland pastures.

So an initiative by the UH School of Engineering to reward their best faculty members with merit raises appears to be a good start in plugging the "brain drain".

As higher education writer Beverly Creamer reports, Wai-Fah Chen, the UH dean of engineering, tweaked his internal budget and tapped other reserves to give merit pay to a half-dozen faculty members who shine.

Chen used a meticulous and inclusive method to evaluate his staff, and we hope that others in his position follow his example.

It's important that any evaluation process be careful not to create an academic superstar system. There are many ways to measure excellence: a brilliant teacher of introductory English can be as valuable as a star winner of grants.

Each applicant, then, must be judged within the context of his or her potential and limitations.

If handled poorly, the art of merit raises can lower the morale of many while elevating that of the chosen few. Indeed, that was the concern during contract negotiations last year when Gov. Ben Cayetano offered UH faculty a 9 percent pay increase over two years that would go solely to merit raises.

At the time, Cayetano said that the days of giving raises for the sake of giving raises are over. But critics said that limiting the raise to merit pay would tear the university community apart because many people would receive no raises.

If handled fairly, merit pay can be used to bring UH salaries in line with better-paying Mainland institutions, retain outstanding faculty and generally raise the level of excellence.

Appendix 13.6
UH Executive Salaries Increase
主管薪俸

By Mary Vorsino
Ka Leo Editor-in-Chief
May 07, 2003
Permission to reprint, 2006

When the Board of Regents at a meeting in March 2001 announced its pick for who would succeed Kenneth Mortimer as president of the University of Hawai'i, the institution's professors — many of whom would walk picket lines for pay raises a month later — stood in the Bachman Hall foyer to welcome the former president of Trinity College in Hartford, Conn.

But shortly into the meeting, many faculty members couldn't help but gasp when the board announced what would be Evan Dobelle's pay, $442,000, an annual salary that was more than two times the size of Dobelle's predecessor and, in some cases, 10 times or more their own.

A little over two years later, some professors say they're not as alarmed about Dobelle's compensation as they are about a system wide trend — whose start coincided with Dobelle's hiring, according to university-provided data — toward markedly increased salaries for UH executives and managers.

This year, the university will spend $20,359,648 on 192 executives and managers, giving each an average salary of over $106,000.

The average salary for the 209 administrators employed in the year before Dobelle arrived, when the university was spending $3 million less on that tier's salaries, was $83,700.

But while the average salary for administrators has grown more than $20,000 since 2001, the average wage for full professors at the university's flagship campus — just over $87,000 — has crept up by less than half that, growing by $7,000 from 2002 to 2003, according to statistics compiled by the Chronicle of Higher Education.

Below is the top 10 University of Hawai'i executive salary wages and their earnings for the current year, 2001 and 1999.

2003

o   Evan Dobelle, UH system president, $442,008

o   Edwin Cadman, John A. Burns School of Medicine dean, $367,776

o   Rolf-Peter Kudritzki, director of the Institute for Astronomy, $260,016

o   Thomas Shomaker, School of Medicine's associate dean of academic affairs, $256,344

o   Walter Kirimitsu, senior vice president for legal affairs and university general counsel, $256,248

o   Deane Neubauer, Interim vice president for academic affairs, $254,040

o   Peter Englert, UH-Manoa chancellor, $254,016

o   James R.W. "Wick" Sloane, vice president for administration and chief financial officer, $227,016

o   Carl-Wilhelm Vogel, director of the Cancer Research Center of Hawai'i, $222,912

o   Joyce Tsunoda, senior vice president and chancellor of the community colleges, $218,520

2001

o   Evan Dobelle, UH system president, $442,008

o   Edwin Cadman, John A. Burns School of Medicine dean, $345,720

o   Thomas Shomaker, School of Medicine's associate dean of academic affairs, $240,960

o   Walter Kirimitsu, senior vice president for legal affairs and university general counsel, $240,864

o   Deane Neubauer, Interim UH-Manoa chancellor, $238,800

o   Rolf-Peter Kudritzki, director of the Institute for Astronomy, $225,264

o   Allan Daniel Robb, director of business and hospitality affairs, $213,216

o   Carl-Wilhelm Vogel, director of the Cancer Research Center of Hawai'i, $209,544

o   Cecil Raleigh, School of Ocean Earth Science and Technology's dean, $197,472

o   Wai-Fah Chen, College of Engineering dean, $193,824

1999

o   Carl-Wilhelm Vogel, director of the Cancer Research Center of Hawai'i, $200,016

o   Cecil Raleigh, School of Ocean Earth Science and Technology's dean, $188,496

o   Wai-Fah Chen, College of Engineering dean, $185,016

o   Kenneth Mortimer, UH system president and UH-Manoa chancellor, $167,184

o   Sherrel Hammar, Interim dean of John A. Burns School of Medicine, $163,560

o   Dean Orren Smith, senior vice president and UH-Manoa executive vice chancellor, $133,968

o   Alan Teramura, senior vice president of research and dean of the graduate division, $133,752

o   Klaus Keil, director of the Hawai'i Institute of Geophysics, $132,440 o Frederick Greenwood, director of the Pacific Biomedical Research Center, $132,072

o Robert McClaren, interim director of the Institute for Astronomy, $130,224 ∎

Appendix 13.7
Publications
論文集
University of Hawaii 夏大
2004 - 2006

Dr. Chen is the author or co-author of more than 595 articles in various refereed Technical Journals (347), Conference Proceedings and Symposium Volumes (248). The following is the list of journal articles during his years at the University of Hawaii as Dean of College of Engineering (1999 to 2006).

Doctoral Student – Ken Hwa (2003)

2004 – Journal Articles (5)

1. K. M. Abdalla and W. F. Chen, Effect of Control Fluid and Surface Treatments of High Strength Bolts, International Journal of Applied Science and Engineering, Vol. 2, No. 1, March (2004) 1-15.
2. H. L. Cheng, E. D. Sotelino and W. F. Chen, Sensitivity Study and Design Procedure for FRP Wrapped Reinforced Concrete Circular Columns, International Journal of Applied Science and Engineering, Vol. 2, No. 3, July (2004) 148-162.
3. K. Hwa and W. F. Chen, Survival Time Prediction of Steel Frames under Elevated Temperature Using Advanced Analysis, The International Journal of Steel Structures. Korea Society of Steel Construction, Vol.4, (2004) 187-196.
4. M. Komuro, N. Kishi, and W. F. Chen, Elasto-Plastic FE Analysis on Moment-Rotation Relations of Top- and Seat-Angle Connections, Connections in Steel Structures V, Amsterdam, June 3-4 (2004) 111-120.
5. N. Kishi, M. Komuro, and W. F. Chen, Four-Parameter Power Model for M-Theta Curves of Eng-Plate Connections, Connections in Steel Structures, Amsterdam, June 3-4 (2004) 99-110.

2005 – Journal Articles (5)

1. C. B. Joh and W. F. Chen, Effects of Concrete Slab on Ductility of Steel Moment Connections, International Journal on Advanced Steel Construction, Hong Kong, Vol. 1, No. 1, June (2005) 3-22.
2. W. F. Chen, Foreword for the Inaugurate Issue of the International Journal on Advanced Steel Construction, Hong Kong, Vol. 1, No.1, June (2005) 1.
3. H. Chen and W. F. Chen, Major Revisions of the 2005 AISC Seismic Code, International Journal on Advanced Steel Construction, Hong Kong, Vol.1, No. 2, September (2005) 3-16.
4. S. L. Chan and W. F. Chen, Advanced Analysis as a New Dimension for Structural Steel Design, International Journal on Advanced Analysis of Steel Construction, Hong Kong, Vol. 1, No. 2, September (2005) 87-102.
5. W. F. Chen, B. Chen and X.L. Liu, Developments of Structural Calculations, Keynote Lecture in Proceedings of the International Symposium on Innovation and Sustainability of Structures in Civil Engineering – Including Seismic Loading, Nanjing, China, November 20-22 (2005).

2006 – Journal Articles (5)

1. W. S. King, C. J. Chen, L. Duan, W. F. Chen, Plastic Analysis of Frames with Tapered Member, Journal of Architecture and Civil Engineering, Vol.23, No.2, June (2006) 9-19.
2. W. F. Chen, Foreword on Recent Work on Advanced Analysis, International Journal of Steel Structures, Korea, March (2006).
3. K. M. Abdalla and W. F. Chen, Base Plate Design in Steel Structures – A New Approach, Journal of Structural Engineering, Chennai, India (2006).
4. W. F. Chen and Y. M. Lan, Finite Element Study of Confined Concrete, Keynote Lecture at the International Symposium of Confined Concrete, June 12-14, 2004, Changsha, China, ACI SP-238, Editors, Y. Xiao and Y. R. Guo, (2006).
5. W. F. Chen and S. L. Chan Advanced Analysis of Steel Frames – from Theory to Practice, Keynote Lecture in Proceeding of the XI International Conference on Metal Structures, ICMS'2006, Rzeszow, Poland, June 21-23, 2006. Editors, M.A.Gizejowski, A. Kozlowski, L. Sleczka, J. Ziolko, Progress in Steel, Composite and Aluminium Structures, Taylor & Francis, London, (2006) pp.20-29.

Reception for Mayor Ma Ying-Jeou (馬英九) with Chancellor Rose Tseng (曾張蘊禮) (right of Ma), Dean Chen (coat, standing), and Linlin (玲玲) (left of Ma), and members of Association of Chinese Scholars in Hawaii (夏威夷國建聯誼會).

Dean Chen, Linlin (玲玲) and his secretary, Fay Horie, at the College of Engineering Banquet held at the Hawaiian Hilton Village in 2006.

Dean Chen, and Linlin (玲玲) with a group of undergraduate students at an exhibition held at the Hawaiian Hilton Village in 2006.

In 2004 Taipei, Dean Chen (center) met UH civil engineering alumnus Lin Wen-Yuan (林文淵) (right), Chairman of China Steel, and, Hou Ho-Shong (侯和雄) （left）, Nation Policy Adviser to President Chen Shui-Bian (陳水扁).

14

On Higher Education Reform in China
關于中國高等教育改革

As a 40-year veteran of academic life and career, I have served as an external reviewer of institutions of higher learning in China and Taiwan, as a panel member for evaluation of major research programs for an academic excellence in US, Hong Kong, and Taiwan, as a member of selection team for appointment of leadership positions in universities and government agencies, I prepared this article for the publication in the 2006 *Journal of Scientific Chinese*（科学中国人）to share my views on the reform of higher education in China. In this discussion I offer some perspectives on the current issues in higher education, lessons about reforms, and some thoughts on how to create a number of world-class universities in the rise of China as a major power in the 21st century.

14.1 Introduction 介紹

As China's economy booms, higher-education also expands rapidly. The government has pumped billions into higher-education reform with the aim of creating a number of world-class universities. But along with these opportunities, challenging problems develop that accompany the growing pains. Money alone is unlikely to fix the problems and attain the goal of developing some world-class universities.

Recent surveys point to the fact that while China has 10 universities ranked in the top 200 universities in the world, not one university is considered to be of world-class quality and reputation. In fact, out of the world's 20 top universities, 18 are American (Newsweek, June 12, 2006). The problem is the out-dated thinking used to develop the education system. This is in large part because of the influence of the past education system the leadership has gone through; and in part because of the thinking of planned economy for higher-education development.

To have some of China's elite universities join the "*club*" of top-tier, world-class institutions, we must depart from planned economy thinking of protectionism and embrace market-driven enterprise by introducing competition from the private sector.   This article examines the background of the challenges to this development and suggests some fundamental solutions for the transformation.

In Appendix 14.1, a vision article entitled "*How to Build a First-Rate Engineering College at UH*" is used as an illustrative example on how to build world-class universities in China.

In Appendix 14.2, the article entitled "*Engineering Education in Rapidly Changing Times: Engineers in Transition*" identifies the basic requirements of engineering education to produce the type of future engineers to address global problems.

14.2 Rapid Development 急速發展

China's remarkable growth rate of more than 8 percent per year for more than a decade cannot be sustained unless it can produce the skilled labor a modern economy needs. Several industries have scaled back their growth because they do not have qualified engineers to produce the products the customers want. Obviously, China cannot separate the reform on higher education system from the country's economy growth and transformation.

The Ministry of Education, with participation from a select group of universities, developed plans for major investment in higher-education. A glance of this rapid development for higher-education in China includes (Mooney, 2006):

o   The government doubled its investment in China's 2000 colleges and universities for a total of $12 billion from 1998 to 2004.

o   The total campus size has tripled in acreage.

o   The enrollment has increased from 3.4 million in 1998 to 19 million today (Minister of Education, 2004).

o   The national education excellence initiative known as 985 Project provided major funding in 1998 to a group of select elite universities for a five-year period. For example, Peking University and Tsinghua University received

$225 million each; and SJTU and Nanjing University received $150 million each to help raise them to a world-class university level. However, a typical operating budget for a public university is only $2 to 3 million which is grossly inadequate. In terms of GDP expenditure on national education, it is only 3.41 percent of the GDP, which is far below the 6 percent proposed by the United Nations for developing countries (Liu, 2005). Budgetary limits for most of these public institutions mean that they cannot build modern campuses, hire the best professors, so they were never able to attract top students. The famous old saying that "*The rich get richer, while the poor get poorer*" applies to China.

There is no clear definition for standards of a world-class university. One trend seems to focus on size by merging smaller universities into a large comprehensive one to compete for ranking in terms of numbers in research funding, paper publications and citations, and patents awarded, among others. Others are focusing more on hiring internationally known scholars, buying advanced laboratory equipment, or developing cutting-edge research programs. The government may be focusing too much on the development of hardware, and not paying enough attention on creating a scholarly environment for intellectual discovery, learning and teaching.

14.3 Bean-Counting Curriculum 填鴨式課程

Since the centrally-planned economic thinking is applied to higher-education development, it leads to a curriculum that focuses on numbers. It focuses too excessively on "*publish-or-perish*" for faculty members. As a result, teaching suffers. The traditional forced feeding students known as "*stuffed duck*" approach to teaching continues. Students are not happy with their education experiences because:

o   The education does not encourage creativity.
o   The professors only teach knowledge, not critical thinking.
o   Some old professors are still using their outdated notes.
o   Many courses have no relationship to the real world.
o   They have to take 8 courses per semester which is too many.
o   They want to learn more about practical applications and teamwork skills.

14.4 From Planned Economy Thinking to Market-Driven System
從計劃經濟到市場經濟

Most universities in China are founded or supported by the government and most presidents of the universities are appointed by the government. To some extent, the administrative mechanism in universities works like a government agency. There is very little flexibility in the administration and educational programs at the universities.

In most cases the curriculum in different departments is fixed by a national evaluation organization, even to the extent that the textbooks are selected by the government, not professors in the field who should be determining what should be taught and how. In general, government officials have the final say on who is hired and who is fired. Professors and administrators are fearful of making decisions and still defer to the government.

In a market-driven economy, the nature of teaching is very different. Students must understand how supply and demand works in the economy. They must be encouraged to argue and to challenge the teachers. Teachers must be up-to-date to gain the respect from their students. With a deep-rooted Confucian teaching, the process can be painful for students, who were all raised with the idea that to question a teacher is not only disrespectful but also shameful.

14.5 From Specialization to General Education 從專業到通材教育

Historically, China's education system follows the old Soviet Union system which puts emphasis on specialization. Most faculty members are knowledgeable in their sub-special areas and lack multi-disciplinary training (Liu, 2005). These faculty members were trained to develop technical skills for special subjects rather than in basic science, computing or physics. This limits their ability to do cross disciplinary work.

In a Confucian society, teachers are always right and students do not challenge them. This tradition discourages open discussion in the class and limits academic freedom. Furthermore, admission is determined by the national entrance examinations with some subjects unrelated to a student's field. This creates mismatch between individual talents, interest and educational opportunity. There is little opportunity for students to transfer to their preferred university or switch to their desired department of specialty once the decision is made by the entrance examinations.

The government knows that higher education in China needs to be completely revamped, but it is not easy to change the thinking of the powerful officials who do not really understand a market-driven economy. Low cost, fairness, and equal opportunity for all are the hallmarks of the higher-education policy in China, not the demand or need of the market. The nation-wide entrance examination has been successful in helping prevent corruption in the admission process for a country with a 1.3 billion population.

With today's environment of cell phones and Internet access, what students really want to learn are English, how to work in teams, and how to communication. English is increasingly required to get a good job. Students prefer that their professors teach their classes in English so they can learn to speak English. Wealthy Chinese are willing to pay for the best education that includes English-language centers and computer-training institutes. Students do not want to enroll in courses having no relevance to their careers.

14.6 Academic Work Plagued by Plagiarism 學術受抄襲折磨

During the Cultural Revolution from 1966 to 1976, the general culture was not to tell the truth. Academic integrity and ethical values were not respected (Liu, 2006). One generation was brought up in that environment. Even today, the impact of that culture still exists. This situation needs to be changed in the coming years with the help of increasing international collaboration and returning the respect to Chinese traditional values (Monaghan, 2006).

When students are forced to take useless or irrelevant courses and learning nothing, cheating to earn their degrees becomes a norm. As a result, cheating is no longer considered the great evil it once was. There is an old American saying, *"You can lead a horse to water, but you can't make him drink."* You cannot force people to learn. Students must be motivated to learn new things that have relevant to the fast-paced world they live in. If we simply continue the "*stuffed duck*" teaching with unrelated subjects, we will continue the wasteful cycle of "*Teacher pretends to teach, and student pretends to learn*".

China is currently facing a strong international competition for outstanding students. For many top engineering students in China, "*going abroad*" is still the first choice for their best and brightest graduate students. Educational system reform will help retain these top-notch students, recruit those foreign-trained students to return home to China, and improve academic integrity and ethical values of higher education in China.

14.7 Higher Education Reform 高教改革

There is growing frustration over the fact that students spend years in universities but graduate with irrelevant knowledge. They are asking: Why we don't have world-class universities? Why isn't there freedom to develop academic programs to meet market needs? Why isn't the higher education system merit-based and market-driven?

China needs to place emphasis on creative, cross-disciplinary research and focus on general education and avoid "*stuffed duck*" teaching. We must change the admission process to recruit quality students with talent. We must give control to university administrations to hire and fire faculty and staff. We must define different missions for different types of universities.

The government should lift restrictions on foreign investments on education and encourage them to invest in China. For example, in 2001 the Royal Melbourne Institute of Technology received permission to offer degrees in Vietnam and now runs popular programs in information technology, engineering and business (Overland, 2006). Similarly, high quality and popular joint M.B.A. programs with foreign institutions are rapidly developing in China. These provide healthy competition and showcases for other institutions to follow and serve as role models for reform.

An Illustrative Example

An article appeared in the 2006 *Chronicle of Higher Education* (Overland, 2006) described how Harvard teaches capitalism to communists in Vietnam. The Vietnamese government planned to build a new massive airport facility for Ho Chi Minh（何志明）City in a neighboring province. The current airport can easily handle the projected numbers of passengers while the new airport will be convenient to no one.

A professor from Harvard called the construction plan a suicide plan, but some students argued that the new airport would enhance Vietnam's image to the world. When the professor asked the class, *"Would you buy the bonds issued to raise money for the project"*, no one raised their hands because no one wanted to risk their own money by investing in this financially suicidal project. That is how the market-driven process works in the real world of engineering. The process should also work for higher education reform in China.

14.8 Summary and Conclusions 簡述與總結

Rome was not built in a day. Higher education reform must evolve but not be a revolution event. It will take time for the transition from planned economic thinking to market-driven enterprise to occur. Simply throwing more money at established universities will not solve the problem.

China needs competition from private universities. They need open-door policies to recruit international students and faculty members. They need less government interference. They need to produce graduates that employers want to hire. They need to make their own hiring decisions. They need to set their own curricula and design their own budget. They need to set up their own standards for admission and accreditation. They need to develop their own strategic plan and to articulate what they need to get there.

World-class universities will eventually emerge if they are allowed to stand on their own, compete freely, and receive their rewards accordingly. Once you enjoy the ability to make decisions and choose your own path, you cannot return to rigidness and no selection. *Openness, diversity, competition, and tolerance* are the fundamental values and traditions of top-tier universities. Once this scholarly environment is created and becomes a tradition in a campus for intellectual discovery, learning, teaching and sharing, a world-class university has emerged. The Nobel laureates today are the product of decades of work in such an environment. This is what we define as a world-class university. This is what we want to build it.

14.9 The Timing for Reform 改革時机

As Charles Dickens said, "*It was the best of times; it was the worst of times.*" This is good time for China to fundamentally reform its higher education system. As students have grown up with reforms, they understand that central planning leads to monopolies, mediocrity and inefficiency.

The public also understands the need for reform. All they have to do is to look around at the booming economy and their rapid changing of standard of living. Telecommunication competition makes cell phones so inexpensive and convenient that almost everyone owns one. There is no reason why a market-based approach will not work for the reform of higher education.

References
參考文獻

1.	Newsweek, June 12, 2006, p. 44.
2.	Paul Mooney, "The Long Road Ahead for China's Universities", the Chronicle of Higher Education, May 19, 2006, pp. A42-A45.
3.	Statistic Data of Education Reformation and Development in China, April 28 of 2004, News Issuance of the Ministry of Education in China (in Chinese).
4.	Paul Mooney, "Plagued by Plagiarism", the Chronicle of Higher Education, May 19, 2006, pp. A45-A46
5.	Peter Monaghan, "Open Doors, Closed Minds?" the Chronicle of Higher Education, May 19, 2006, pp. A14-A16.
6.	Xila Liu, "Challenges of Engineering Education in China", Presentation at the 7th WFEO World Congress on Engineering Education "Mobility of Engineers", March 4-8, 2006 Budapest, Hungary, pp.200-202.
7.	Martha Ann Overland, "Higher Education Lags Behind the Times in Vietnam", the Chronicle of Higher Education, June 6, 2006. pp. A36-A39.

Dean Chen gave a speech on higher education at the Association of Chinese Scholars of Hawaii.

Appendix 14.1
An Illustrative Example on How to Build World-Class Universities in China
在中國建立世界級大學的列範

How to Build a First-Rate Engineering College at the University of Hawaii 如何建立一流夏大工學院

W. F. Chen
Dean, College of Engineering
University of Hawaii at Manoa
Honolulu, HI 96822

Introduction 介紹

There are three ingredients to a great engineering college: it must have a first-rate faculty, a highly selected student body, and very good facilities. First-rate faculty attracts quality students and quality students attract quality faculty. Each is dependent on the other and together they bring excellence and prestige to the University. When I first arrived at the UH on September 1, 1999, I noted that our operating budget's purchasing power was reduced by about 30%, our total number of faculty positions was reduced by about 20%, and our total enrollment has dropped by more than 30% over the last five years. Some of our critical facilities have not been upgraded for many years. To build up our college, all three of these elements must be excellent.

Renewal 更新

Our highest priority must be to strengthen the faculty, as it is their quality that will attract the excellent students and help to enhance our facilities. Without a strong faculty, there is no strong college of engineering. Period. During the next few years, the College will have to recruit more than 30 percent of our faculty, more than 20 positions, and have to do this at a time when Ph.D.'s are in short supply in some critical disciplines and competition for the best talent is fierce. To this end, we must secure funds for start-up packages and use them to recruit the best and brightest young faculty, integrating them with our current faculty in our focus areas of research so they can grow and flourish with the College.

Crossing Boundaries 誇領域

Because our resources are limited and the size of our College is relatively small, we cannot constantly add new faculty positions in emerging areas of technology and to compete at all national levels for excellence. When we recruit, we must look for individuals with multiple interests who can teach and do research by crossing the specialty boundaries in the same department or even crossing the boundaries with different departments. We must concentrate our effort to strengthen the focus areas of research in our College, rather than just building one department at a time. We ask and encourage our faculty to work as a team, integrating their specialties in the creation of focus areas of research in the College, and they are starting to do so. Our Electrical Engineering faculty focus on the area of Advanced Wireless Communication, our Civil Engineering faculty cooperates in the area of Modeling and Simulation of Civil Infrastructure System, while our Mechanical Engineering faculty concentrates their efforts in the area of Biomedical Engineering.

Measuring Success 量度成功

How do we know we have succeeded in renewing our faculty and strengthening our research base and academic programs? One important indicator is how many dollars our faculty is able to secure via research grants from federal and state governments as well as from the private sector. Another indicator is the support we will receive not only from UH alumni but also from national corporations and foundations. One more indicator is whether the College is able to recruit and attract excellent, promising faculty members as well as top-notch students. All the major research universities are in a marathon race to maintain their excellent faculty and academic programs and attract high caliber students. We want to join this "club" of excellence. This is our goal.

Conclusions 結論

To join this "club" of excellence, we must have faculty members with national or international reputations. To keep and recruit quality faculty members, we need attractive salaries, strong colleagues, access to equipment and resources, and opportunities to work with top-notch students. Nothing can be taken for granted. We must continue to pay attention to our current faculty while working to attract new people. Recognition is a critical part of the process. It is very important that we demonstrate how much we value our colleagues' work, achievement and success. We can highlight our faculty's work in the five areas: Teaching, Research, Scholarship, Mentoring, and Service, and recognize their success with support and rewards.

Excellence is not something we achieve and then go on to do something else. It is a daily struggle, an annual struggle, and endless struggle. The struggle is not just for dollars. Dollars are necessary but we must have the vision. We must use the resources and potential opportunities to create an environment that allows us to recruit talented faculty and retain our good faculty, whose reputations will attract the quality students. To recruit top faculty and keep them happy is not only the responsibility of myself as the dean but also that of the University and the State of Hawaii.

Appendix 14.2
Engineering Education in Rapidly Changing Times
在急變世界中的工程教育
Engineers in Transition
工程師日新月異

Introduction

As a dean, I presided at convocation ceremonies for the College of Engineering. To the graduates, I always started with congratulations on successfully completing their course of study. They were happy to hear that their long study sessions, many homework problems, laboratories, and reports had been completed. But then I reminded them that they were not quite through yet. I told them that in order to compete successfully in world markets, they would need to learn continually in order to keep up. They need to be lifelong learners as engineers in the 21st century.

I believe that the 21st century is the engineered century and that we are at a particularly exciting time. We are fortunate because the industrial world is making the transition from the industrial age to the high technology age. The high technology age will be just that – *an age of technology*. It is a time of change and indeed it is a time for change. This paper, based on my 40 years of experience as a teacher, researcher and administrator in major private and public universities in U.S, provides my perspective on a sustainable engineering education curriculum to meet the needs of a high tech economy and global competition in rapidly changing times.

High Tech and New Economy

We use computers and communications gear to work smarter and more efficiently to increase our "*productivity*". After World War II, according to available statistics, productivity in U.S. increased 2.7 percent a year for 25 years. From 1995, it has been growing at 3 percent a year. This increase has to do with information technology. Will the current productivity growth hold up? Yes, we have only implemented 20 to 30 percent of what technology can do for business, if workers and consumers are willing to change their ways of doing business. This is known as *"new economy"*. In the new economy, we know that technology changes rapidly, but people do not. The most important thing is not technology. It is the thinking and redesigning of one's business.

The reasons for rapid changes are due to computers and communications. There are three key developments for communications in the 21st century: *Internet*: connect more people together from around the world. *Wireless*: free people from local physical infrastructure—use wireless local networks, mobile phones, and satellite communications. *Multimedia*: provide broadband contents—graphics, video, music, and voice.

As a result, science and engineering are in transition on two major elements:

1. Increasing globalization of R&D—contribute to stronger economy.
2. Prolific growth of Information Technology (IT)—impact on all facets of society.

Engineering Education in the Information Technology Age

In a recent survey made by the American Society of Engineering Education of its members, the first question was *"What are the three most significant issues facing engineers today?"* The second question was *"What will be the three most significant issues facing engineers in the next ten years?"* The majority of members identified the following three most significant issues for both questions:

1. Engineering education in the 21st century.
2. Engineering and public policy.
3. Managing complexity/information explosion.

As for the engineering education in the 21st century, four key issues were identified:

1. Lifelong learning.
2. Increasing the interdisciplinary nature of engineering.

3.　Preparing at the K-12 level for engineering careers.
4.　Teaching and learning engineering fundamentals.

Regarding engineering and public policy, three issues were identified:

1.　Communications between engineers and the public.
2.　Involvement in public policy decisions.
3.　Support for R&D.

With respect to research and development, we see federal government research missions and goals changing steadily over the years: from basic research to strategic research, then to applied research, and now to product research. Since around 70 percent of federal support of engineering research in U.S. universities comes from mission agencies such as DOD, DOE, and NASA, we must prepare our graduate students for the real world of research and development funding and work closely with industry toward new partnerships.

On the third issue on managing complexity information explosion, four issues were identified:

1.　Implication for education.
2.　Need for systems/multidisciplinary approach in engineering.
3.　Effect of globalization.
4.　Industry/university partnerships.

In short, the new economy is a knowledge-based economy, with education as the key element. Buying computers and software is easy, but rethinking and redesigning the way we do business is not. Technology changes rapidly, but people do not, and we must properly educate and prepare our workforce for the new economy. We must address the fundamental issue of K-12 educational, and build a quality engineering and science program at universities to meet societal need.

Engineering Education in Transition

According to recent MIT statistics, from a job market point of view, 45 percent of its recruiting companies were from the service sector, and many of them were in financial services. This means that engineering students were hired not just by traditional engineering firms but also by Wall Street companies. To some extent, engineering education has provided a foundation upon which to build other careers. For example, engineering today is becoming a popular foundation for careers in business, financial consulting firms and high tech sales.

Recent statistics show that less than one-half of the engineering graduates stay strictly in engineering jobs after a five-year employment period. Many of the graduates who do start in technical positions eventually move into management positions. World-renowned financial consulting firms hire engineers to manage their complex computer systems. As career vistas broaden for engineering students, the engineering curriculum must be changed to prepare for new career opportunities.

From a student background point of view, 35 percent of the MIT freshmen grew up in two-language home. It showed the drastic change in diversity of the American population with more Asian and Hispanic immigrants. Furthermore, 42 percent of entering freshmen at MIT were now women. There were very few civil and mechanical engineering women engineers during my time. Now, almost one-half of our engineering students are women. It is going to be a very diverse group, racially and culturally, compared to the population with which most of us grew up. These statistics from MIT reflected the general trend and changes in job market, family background, and the rapid increase in the number of women entering the engineering profession. As a result, the engineering curriculum must be adjusted accordingly to meet these societal changes.

Another important change for engineering education is globalization. Globalization is not something that is coming, it has already happened. Corporations can now purchase goods and services from anywhere in the world and international competition is keen. For example, to be promoted in American industry today, you need overseas experience and foreign languages. Engineering curricula must help prepare the students for such career advancements.

Since the industrial world is making the transition from the industrial age to the high technology age, it is changing the way we work and the way we educate students. As a further example, books and journals are now being published and courses taught on the World Wide Web.

Engineering education and curricula for the 21st must therefore meet the following three criteria:

1. They must be *relevant* to the lives and careers of students.
2. They must be *attractive* to the excitement and intellectual content of engineering.
3. They must be *connected* to the real world of engineering.

AISC Lifetime Achievements Award ceremony in Baltimore in April 2003.

Relevant to the Lives and Careers of Students

To the end, we must reform our undergraduate engineering education and curriculum to support two classes of career aspirations:
1. Those students who have a motivation to practice engineering.
2. Those students who desire curricular with significant technical content, but focus on various non-engineering career objectives.

To achieve these two classes of career objectives, we must:
1. De-emphasize narrow disciplinary approaches in our curriculum and the way we teach students to think.
2. Prepare them for a broad range of careers as well as lifelong learning.
3. Reduce the number of technical requirements, thus allowing students more flexibility in taking electives such as finance and economics.
4. Integrate teamwork and communication skills into engineering courses.

Similarly, we must reform our graduate engineering education and curricula to support these two classes of career objectives by:
1. Integrate bachelor-master degree into a 5-year program with practical orientation as the first profession degree.
2. Combine engineering-management into a new degree program with practice

orientation.

3.   Orient Ph.D. thesis toward industrial practice with research orientation.

4.   Hire senior faculty from non-traditional academic careers.

The challenges of a rapidly advancing, technology-base society create immense opportunities for engineering students of the 21st century. The goal of the sustainable engineering curriculum is to prepare the students to lead rewarding lives as leaders, who create waves of innovation in industry, academia, and government.

Attracted to the intellectual content of engineering

There is a general consensus that the undergraduate curriculum should teach only the fundamentals. The difficulty comes on the definition of fundamentals. In my own experience, the fundamentals are known as "*engineering science*". The fundamentals are continuous mathematics, applied physics, materials, and computing. In the information technology age, the new fundamentals are now changing to discrete mathematics, not continuous mathematics. In addition, the chemical and biological sciences are fast becoming fundamentals to engineering because of energy demand and bioengineering development.

This leads to the problem of background and training of our current faculty members. Most of them were trained in specialties and were narrowly focused on their research subjects. Our current reward system does not encourage our current faculty to gain experience in these emerging fields. Since these faculty members have the largest say in the engineering curriculum, it is difficult to reform the current curriculum without their participation. The pressure must come from industry leaders as they become increasing vocal about their discontent with current curricula and engineering graduates.

The fundamentals the students learn from school should be adequate for them to become lifelong learners. In business schools, for example, continuous education has been a part of the culture as the best of the best embraced executive training programs with the best teachers to teach these courses. In the engineering community, the notion of lifelong learning has not been a part of the engineering culture. Most will enter the MBA program to gain more managerial skills as their careers advance to higher levels with more responsibilities. Few will attend schools for simply upgrading their technical expertise in this very rapidly changing profession.

Arnold (中毅) and W.F. (惠發) at the front door of National Academy of Engineering's Einstein stature 1998.

Connected to the Real World of Engineering

We need to educate students to work better in groups and pay more attention to the context in which engineering is practiced. We need to give students more hands-on engineering experience, or *"design-build-operate,"* as we usually call it. This new teaching known as *"Project-Based Learning Approach,"* should be emphasized.

As an example, the Engineering Clinic Program is being developed at the University of Hawaii for our undergraduate students. Through this program the College of Engineering attracts industrial support for projects to be performed by undergraduate students. The program is aimed at bridging the gap between academic education and real world engineering practice. Projects are suggested by industry and are supervised jointly by faculty and industry liaisons. The College charges each company for each project; and the fund is used to provide scholarships for participating students. Students get academic credits for working on these projects. With the participation of industry liaisons, the faculty-student relationship extends far beyond assignment and grading; it becomes more of mentoring, coaching, and providing tight guidance to help achieve successful outcomes. This is the project-based learning program we are developing to provide the level of individualized engineering education that we envision for a research university.

Left: Visited the construction site of the "big egg" (國家大劇院) in Tiananmen Square (天安門廣場), Beijing, China.   Right: Chaired a conference in Beijing in 2005.

I am quoting the following Chinese sayings as my perspective on this new approach to teaching:

- o   Tell me, I will forget.
- o   Show me, I may remember.
- o   Involve me, I will understand.

- o   口到　眼到　手到　心到
- o   百聞不如一見，百見不如一試
- o   聽不如看，　看不如做
- o   耳聞是虛，眼見是實
- o   不經一事, 不長一智

Shifting Education Goals

The goals of traditional education are (past and present):
1. Working with one professional group such as civil engineers.
2. Rational analyses leading to quantifiable results.
3. The more mathematics, the better the results.

Goals of education in the future are:
1. Working as an interdisciplinary group including natural sciences and economy and public policy.
2. Learning to include subjective arguments, decision-making and being sensitive to values, ethical questions, and beauty, fear, danger etc.
3. Don't try to be exactly right; you do better directing your efforts to be roughly right.

English as a Second Language

In the Internet and information technology age, it is critically important for non-English speaking countries to dramatically strengthen their English education including the possibility of making English, say, in China, a second language in the future. The increasing globalization of the world in business, R&D developments, and finance will surely lead to an increasing mobility of people, which in turn will lead to more contacts and culture exchanges. English will soon become, if not already, a world language as an effective communication tool for interactions in the 21st century.

Concluding Remarks

We have described the important attributes that engineers must possess to be effective in addressing global problems. In essence, an engineering education must prepare graduates to:
1. be lifelong learners,
2. be flexible in career opportunities,
3. be team workers and good communicators,
4. be knowledgeable about solid engineering fundamentals, and
5. be sensitive to cultural differences.

When we discuss new requirements for engineering education brought by sustainable development, both what we teach and how we teach have an important impact on what students can learn.   We must keep reminding ourselves the famous saying: "*I hear, I forget; I see, I remember, and I do, I understand.*"   The central theme here is how best to engage students in active and problem-based learning processes, that are relevant to the life and career of student, that are attractive to the intellectual challenges of students, and that are connected to the real world of engineering.   This is the new criteria of future engineering education to produce future engineering students who can truly make a difference.

I think it is appropriate and informative to mention the following two recent reports and one book in U.S:

1. *Rising Above the Gathering Storm: Energizing and Employing America for a Brighter Economic Future*, Washington D.C.: The National Academies Press, 2006, 509pp.

2. *The Engineer of 2020: Visions of Engineering in the New Century*, Washington D.C.: The National Academies Press, 2004, 118pp.

3. Freidman, Thomas L., *The World is Flat: A Brief History of the Twenty-first Century*, New York, New York: Farrar, Straus, and Giroux. 2005, 496pp.

These publications identify the critical role of engineers play in advancing "*productivity*" in our global economy.   The two National Academies reports also identify the basic requirements of engineering education to produce the type of future engineers able to address global problems, including the issues brought by the sustainable development as the theme of this Workshop.

Acknowledgement:
This paper was presented at the International Workshop on Engineering Education for Sustainable Development held on October 31 to November 2, 2006 in Beijing, China.

15

My Adviser and I
导师與我

By Xila Liu 劉西拉

应该说，在陈惠发教授指导的众多研究生中我是比较特殊的一个学生：其一，我是唯一的一个毕业于北京清华大学的研究生，其二，我是年龄最大的一个研究生，跟陈教授读书时已经 41 岁。实际上，陈教授只比我大三岁。陈教授是我的老师，在二十多年的相处中，我们之间的关系是 "师生关系"但更像"朋友关系"。在陈教授的这本"*我的生涯與省思*"中，我愿用这一章有限的篇幅，作为一个学生，来感谢老师对我的教导；作为一个朋友，和他一起来见证中华之复兴，因为我们都为此付出了艰辛。

15.1 清华的校训和清华的精神 The Tsinghua Spirit

2007 年对我来说真是一个值得回忆往事的年头。50 年前，也就是 1957 年，我考上了北京清华大学。回想起来，那年的高考特别难，由于国家各方面条件的限制，全国的高校招生总人数比 1956 年下调，全国仅招 10 万零 7 千名大学生。我记得，从那年开始，中国和前苏联的关系开始恶化，大量已经决定派往前苏联留学的学生也停止前往，改为在国内读大学。他们中间的大多数因为家庭政治背景好，加上学习相对比较优秀，所以国家给他们一些特权，他们可以在国内挑选最好的学校求学。这样，国内已经为数不多的名牌大学变得更加拥挤。我所在的中学就有一些同学对此表示不满，组织起来开会、抗议，后来就在随之而来的"反右斗争"中被划成了"右派"。我还算比较走运，我不大相信靠开会、抗议能解决什么问题，我相信"实力"，有时间还是把课程复习好，最后还是靠"分数的较量"。

　　那年我是在南京参加全国入学考试的，地点就在南京工学院，解放前叫"中央大学"，现在叫"东南大学"。考试第一天就不顺利，我是骑自行车去考场的，半路上自行车胎突然漏气，只好临时改乘公共汽车。那时南京的"江南汽车公司"总共也没有多少公共汽车，加上没有汽油，每个汽车都是背一个烧木炭的再生煤气炉，行驶速度很慢，上坡的速度和人行的差不多。等我进入南京工学院大门，开考的铃声已经响起。我记得，我是飞跑进入考场的，还好老师没有取消我的考试资格。应该说，考试开始发挥得比较正常，几门课考下来，心中十分得意。谁知最后一门数学出了问题，那年的数学题目出奇得难，其中有两道题都很困难。后来我才知道，如果再多做半道题，我的数学成绩就不错了。我当时因为是班上的第一名，填报的十个志愿都是名校、名专业，这样一来，我前面的一批志愿都要垮。考完以后，我基本上可以准确地估算出每门考试的得分，我的父亲用概率分析了一下告诉我说："你可能上第七志愿，去清华土木系了。"父亲算得真准！就这样我是以第七志愿上了清华土木系。

　　刚进清华时，虽然觉得清华不错，但还是认为"土木"不大好，有点"又土又木"的味道，不如人家学"核工程"和"电机工程"的。然而我这个人总是很乐观，干什么都有兴趣，所以很快就适应了。走过这 50 年，回想起来，我很庆幸自己学了土木工程。土木工程领域里既有自然科学，又有社会科学；土木工程既是技术，又是艺术，合理的结构才能创造真正的美。土木工程讲究个性，世界上没有两个工程是完全一样的；土木工程又讲究综合性，能包容许多不同的学科。所以，我现在一直对土木系一年级的新生说，"如果我还有一辈子，我仍会选择土木工程"。

　　从 1957 年秋天开始，我在清华一呆就是 11 年。开始是五年本科学习。当时全国一般的工学院，本科是四年，但是清华特殊，本科是五年，而且还要延长到六年。我们那一届，正好处在五年改六年的过渡期，结果读了五年半，一直到 1963 年的春天才毕业。我大学的成绩不错，在 120 名同年级的学生中有两名获优秀毕业生金质奖章的，我是其中之一。随后我参加了全国的研究生联考，又被录取为清华大学结构工程教研组的研究生。当时的研究生属于教师编制，所以我们佩带的校徽是红色的，而学生的校徽是白色的。研究生由国家计划委员会管理，整个清华大学一共才 200 多名，都是最优秀的学生。我们按计划应该在 1965 年毕业，但是 1965 年毛泽东主席下了一道指示：“阶级斗争是主课，主课不及格不能毕业。”于是，我们这些即将毕业的研究生都下到农村去参加当时的“四清运动”。这一去直到文化大革命爆发，才回到学校。真正到工作岗位已经是 1968 年的四月。这样算下来，我在清华不是 11 年吗？

清华有一个国内外知名的校训，即“*自强不息，厚德载物*”。这是出自 1914 年著名学者梁启超先生莅校作“君子”为题的讲演时，以“自强不息”、“厚德载物”勉励学生的一段话，后被铸入校徽，高悬于大礼堂的上方，成为师生共同遵守的校训。这段话出自《周易》，即：“天行健，君子以自强不息”（乾卦）、“*地势坤，君子以厚德载物*”（坤卦）。意谓：天（即自然）的运动刚强劲健，相应于此，君子应刚毅坚卓，奋发图强；大地的气势厚实和顺，君子应增厚美德，容载万物。作为一个真正的人，在做事方面应不屈不挠，永远向上，永不停步，在为人方面应胸怀博大，团结众人，共同奋斗。这里既谈到“天”又说到“地”，既讲“做事”又讲“为人”，这是中华文化中多么精彩的一段！可能很多人不愿提及，这样一段精彩的校训，在清华园里曾被封闭了 30多年！从 1957 年进入清华，到我 1986 年从美国回校任教，在长达近 30 年的时间里，我在清华从来没有听过这个校训。

我第一次听到“自强不息，厚德载物”还是在后来的一次校庆活动中听当时土木系的老教授卢谦先生讲的。在 30 多年的时间里，我们根本看不到高悬于大礼堂上方的这八个字，看不到清华的老校徽，那块地方很长时间里是被一个红五星覆盖着。随着时间的流逝，历史终究恢复了它原本的面貌。现在所有的清华人都可以在大礼堂正中拱圈的中央看到清华的老校徽和“自强不息，厚德载物”这八个大字。它们的恢复真来之不易！梁启超先生的长孙梁从诚先生是我的好友，我们在全国政协任常委时有很多交往，彼此有许多相同的观点。梁启超先生的公子梁思成先生是 1958 年清华土木系和建筑系合并后我们的系主任，他曾亲自给我们上过建筑历史课。想到梁思成先生在文革时被拉到清华学堂（当时的土木建筑系馆）阳台上被红卫兵批斗的情景，不免又要出一身冷汗！毕竟，历史已经翻过了这一页。

清华传统的校徽和校训
The Tsinghua and Value

清华的人都有一种说不出的味道，现在我也常常和清华的老师、同学们说，希望大家来琢磨一下，什么是清华的"味道"？为什么有些人到清华任教十几年、几十年，就是没有清华的味道？为什么有些人一进清华就让人感到他们像土生土长的清华人？这里有一种说不出的清华传统，一种清华的精神。什么是清华的传统？什么是清华的精神？很长一段时间，我得不出结论。上届总理朱镕基在清华经济管理学院任过 17 年院长，他离任前在清华有一个演讲，谈到他认为的清华精神就是"追求完美"。我认为他归纳得很好。前几年在香港参加了一次香港清华校友的聚会，北京清华和台湾新竹清华的校长都来了，会上有一个从美国加州回来的老学长发言，他说解放前清华的体育代表队参加比赛时，清华的拉拉队喊的是英文，他们的口号是："We do the best, you do the rest." 准确！这就是清华的精神！

15.2 清华求学和清华教我做人 The Tsinghua Education

15.2.1 清华的学习生活 The Academic Life at Tsinghua

我非常庆幸自己能在清华读书，清华的环境、氛围使得每一个学生都要努力读书，丝毫不能懈怠。我从进清华起，就一直没有间断过自己的小提琴学习。那时除了每天要保持一个多小时的练琴外，每个星期天还必须从郊区清华园赶到城里的中央广播乐团去上课。每个星期天，别的同学可以好好复习一下一周紧张学习下来的功课，我却要步行到南门外，乘上慢腾腾的 31 路公共汽车进城，在中央广播乐团一呆就是一天，回来时已是傍晚，等回到宿舍，其他的五个同学已经围着一个大桌子开始晚自习了。他们和我一样，都是中学里出类拔萃的学生，这对我就是一种无形的压力。我不想放弃学小提琴，如果还要想和他们竞争，唯一的办法就是提高效率。不少人说："后退一步自然宽。"而我在清华的指导教师籍孝广先生却告诉我："一步也不能后退"。我欣赏后者，成功往往产生于最后努力一下的坚持中。后来在清华本科毕业时，我不仅拿到了优秀毕业生的金质奖章，而且还当上了清华大学管弦乐团的首席小提琴手，成功举办了个人小提琴独奏音乐会。上个世纪 60 年代的清华校友中，知道我拉小提琴的人很多。记得 2003 年元旦，我在北京参加全国政协的团拜会时，遇见总书记胡锦涛。已经有 40 年没有交谈了，我忍不住上前主动打招呼说："我是刘西拉。" 胡锦涛一见如故说："我知道，我知道！"他握着我的手说："怎么样？还拉琴吗？"我说："还拉。"他说："这样很好，你有终身爱好啊！"我说："您不是也有终身爱好吗？""什么？"他疑惑地问我。我说："您可以跳舞啊。"我有意让他想起 60 年代他在学生文工团舞蹈队的日子。他恍然大悟："啊，跳不动了，跳不动了！"于是大家都大笑起来。胡锦涛讲得很对，小提琴的确是我的终身爱好。

清华的本科学习十分严格。当时是学苏联，本科的主要课程，如数学、画法几何都要口试。学生不仅要在规定时间内完成笔试答卷，还要拿着答卷被老师一个一个地单独"盘问"。这种考试是世界上最严格的。我们必须在很短的时间里，把中学那种"教三、做二、考一"的习惯改成适应"教一、做二、考三"。记得大学三年级上材料力学，老师讲"薄壁杆件的约束扭转"。现有的教科书内容不够，老师就介绍了一本俄文的原版书，作者是符拉索夫（Власов）。这是一本很薄的书，关于"扇形正应力"的概念讲述得十分清晰。为了保持概念的准确，我记得自己当时的笔记是用俄文记的。20 多年后，我来到美国 Purdue 大学读研究生，由于报到晚了，只能旁听我的指导教师陈惠发教授讲授的"高等材料力学"。正好陈教授在讲"薄壁杆件的约束扭转"，他介绍了美国用的一种简易法，这使我回想起在清华学习的理论，仍然感到原来的概念十分清晰。几天后，陈教授要考试了，我就向陈教授申请参加这个考试。陈教授很吃惊，我怎么刚来美国就敢参加考试？我其实是想测评一下自己的实际水平，反正是旁听，也不算成绩，只是请老师多改份考卷，陈教授同意了。结果我的考试成绩是全班第一，陈教授很高兴，我也开始对自己在美国的学习有了一点信心。后来在 Purdue 陈教授给我们讲弹塑性力学，是用张量的形式讲的，这对清华的学生也不难，因为我们在本科时上的弹塑性力学就是用张量讲的。那时给我们上课的古国纪先生，是清华土木系第一个用矩阵做结构分析的（1954），也是清华土木系第一个用第一代计算机分析潜艇外壳失稳的，那时还没有程序语言（1962）。古先生讲的平衡方程，在没有体积力时就是 $\sigma_{ij,j} = 0$，我在上课时非常赞叹这种简练，这真是一种美！这些事至少说明，清华的本科教育是很扎实的。

清华的本科工程教育也非常重视实践。我们在大学二年级就直接参加学校一个通用车间的施工，我被分配学瓦工，跟着工地上的老师傅当学徒，学砌砖。没有多久我不但学会了如何放线、立匹数杆、把砖砌得横平竖直，而且能跟着师傅砌砖拱吊车梁了。我也学会随手捡一块砖，用一只手让砖在手中打转，如何尽可能把砖的好面放在外侧。现在，我是一个土木系的教授，但我很为自己有这段经历自豪。现在有几个土木系的教授会使用瓦刀、大铲砌砖的？三年级我们就跟着高班同学直接参加清华主楼的设计和施工。那时，我们分析结构框架用的是电模拟系统，也就是把一些的电工器件用导线拼成一个系统，使它的电阻、电容设置和电流、电压输出与结构分析中的弯矩、轴力和剪力一一对应。通过改变参数模拟荷载作用，通过量测参数求得内力分布。我们在清华主楼基础施工时用的是"井点降水"法，白天在工地，晚上听梁思成先生的课。我们就这样在上个世纪 60 年代初，把一个北京最高的清华主楼建起来了。

15.2.2 清华教我如何做人 Tsinghua Tradition and Value

学校一般是传授知识的场所，但在清华更强调能力的建设。我们当年的蒋南翔校长就多次讲过，教书是给学生"猎枪"，要教会学生如何打猎，而不是给学生"猎物"。这就是目前国际上普遍接受的"能力建设"（Capacity Building），看来"能力建设"比知识传递更重要。但是回想我的人生轨迹，却感到我在清华最大的收获是清华教会了我如何做人。

　　我是一个很不希望受约束的人。从小学六年级开始，我在班上就一直是第一名，但是上课很不安分，喜欢和周围同学讲话，喜欢提各种问题，有时把老师弄得非常尴尬。进清华以后，正赶上学校一个又一个的政治运动："红专辩论"、"处理右派"、"双反运动"、"反右倾运动"。我常常在这些运动中直率地谈自己的观点，而那些观点在当时多是"反面"的。但是我那时很自以为是，在运动中不但不认错，还组织一些同学成立"辩论团"，敢在会上公开地为自己的观点辩护，有时还能获得一片掌声。这个局面不是学校所希望的，不久我就遭到全年级有组织的批判。　那时针对我的批判会很厉害，有好几个重点发言，都是集体专门准备的。同学批判我的大字报贴到我的宿舍，甚至贴到我的床上。在那种压力下，我必须低头。现在想想，那些运动多半不是我的错，但我不喜欢埋怨历史，我觉得那些运动至少教会了我"不应该太自以为是，一个人应该有点自以为非的态度。"

　　我出生于一个教师家庭，父母都是教师。我从小比较娇生惯养，很少接触广大的平民百姓。这样成长的年青人有如"无源之水"、"无本之木"。我真感谢清华给我们这些学生有更多的机会到社会的基层去。比如说 1958 年的春天，我们全班参加了十三陵水库的建设，一去就是 10 天。我们班，晚上 11 点开始上坝挑土，到早上 7 点才下工。挑土很简单，就是用一根扁担挑两筐土。当时我们大家都是互相比的，筐里装的土要越多越好，要冒尖，这叫挑"窝头"。我们挑着两筐土往坝上跑，跑不动了再由另外一个同学接着挑着跑，我们一般要换三次才能把两筐土挑到坝顶。晚上劳动的工地上，人山人海，灯火辉煌，十几万人，来往穿梭，这时你会觉得自己完全置身于大众的熔炉当中。当时广播里面播放的歌是黄河大合唱，其中有一段让我特别震撼，那就是"五千年的民族，苦难真不少…"，每当我听到这两句，眼泪就要夺眶而出。中国近百年的苦难太多了，我们一定要努力奋斗！我们劳动的时候，旁边农民见到我们就喊"向大学生学习！"可是我觉得我们没什么值得他们学习的，在这样浩大的场景里，个人实在是太渺小了。我们当时干得非常累，因为都是强体力劳动。

　　中学时，我看过一本前苏联的小说《勇敢》，是讲西伯利亚开发的事情，他们讲共青团员在开发西伯利亚时晚上排队往工地上走都会睡着，前面的人一停，后面的人就撞到前面的人。这种事情过去是在小说上看到的，可是在十三陵水库工作的过程中我还真体会到了。这次劳动真是给我们带来了无穷无尽的享受，我们在挑最后一次土的时候，每个同学都挑四筐，从第一站开始就不休息，一直把土挑到坝顶。等到了坝上以后我们连人带土一起都倒在上面。从十三陵回来的时候，我在工地拣了一块石头，到现在还珍藏着，我觉得这次劳动对我是一次"洗礼"。

　　我们那时候也会在假期安排到贫困农村劳动。我记得 1961 年国家困难时期，我们到当时长城脚下有一个叫三堡的地方，在那里要建一个清华教师的活动中心。我们在那里劳动，每个学生一天只有一斤粮食。像我的个头比较大，一个月给 34 斤粮食不敢都吃完，要留一点粮票，以防出现更困难的时候。我们早上 2 两，中午 4 两，晚上 4 两。为了在抬石头的时候两条腿不打颤，一下工我们就都跑到山上爬上树摘"山里红"（山楂）吃，吃得差不多有点饱了，再回到食堂吃饭。那个时候，我们还利用工歇的时间访问周围的农村。我记得，当时已经是深秋季节，11 月底了，我们到附近农村里去看望一位老大爷，到了老大爷的家里，发现他整个房间的窗户纸都还没有糊。我问大爷为什么不把窗户纸糊上，老大爷说没有粮食熬浆糊。那天，我跟老大爷谈了一个晚上。他说，你们城里大学生总以为上大学是父母花的钱，你们有没有想过，如果没有我们种粮食怎么办呢？这个国家谁能承担呢？我认为老大爷给我上了很好的一课。我在想，我们能够上大学的条件是谁提供的？我在那天晚上才明白，没有广大农民的劳动是不可能支撑起我们这个社会的，我们是吃不上饭的！今天，我们在这里享受的时候，有没有想到还有一批人盼望着我们好好工作去改变国家的面貌？所以，等到我们毕业的时候，全班同学都要求到祖国最需要的地方去。那时刚得知大庆发现石油了，我们国家再不是缺油的国家了。所以当时没有一个同学填的工作志愿是大城市，大家填的都是大庆。第一志愿去大庆，第二志愿去西北，西北什么地方？随便。当时如果谁想去上海和北京，都下不去这个笔，觉得如果这样就对不起国家的培养，所以最后我们的同学都到很边远的地方去了。

胡锦涛当年也是到了西北刘家峡水库工地。我们班有一位同学叫孙勤悟，他分到了新疆伊犁。毕业后没有一个同学见过他，后来知道他还在伊犁。1997 年校庆，我们希望他能来北京聚会。打电话给他，他说很高兴，非常想见大家，但是如果坐火车从伊犁转到乌鲁木齐再过来，时间太长，他现在身为总工，还没有退休，没有那么多时间。如果让他坐飞机到北京，他又没有那么多钱。后来全班同学捐款，买了来回的机票送到他的办公室桌上。他看到机票，热泪盈眶。后来在北京我们见到了孙勤悟，他已是一位白发苍苍的老人了！四个班一百多人在一起聚餐，中间有一个主桌，主桌中间那个最重要的位置谁坐？要说当官的，我们班里有中国贸促会的会长，要说做学问的，我们有中国工程院的院士，但是我们没有一个人敢坐主桌中间的位置，后来我建议请孙勤悟坐，大家都同意了。为什么呢？因为当时我们说了要"到祖国最需要的地方去"，他不但去了，而且一干就是 35 年。他是一个浙江人，35年间只有他母亲病故才回了一次浙江，剩下的时间都在伊犁。这就是我们那一代清华人，一代决心奉献的人。所以我们那代人经得起考验，永远满腔热血，总是勇往直前。

15.3 祖国的"大三线"建设和改革的春风 The Defensive Strategy and Reforms in China

15.3.1 祖国的"大三线"建设 The Defensive Strategy in China

1966 年到 1976 年，在中国发生了文化大革命，现在这场革命已经过去 40 多年，它究竟是怎么发生的，绝大多数人仍然弄不清楚；它给中国造成多大的危害，大多数人也弄不清楚。1968 年初，整个学校一片混乱，校园几乎处在无人管理的状态。学生分成了两派，相互攻击、吵骂，几乎快开始"武斗"了。我和当时的女友（后来的妻子）陈陈就在这个时候离开了清华大学到四川任职。我们选择四川，主要是因为那里被称为祖国的"大三线"。所谓"大三线"就是说，如果打仗，东南沿海就是"一线"，而西南内地就是"大后方"，那里是今后国家建设的重点地区。我和陈陈总希望能多争取些时间工作，为国家多做些实事。我和陈陈的工作地点并不在一个城市，相隔 71 公里。我在四川省的省会成都，而陈陈在成都附近一个工业城市德阳。我在当时的国家建委西南建筑科学研究所，她在当时全国的三大电机制造厂之一的东方电机厂。我所在的研究所是当时国家建委研究任务最繁忙的研究单位，她所在的厂是当时国家机械工业部最受重视的大电机厂。我们虽然远离了北京、上海、南京这些大城市，远离了我们的父母亲人，但是我们非常满足，因为我们都有机会在国家最忙碌的岗位上发挥我们的一技之长。我们那代人脑子很清醒，有机会为国家服务和做事情就是我们的乐趣。

　　我来到西南建筑科学研究所以后，在同事面前有些特殊，一是来自名牌大学清华大学，二是学历最高，是毕业的研究生。那时候所里一个非常好的老总工程师叫曹居易，他建议我不要总呆在所里，要下到基层去，他说"越往下"对我越有好处，他对人非常真诚，我就相信他，按他的意见办了。 1968 年 12 月 1 日我带着两个同济大学的本科生下到四川内江的一个构件加工厂去，现在这两个大学生都当上了总工程师。当时我们三个一起下去，任务是生产一种不用木模的离心管柱。接待我们的工长是个上海浦东人，赵师傅，他是一位非常负责的老师傅，后来我在他身上学到了很多。但是我们第一次见面的感觉并不和谐，我们希望能和工人们住在一起，他说没必要，说我们可以住到加工厂附近的宾馆去。后来我跟两位同济大学的同学商量，我们下来就是要吃苦的，不能住宾馆，于是就在钢筋工工作的草棚里面睡下了。我们不住宾馆住草棚是表示我们的一种决心，我们一定要享受工人们的待遇，跟他们同吃、同住、同劳动。

1968 年 12 月 1 日的午夜，我躺在钢筋棚的工作台上，透过茅草顶上的破洞还能看到外面的月亮，已是深秋季节，不免感到寒冷。就这样，我们在钢筋棚的工作台上睡了一个礼拜，工人师傅们都很感动，后来天更冷了，他们就主动把他们的房间腾出来，让我们跟他们一起住。当我们跟工人们睡在一起的时候心里十分高兴，因为工人们已把我们当成自己人了！一个知识分子，能被老百姓当成自己人，这是一种荣耀。后来我们就在工地上出方案、做试验，随时随地请教那些工人师傅，从他们那里我们真学到许多在学校里学不到的知识。我们和他们一起背水泥、弯钢筋，和他们一起"打牙祭"、喝老酒，在他们身上我们学到了工人的许多优秀品质。他们那种诚实、正直深深地影响了我。整整 13 年，我在四川跑了许多工地，在现场做了许多设计，我们和师傅们合作的"钢筋混凝土离心管结构"，在全国推广了 200 多万平方米，得到了 1978 年全国科学大会的奖。"下基层"下对了，老曹工确实说得对。现在我和陈陈回忆起来，清华毕业后，我们到过祖国的西南，下过基层做工程，到过美国留学，年过 40 再去西方重修美国的 16 门课，这是我们的幸运。这两件事，构成了我们能成为知名教授的坚实基础。这两件事，不是我们所有的同龄人都可以享受到的。如果一定要问我们，"下基层"和"出国"哪个收获更大，我和陈陈的意见是一致的，那就是"下基层"。

现在大家在谈到文化大革命的时候，都不可否认那是一场灾难。我不喜欢一味地埋怨，我欣赏一句话，那就是"迎着太阳走，把黑影留在身后"。可以想象，如果我们当时仍然留在北京、守在清华，那将是什么日子？那样，我们要浪费近十年的时间，失去在工程第一线锻炼的机会，天天在学校里"开不完的会，排不完的队（根据每人政治上不断的表态评价每人的政治态度），写不完的检查，流不完的泪。"当然在那种日子里，对知识分子总的环境是压抑。我所在研究所的气氛几乎不是一个可以工作的环境。没有正式的上下班制度，晚上大家几乎是清一色地打桥牌。打牌，我认为是浪费光阴。所以我在所里的时候，晚上常常一个人到机关顶层的房间里去学德文、拉提琴。我宁可在四川的各个工地跑，多做一些实际的工程技术工作，也不太愿意回到成都。我当时管自己的做法叫"不着陆飞行"。到了周末，我就沿着川陕公路骑 71 公里的自行车到陈陈的厂里去，在那里可以真正休息一下。我自己满意的是，在文化大革命那样的日子里，我基本上没有浪费时间，而且把精力集中到专业工作上。

这场文化大革命是以毛泽东主席的去世告终的。我还记得那一天，我和我的科研小组在安徽马鞍山的工地上制作钢筋混凝土管柱节点的试件，准备运到南京工学院去试验。早上广播里开始还在骂我们这些知识分子是"臭老九"，后来突然宣布毛主席逝世了。这个消息并没有给我带来什么悲哀，只不过是一种震惊。人们普遍兴奋的是"四人帮"的垮台，那真是全民欢呼的日子！冬季，终于过去了，春天来了！

15.3.2 改革的春风 The Reform

从 1976 年开始，我就直接参加了国家新的钢筋混凝土结构设计规范编制工作，开始做一些有关钢筋混凝土结构基础理论的研究工作。从那时就开始接触我完全不懂的计算机，利用计算机模拟混凝土柱的破坏过程。我们那时使用的计算机语言是 Algo60，现在恐怕在计算机教科书里都不提这种语言了。那时输入还是使用纸带，用手工打孔。一个拥有 200 万人口的成都市只有一个为公众服务的四川省计算站。如果要去计算，需要预约。我有时被安排在夜里两点上机，这就必须先住在省计算站的招待所过夜。在那里其实是不敢睡的，因为过了时间，不但失去了上机的机会，而且还要付上机的费用。我就是在这种条件下第一次自己动手，编写了描述钢筋混凝土柱破坏全过程的大程序。为了模拟到柱子在破坏以后的软化行为，真不知度过了多少废寝忘食的日子，这对我又是一次"洗礼"。

很快我们就应该在计算机分析的基础上开始试验了。为了这次试验我几乎跑遍了全国去向老教授们请教。对我帮助最大的是西南交大（原唐山铁道学院）的黄棠教授，他是美国 Lehigh 大学黄棣教授的哥哥，出身于一个教授世家。他要求极为严格，对国外同类文献的分析十分透彻，对我的影响很大。我在清华的教授们也给我的试验规划出了很多主意，上面提到的籍孝广教授再三嘱咐我，在试验的要求上"一步也不能后退"。我是牢记这句话开始我的试验的。我制作了近 20 根离心混凝土的管柱，每根都是 7～8 米高，准备在我们研究所的 500 吨长柱试验机上进行试验。这批柱子从钢筋制作、贴电阻应变片到离心、养护，我都亲自参与。帮助我的小青年都是研究所里一些小学文化的学徒工，他们在我的示范和严格训练下工作十分出色，预埋在钢筋上的几百张电阻应变片，通过混凝土浇注、离心后居然 99% 都能正常工作！试验的时候，为了同步记录混凝土柱破坏过程中的全测点数据，我几乎把整个成都平原各单位能借到的 X－Y 记录仪都借来了。整个加载过程中，除了几位老教授和老工程师帮助外，其他十几个都是学徒工。有几天我的腰伤发作了，我就躺在试验机旁的一张板凳上指挥这十余人加载。现在回想起来，这批成果中的每一个数据还真都是按着"一步也不能后退"的要求得来的。当时怎么可能想到，就是这批数据成了我们后来在美国获得美国 ASCE1985 年 Raymond C. Reese 研究奖的重要基础工作。

文革的结束解除了我所有思想上的禁锢，繁重的科研任务又给了我施展自己能力的广阔天地，我的心情从来没有这么舒畅过。可能是因为我工作努力，得到了各级领导的重视。不久我被晋升为整个四川建工系统最年轻的工程师，经常代表省里的领导单位去处理一些重大的工程问题。我花费了近两年的时间，随一批资深的总工程师在全四川跑了几十个工地进行调查，主编了一份详尽的"四川工程质量调查报告"，受到国家建委的表扬。那时我仅三十几岁，我的业务水平已渐渐被工程界认可了。我所在的研究所，长期科研规划是我参加制定的，全所的科研规划报告是我做的。在全国的钢筋混凝土结构规范会议上我的发言常常受到大家的重视，甚至有一次全国的规范会议就因为我提出的问题需要讨论而延长了一天。

70 年代末，我已经被省里推荐为四川省青年联合会的委员，被四川省建工局和研究所的领导考虑要提拔为研究所的后备领导干部。按这个发展趋势，大概在四川干一辈子也不错。但有一个问题仍然摆在我们面前，那就是：离开清华十几年，我和陈陈仍在两个不同的城市里工作。这种分离，在我们当时的同龄人中已经非常罕见。四川省青年联合会为此专门给省委写了"情况反映"。后来听说，当时的省委书记赵紫阳（后来成为中共中央的总书记）亲自为这件事做了批示，大概的意思是：要尽快解决我和陈陈分居两地的问题，但是要保住这两个人才，不能让他们离开四川。按照这个指

示，无论我们两人都到成都或者都到德阳，我们之间必须有一个人
牺牲自己的专业。这对我和陈陈两个都对专业十分专注的人来说，
是不可能接受的。正在这个时候，邓小平提出要大量往国外派遣留
学生。在这种情况下，我和陈陈想出了唯一解决两人分居的办法就
是：两人设法出国留学，学成回国后再设法调到一起。这是我们当
时的"战略"决定，时间是 1978 年。

1978 年陈陈参加了当时一机部公派出国留学生的考试，总分
是西南地区第一名，顺利被录取了。她的英语很好，小时候就读的
中学是上海知名的"中西女校"，是当年宋霭龄、宋庆龄、宋美龄
三姐妹就读的中学，那里有很好的英文教育传统。经过一年的英语
训练，她就踏上了到美国 Purdue 大学的旅程。应该说，陈陈他们
是中国改革开放以后最早赴美的几批留学生。那时从中国大陆直飞
美国的太平洋航线还没开通，他们是从巴黎转机的。然而，我就没
有那么幸运。我小时候是学俄文的，英语是我的第二外语，所以我
的英文只能看看技术书籍，听说写都不行。那时也没有条件补习，
只能自己下功夫。那时我自己买了 New Concept English Course
的第二册，每天背一课、默写一课，不完成不睡觉。每天早上工作
之前，先把一课英文完成。就这样用了一年的时间，把整本书的 96
课背了四遍。同时，我又买了一台录音机不停地练我的听力。这就
是我到美国前的全部英文功底。我能够去美国留学，除了国家的政
策开放以外，关键还是我在普渡大学的导师陈惠发教授。

15.3.3 跨出国门 Going Abroad

我的出国没有陈陈那么顺利。那时候我从事的计算分析和试验都是
围绕着钢筋混凝土的纵向弯曲问题，在查阅文献时发现一个叫
"W. F. Chen"的教授在国际上有大量的文献，Lehigh 大学黄棣教
授、方晓阳教授到四川访问时，我曾和他们有接触，就打听了有关
"W. F. Chen"的情况，原来他叫陈惠发，已经到 Purdue（普渡）大学
任教了。黄棣教授曾详细给我介绍陈教授在梁柱理论方面的贡献。
后来同济大学的蒋大骅教授、朱伯龙教授和清华大学的沈聚敏教授
都跟我谈到陈教授在塑性理论方面的贡献，他们在丹麦 IABSE 开会
时见到过。于是陈陈去普渡大学时，我就请她把我的有关钢筋混凝
土离心管结构的著作和几篇中文论文送给陈教授，想看看他对我是
否有兴趣。我开始只是抱着试试看的想法，因为预先一点直接联系

都没有。没想到，陈教授给我的答复是积极的，他不但接受我去读研究生，而且同意给我资助，这真叫我喜出望外！如果我真的实现了出国的愿望，不但可以使我在工程界工作十余年后有了再度深造的机会，而且又可以和陈陈团聚了！

但是事情并不是那么顺利。当我开始正式向研究所的领导提出出国申请时，所里的领导没有把握。像我这样的人，妻子已经在国外，从政策上讲，能不能出国呢？以往夫妻两人一起出国至少应该是外交参赞一级的官员，一个技术人员是否符合这个规定呢？为此，研究所专门派人去国家建设部，向外事司司长请示。后来我才知道建设部同意了。随后出现的问题是，我已经是一名工程师了，"工程师"当时属于国家的业务骨干，按照规定必须由国家"公派"出国，而不能"因私"出国。可是当时在整个四川省，还没有任何"公派"出国、外国"资助"的先例，可以想象这个申请的过程有多么困难！最后我的政治审查是中共四川省委组织部进行的，我的护照是四川省的副省长亲自签署的。现在想想，真够特殊的，这真叫"好事多磨"。

为了出国要做些业务准备，1981年春，我专门到南京大学进修了半年的数学，主要是"数值计算方法"、"最优化方法"和"样条函数"。意想不到的是，就在这年5月，我在南京第一次遇见了陈惠发教授，他正好应东南大学丁大均教授之邀到南京讲课。那时的气氛还不开放，因为陈教授是美国客人，住在江苏省委的招待所，一般外面的人不能随便接触，要见面一定要有组织的批准。我那时就不顾这些，通过我父亲的中国药科大学，给省委招待所的门卫打了一个招呼，就直接闯到陈教授的房间里。现在想想，这也太没规矩了。因为是私自会见一个美国教授，怕引起一系列政治麻烦，所以见面的时间不能很长。我的第一个印象是陈教授没有架子，和我谈话不像是一个美国教授对中国学生，简直就像两个朋友。我们谈话时，他的大儿子 Eric 也在身边。我这才知道，陈教授小时候也在南京住过，是30年前从大陆去台湾的。他离开大陆的时候是他的父亲雇了一条小船把他们全家接出去的。他当时还是一个十岁刚出头的孩子，不知道离开大陆以后能做什么，他就带了几件干活儿的工具，准备靠自己劳动活下去。这样一个孩子，在他幼小的心灵中就很务实。而我是另外一条路，1945年从内地到南京，后转到上海一直到大陆解放。那时我父亲在美国，我母亲差一点也要离开大陆到英国去了，只是我和弟弟还小没有人带，这才暂时留下。历史把我们一隔开就是 30 年！但是无论多久，中国人还是中国人，同样的中华民族的热血仍然在我们全身流淌着。我们身处不同的国度，但从内心都一样地期望中国能够早日强盛起来。在当时的环境下，一个美国的教授决定要招收中国大陆的研究生要顶住多大的政治和舆论的压力？这种压力，在我到达 Purdue 大学以后才真正明白。我非常感谢陈教授。

15.4 普渡大學校園里的清華學子 Tsinghua Students at Purdue

15.4.1 留学的艰难和我的指导教师 My Challenges and My Adviser

1981 年 9 月，我也踏上了赴美的旅程。我记得，临行前，四川的研究所预支给我两年的工资，换成大约 1500 元美金，请专人送到北京给我。我怕到纽约后会遭抢劫，于是把这些现金都藏在我穿的皮鞋里。我记得，离开北京的前一晚，我一个人骑车到北京建国门外的立交桥上兜了一圈，感受了一下第一座立交桥是怎样的。我真感到一种自豪，国家变样了，虽然当时国家经济正在"调整、巩固、充实、提高"。我下定决心一定要在 1985 年赶回来，迎接下一个建设高潮。我还记得，陈陈出国之前，我和她 起去看了清华的老校长蒋南翔，他当时给我们提出了一个问题，这个问题也很值得大家现在去思考。他说，50 年代中国派了大量的留学生到苏联去，这批人学成后都回来了，跟清华大学自己培养的学生相比，发现他们如果不是搞新的专业，常常比不过清华大学自己培养的学生。这是什么原因呢？可能是语言问题。因为刚到苏联要学高等数学等一些基础理论课，这些课又不能用中文讲，这样对他们的基础就会有一定的影响。蒋校长希望我们："你们出去试试，看看我们国内大学的学生出去拿美国的博士难不难？"那时，我们这批留美的学生，心中总感到有种责任。

刚到普渡大学时生活相当艰苦。我们没有汽车，在校园里捡了几部别人丢弃的自行车，花了一个上午自己拼成了一部不错的自行车，这辆车我一直用到 1985 年 11 月离开普渡。周末，我一般是和陈陈骑着自行车去买菜，大冷的冬天，气温零下摄氏 40 度，弄不好手会粘到自行车把上。骑车行驶在路上，由于马路没有专门的自行车道，要么在路边扫出的雪堆中穿行，要么在扫出的道路上与汽车抢道，十分危险。这时才体会到，要尽可能买一辆汽车，再破的车也要买，这是为了生活，为了安全。我们几乎整天都是规律的生活：早上八点多到办公室，中午在办公室啃早上带出来的三明治；下午五点多回家，和陈陈做一顿晚饭，看一会儿电视，实际是打瞌睡；晚上八点又回到办公室，一直工作到第二天凌晨一时许才和陈陈电话联系，开车回家。两天的周末，只有星期六上午买菜，其他时间全部在学习、工作。现在回想起来，这段生活是我们一生中难得的可以集中精力的日子。

我那时最大的困难是英语。上课时老师开始布置作业都听不懂，到了晚上看着大家都在忙，才知道明天要交作业，只能连夜起，　干就是一个通宵。天亮了，我在回家的路上非常难过，心想这辈子读书从没读得这么被动，不就是英文没听懂吗？但是英文又不是突击一下就可以掌握的。我的英语最后是靠修了一门英语作文课（English 002）勉强过关的。这门课只要求写十篇作文，十篇作文合格，英文就不用再考了，也不用考 TOEFL 了。我的英文作文做得很好，有好几篇文章被我们的英文教授留下来做为范文，介绍给以后修课的学生。其实，英文作文做好的窍门就是要先把中文编好。

1981 年和陈陈在普渡团聚

在普渡，我的指导教师陈惠发教授给我的帮助是时时刻刻的，他的工作风格无形中对我影响极大。他很珍惜时间，我们之间很少有长时间的闲谈，谈话就要直率、简明。直到现在，我对自己的学生也保持着这种习惯。他很严格，我还记得他第一次请我帮他校对一篇论文，我草草看了一遍就退还给他了。他看了以后，发现有几处漏校，就气冲冲地到我的办公室批评了我一顿。这件事让我一辈子都记得住。后来我在学术界也有"严谨治学，一丝不苟"的声誉，我想这与陈教授以及上面提到的黄棠教授对我的教育很有关系。

　　陈教授很注意学术前沿的变化，而且能以极快的速度抓住前沿。有一次他在课上讲到钢筋混凝土板的 Smear **Cracking** 模型，当时加拿大刚刚做了一批试验，他就把这批试验作为家庭作业让我们全班同学用不同的模型进行分析。事后，大概因为我的作业分析得比较全面，他就建议与我一同发表一篇论文。不久加拿大做试验的这位教授到 Purdue 来讲学，发现我们已经把他的工作当家庭作业做了，大吃一惊！所以我把陈教授的这个风格称为：做事要"Right on time"。陈教授对我最大的教育还是"勇于开拓"。陈教授擅长的是塑性力学，他能用他的所长在钢结构、土力学和混凝土结构各方面全方位地拓展，能有机会在这样宽的领域里工作，是非常难得的。所以在博士论文的写作期间，我常对陈教授说，如果他还有什么新的课题，我可以帮他完成。有一次，他真要我完成一个 EXXON 公司关于北冰洋巨大冰块冲击海洋平台的课题，这里要用极限方法分析冰块的极限破坏荷载。大概因为我完成得不错，他就希望我能和他共同出版一本书叫"Limit Analysis in Soil Mechanics"。我其实没有做过什么土力学的工作，但是在陈教授和他前面几个学生工作的基础上居然也把书写成了，并于 1990 年在 ELSEVIER 出版。这件事与其说是增长了我在土力学前沿和用英文写作方面的知识，倒不如说是给了我一种开拓的胆量。

　　陈惠发教授在一些关键时刻的帮助使我永难忘怀。一件就是我 1982 年秋天的硕士学位答辩。因为出国手续的麻烦，我到学校报到时，已经错过了学校的选课时间。但是我不希望浪费时间，总想争取时间多修一点学分，这样陈教授就同意我先开始硕士论文的写作，因为这也算学分。我论文的题目就是关于钢筋混凝土离心管柱的纵向弯曲，依据的就是我在四川做的试验。我对这篇论文很有信心，因为详细的试验资料告诉我，钢筋混凝土柱的破坏，不是单纯的材料破坏，而是在材料破坏过程中的失稳。这个发现可以完整统一地解释我所进行过的全部钢筋混凝土柱（从 15 米的高压输电杆到 1 米高的短柱）破坏的现象，似乎在人们非常熟悉的规律中看到了新的内涵。但是，这篇论文是我在没有选修普渡大学的钢筋混凝土课程的情况下写的，这可能会让教这门课的教授很不高兴。果然，这位教授在看到我的论文送交审查后，就表示要给我一点厉害瞧瞧。答辩的时候，他真就这样做了。他不断打断我的发言，提一些我根本听不懂的问题，好在我当时并没有害怕，我给自己立了一个对付他的原则，就是"你讲一句，我讲十句"。答辩之后，这位教授要求重审我的论文，暂时不给学位。

当时我的指导老师陈惠发教授就在答辩现场，他理解我的难处，我的问题不在于技术内容而在于语言交流。答辩之后，他把我叫到他的办公室，告诉我答辩委员会的这个决定。这时他突然说了一句："这个教授是搞设计出身的，在理论上他难不倒你。"这当时对我是一个十分关键的支持。一个人的关键时刻没有几次，这个时候陈教授的这句话就等于告诉我，不要怕，要对自己有信心。后来一直等到学期快结束了，这位教授还没有审查完。我急了，就去找陈教授，我说：如果到年底拿不到硕士学位我的博士生计划就无法开始，我在美国的学习时间全要延长。陈教授马上帮我催了。过了几天，一个星期天的上午，这位教授终于来找我了，他来到我的办公室开始一页页地对我的论文提意见。我看到他在我的论文上面有很多批注，批得最多的是"Poor English"。我看到以后一点也不慌，一直面带微笑，因为我想：我英语再破，总比你的中文要好。后来他问到了核心问题，他说你对钢筋混凝土柱的失稳的看法有问题。这时我才严肃起来跟他讲："对不起，这个地方我一个字也不会改，因为我论文的精华就在这里，如果改了的话，就不是我的论文了。"他没有坚持，说可能是他没有看懂，我没有说话，还是面带微笑。

1985 年美国 ASCE 给的奖，就是因为这篇论文的内容。1985年的秋天，我和陈教授在芝加哥 500 多人的宴会上接受颁奖。一进大厅就可以看到我和我的导师陈惠发教授的照片挂在了大厅中间的展板上。就在这个时候我碰到了给我难堪的那个教授，他也上来向我祝贺。我说我也要感谢你，因为你帮我改了英文，他的脸当时一下子就红了。颁奖时，我的名字排在第一，我的指导老师陈教授的名字排在第二，因为这是按我们发表论文时的作者排序发奖的。记得在发表论文时，陈教授一直是把学生放在第一作者，自己的名字居第二。在荣誉面前，他没有像有些教授，喜欢把自己的名字放在前面，这对我又是一次深刻的教育。一个教授在名誉面前不计较，这才是真正的学者风度。回国后，我一直效仿陈教授的做法。在芝加哥得奖之后，人们排着队来向我们祝贺。我记得，有一个美国的老先生来和我握手，问："您是清华大学毕业的？是哪个清华大学？是台湾的，还是北京的？"我说："是北京的。"他说：

"啊，那就是中华人民共和国的。"这真是一种喜悦，这种喜悦有如在国际比赛中看到自己国家的国旗升起一样！

使我难以忘怀的另一件事是在我 1985 年回国以前，当时因为我得到了美国 ASCE 的奖，不少单位、公司和学校来信邀我去面试，可以考虑留我在美国工作。甚至在普渡，也有教授到我家里和我谈到深夜，劝我能和陈陈都留下来。他们认为，中国的现状还没有条件能使我们回去好好发挥作用。我这个人喜欢有计划，在出国前我就计划要回国，而且自己定的时间是 1985 年。我始终认为，我回国是自然的，不回国是不自然的。但是，的确有不少好朋友劝我留下来。有的甚至说："西拉你还有几天了，要改变主意留下来，还来得及。"为此，我去问过陈教授，想听听他的看法。他的看法非常干脆：应该回去。他说：你在国内很快会被提拔起来。他又说，回去要抓紧多搞些业务，否则提起来以后就没有时间了。在这个关系重大的决策面前，陈教授给了我有力的支持和明确的指导。

1985 年与陈惠发教授获 ASCE Raymond C. Reese 研究奖.

15.4.2 留给 Lafayette 的琴声 Sharing Music in Lafayette

在普渡，陈陈在中国大陆的留学生中名气很大，很多人知道，她不仅有数理化三门均为 100 分、华东地区第一名的成绩考入清华大学的记录，还有大学全部功课都是优秀（5 分）的优秀毕业生的成绩，更突出的是，她毕业时同时拿到清华大学电机系和中央音乐学院钢琴系的两张文凭。1981 年普渡成立了美洲大陆第一个中国访问学者和留学生的联谊会，陈陈被选为第一届联谊会的副主席。我们的联谊会组织得非常活跃和成功，在全美、特别是中西部是有名的。中央和教育部的领导到美国视察留学生，几乎都要到 Purdue 来。第二届，我被选为联谊会的主席。从那以后，我们每年中秋节时都要在校园里组织一次 China Night，邀请华人教授、美国朋友参加，共同度过中国人团聚的时光。作为联谊会主席，每年迎新、毕业时都有许多活动；有时候为了维护中国学生的利益，要与普渡的 Student Office 打交道，甚至要到法院打官司；常常要开车到芝加哥国际机场接送中国的留学生，帮助安排他们的临时住处，甚至要到芝加哥 China Town 帮大家采购点食品。为了不耽误我的研究工作，我做这些事时，特别强调效率，对我的合作同伴也特别强调效率和纪律。时间长了，反倒培养了我一种很好的工作习惯。直到今天，和我合作的同事都知道我的这种习惯。

　　一般刚到美国的留学生都有很长一段时间不能适应美国人的文化环境，但是我和陈陈却是例外。我们可以演奏小提琴和钢琴，交友的范围就比较广泛。从纪念会到联欢会，从养老院到礼拜堂，我们常被邀请去演出。开始时，美国人有些吃惊，由于他们长期没有接触中国大陆来的学生，就奇怪："怎么你们也懂贝多芬和莫扎特的？"也有些美国教授，他们保留着一些中国音乐的录音，但多半是一些丝弦古曲。当他们听到我们演奏的几首新中国的曲子："浏阳河"、"新疆之春"后，马上就感到一种生机，可以少说许多话，大大拉近了彼此的距离。时间长了，陈陈和我甚至被邀请到 Purdue 大学的音乐课上介绍新中国的各种钢琴和小提琴曲目，直接给学生演奏。那时有位在 Lafayette 市交响乐团拉第一小提琴的老太太劝我去考市交响乐团，我也十分希望自己能在一个专业的乐团里试试身手，于是就大胆去尝试了一下。考试那天，乐团各个声部的首席都来了。先考识谱，让我拉了一段"卡门序曲"，后来要我拉一段自己喜欢的曲子，我就拉了"新春乐"，拉得大家眉飞色舞！就这样，乐团指挥通知我，我被录取了。因为我的琴质量不好，乐团专门借给我一把琴。从此，我每星期都有一个晚上在市交响乐团音乐厅参加合练，正式成为一名市交响乐团的第一小提琴手。我们经历了许多演出，几乎每个月都要演出。每次演出，我还可以得到一些酬金。后来，我又被 Lafayette 市交响乐团借到 Illinois 的 Danville 交响乐团去拉琴，每周要加一个晚上开车到 Illinois 州去。这些演出使我有机会接触到一些世界知名的钢琴家和小提琴家，并给他们伴奏，这真是一种享受！直到现在还常常回忆。

1984 年底，陈陈顺利完成博士论文答辩，开始准备回国。Lafayette 市交响乐团正在准备迎接成立 35 周年纪念音乐会。指挥找到我说："这是我们乐团历史上的一件大事，我们希望能演奏世界各国的名曲。我知道，您夫人的钢琴弹得很好，有没有可能请您的夫人来演奏一首中国的钢琴协奏曲？"这真是一个挑战！自改革开放以来，那时在美国没有任何人弹中国的钢琴协奏曲，不但业余的没有，专业的也没有。但是天助人也，陈陈就有这个可能，因为她大学毕业时就弹过刘诗昆的"青年钢琴协奏曲"。和陈陈商量之后，我鼓励她一定要抓住这个机会，接受这个挑战。我觉得，在这种时刻，表现个人的才华并不重要，重要的是要让更多的人通过音乐了解中国大陆，了解大陆的学生。于是我认真地回答乐团指挥："我们很高兴参加这个演出。"指挥也答应会对陈陈的练习给予更多的指导。随后是紧张的准备。

首先我们把这个决定用最快的方式告诉北京中央音乐学院的招翠馨先生，她是陈陈当年的钢琴老师，很快招先生就把 50 年代的全套总谱寄来。打开一看才知道这份总谱是为民乐写的。马上要解决的问题就是要把这份总谱改成西洋管弦乐谱并写出各种乐器的分谱。也多亏清华的培养，在"大跃进"时代，这类工作我们在乐队干过。那时学生自己创作的一首管弦乐"劳动赞歌"居然能到天桥剧场去演出。面对眼前的困难只有自己动手了。其中自然还有些问题：例如洋琴的分谱由什么代替？后来决定用古钢琴。又如中国的小鼓等打击乐器很难找到，也只好用西方打击乐器代替了。乐团指挥大概每隔几天就来指导陈陈一次，那段时间陈陈的手指都弹裂了，只好贴上胶布继续练。与交响乐团合排期间，又正遇美国中西部的大风雪，钢琴与乐队的合作时间只能减到最少。

终于到了演出的那天，Lafayette 市交响乐团的音乐厅坐满了观众，特别是许多中国留学生，包括香港、台湾的同学都来了。音乐会开始，先宣读了里根总统专门为 Lafayette 市交响乐团成立 35 周年写的贺信，然后开始演出。我还是像往常一样，坐在小提琴的第一声部，陈陈的节目被安排在上半场的最后一个。在她的节目开始前，乐团指挥专门对观众说了一段话："下面请你们欣赏新中国的音乐。中国人认为，音乐应该是'古为今用'、'洋为中用'，请你们听听他们是怎样做到的。演奏者是 Purdue 大学电机工程博士陈陈。"演出进行得非常顺利，在全队齐奏歌剧"刘胡兰"的主题曲时，我看到这些金发碧眼的提琴手们陶醉在音乐中的样子，心中十分自豪。原来不仅是西方的音乐，东方的音乐也是动人的，而东方音乐最动人之处在于美丽的旋律。当陈陈的最后一个和弦与整个乐队的最强音一起结束时，音乐厅里爆发出雷鸣般掌声，我看见台下的观众都站起来了，掌声久久不绝……。我们一直珍藏着这个现场的录音带，它给我们带来一生中一个难忘的回忆！

15.5 清華園裏的普渡人 The Boiler Maker at Tsinghua

15.5.1 重归清华园 Returning to Tsinghua

1985 年我和陈陈回到祖国，后来从中国驻美国大使馆知道，我们是中国改革开放以后第一对双双获得博士学位回国的夫妻。因为我在美国获奖，一回国就被请到中南海谈话，那是我第一次进中南海，在一个平房里还见到了当时的中共中央总书记胡耀邦。中南海和我谈话的领导希望我能到国家科委任职，安排的职务不低。第二天国家科委领导就找我谈话，可是我这个人有些"不识相"，当着那位领导的面就说，我想教书，不想当官。我的父亲是老师，我的母亲是老师，我还是喜欢当老师。就这样，我选择了清华大学，离别 18 年又回到了我的母校。在美国中西部，大家给普渡的学生一个绰号，那就是 Boiler Maker（锅炉工），一个憨厚、踏实苦干的形象。在陈惠发教授的手下工作，大家也是凭实力、讲实干，自觉遵循的都是一种 Boiler Maker 的作风，这有如当年我在四川工地上的习惯一样。我期望回到清华能用我诚实的劳动多教一些学生，也给自己的母校增添容光。

应该说清华对我非常宽厚，给我安排了足够的研究生，他们几乎都是年级第一名的学生，让我用英语给研究生讲授弹塑性力学

和结构矩阵分析。开始，我是以讲师应聘的，两个月后，晋升为副
教授，一年以后被破格提升为正教授，随后又成为学校特批的博士
生指导教授，像这样的特批教授在全校不足十名。随后各种荣誉接
踵而来，令人应接不暇。最使我高兴的是，1987 年我能主持起草国
家自然科学基金委结构工程学科发展的战略研究报告，这个报告是
国家自然科学基金委的第一份学科发展研究报告，为这个报告组织
了国内百名专家写了四年，直到现在，看这份报告的内容也不觉得
过时。另一件令人高兴的事是，1987 年和中国科学院刘恢先院士一
起主持全国八个部、局、委共同资助的国家自然科学基金重大项目
"工程建设中智能辅助决策系统的应用研究"，组织国内 20 个单
位、200 多科技工作者合作了五年，取得了 32 项成果，仅提供的知
识工程软件就有 49 个。回国不久就能做这些国内顶级的工作真是
我的幸运。

15.5.2 天安门事件 **The Event at Tiananmen Square**

1989 年夏天，一件预料不到的政治事件发生了：为了反对官员的腐
败，学生开始罢课、游行示威。我是经历过文化大革命的人，对这
样的群众运动并不热心。到上课时间，我还是去上课，但课堂里空
无一人，作为一个老师，我也要在教室里呆到下课。没过多久，学
生和政府的对立越来越尖锐，天安门广场开始聚集越来越多的学
生。不巧，陈惠发教授就在这个时候访华了。他原定是到清华讲
学，然后去长沙在全国的一个学习班上讲课。我到机场去接他，结
果运行李的工人也罢工，行李等了两个多小时才拿到。等我们见面
握手时，我半开玩笑地告诉他："老师这次的访华，可能不是技术
访问而是政治访问。"我还告诉他："老师要有思想准备，长沙去
不了了，可能要从北京直接回美国。"正如我判断的，我们被困在
清华园里什么地方也不能去，最多带他到学生食堂看看大字报，大
部分时间只能坐在甲所（原来的校长住宅）的电视机前看新闻。那
时，国家的电视基本失控了，全天在直播天安门前的画面。陈教授
看了很吃惊："这哪里是中国的电视，这简直成了美国的电视了！
怎么什么都播？"我也真无法解释，究竟发生了什么？
　　一天晚上，我问他有没有兴趣去天安门广场看看？因为出事
后我一直也没去过天安门广场，很想去看看学生们。我说，只能骑
脚踏车去，他表示很感兴趣："骑车就骑车"。于是，我们就骑车

从清华园出发，骑了大约 20 多公里，才来到天安门广场。当时很难想象，一个美国的大教授，在美国开汽车开惯了，居然有这个体力和我骑了这么远的路程。天安门广场果然是人山人海，各个大学都有自己的圈子，不能随便进入。我们来到清华学生集中的地方，自我介绍了一下就进了圈子。陈教授兴致很高，和学生自由交谈。学生可以说是满腔热血，但是并不知道会有什么后果。几天后，陈教授按原定计划飞往长沙。我们完全没有估计到，他的这架飞机到长沙没有着陆就又飞回来了。等回到北京机场，他和我联系不上，自己临时叫了一个非法营运的小汽车，几经周折才回到了清华园。这下大家才明白，局势严重了！陈教授的夫人也不放心，最后他只有放弃长沙的计划提前飞回美国了。

学生开始在天安门采取绝食行动了，我看着局势越来越严重，心急如焚。我不相信对立下去是唯一解决问题的办法，于是就和当时也是刚从 Stanford 回国的余志平博士商量采取了另外一种行动。我们利用一个下午紧急起草了一封直接给中央的信，要求当时的中共中央总书记赵紫阳和总理李鹏亲自到天安门广场看望学生，直接对话，缓和矛盾。同时在清华园紧急联系了 100 多位知名教授和全部 10 名学部委员（院士）在信上签字，当晚我和余志平亲自送到中央统战部，要求把这封信直送国家最高领导人。果然这个建议生效了，第二天赵紫阳等国家领导人到了天安门广场。其实，这就是赵紫阳的最后一次公开露面。

天安门的事件最后导致了 1989 年 6 月 4 日的一场流血的悲剧，我作为一个老师，是不希望看到这个结果的，我相信学生和政府也不希望是这个结果，但是它毕竟发生了。事后，大家唯一可以做的，就是如何避免这类事件再次发生。作为一个老师，我了解学生。他们有一腔热血，能对那些贪官污吏表示自己的不满，这正说明他们是真正的青年。当老师的应该保护他们这种政治热情。但是他们不成熟，不知道中国的改革还有一个漫长的过程，不可能在几天里就完成。当老师的也应该教育他们尽快成熟起来。

15.5.3 出任清华大学土木工程系主任 Leading the CE Department

1992 年我被学校提名为土木工程系主任。这是学校对我的信任，当时清华的 30 多个系主任中不是共产党员的只有两名，一名是外语系的，一名就是我。我开始组建我的副手，结果商量下来，几个副系主任年龄都比我大，其中一位常务副系主任就是我读本科时的班主任。我作为一个相对年轻的系主任领导几位年长的副系主任不是一件很容易的事。但是我的习惯决定了，我不是一个循规蹈矩的系主任。我从 1957 年进入清华土木系，对清华土木系已经了解了 35 年，无论从工业界还是学术界看，这个系的确需要改造。从哪里入手呢？这是我上任前反复思考的问题。

我的第一个决定是针对自己的，我一定要立规矩，对自己有约束、有监督，不能在系里形成我一个人说了算的局面。一个领导能力可以有大小，但是有三点必须做到：一是，必须保证系里群众中的诚实劳动者能得到切实的好处，这个好处就是"利益"。"利益"就是指"名利"。"名"就是提职称，"利"就是发奖金。二是，要求系里的领导在"名"和"利"前面后退半步，我没有要求"全退"，而是"退半步"。因为领导也有他们个人的"利益"问题。这"半步"反映了一种觉悟。这"半步"会大大增强全系的凝聚力。所以一个受欢迎的领导，要多考虑群众的"利益"，对自己要多讲一点"觉悟"，千万不能反过来。有一些领导在群众中威信不高，问题常常就出在把"利益"和"觉悟"的要求反过来了。三是，要有远见。这个远见至少要比自己的任期长。在卸任的时候，能让你的后任有一个更好的发展环境。

我考虑，约束的办法就是成立"教授委员会"。在清华园，"教授委员会"是个非常敏感的话题，1957 年不少教授，像钱伟长教授，就是因为讲"教授治校"而当了"右派"。我认为这个"教授委员会"主要管三件事，一是教师的升迁，二是进人的筛选，三是重大的科研方向和课程改革。为了这件事，我又回到了 Purdue，见到了陈惠发教授，通过他见到退休的原台湾新竹清华大学的校长徐贤修先生。徐老当年在清华是我的老师们的数学老师，他以前的学生，凡留在清华园的，都已经是资深教授了。他离开清华园 40多年，对清华仍然一片赤诚。那天，陈教授带我到了徐老的家里，我详细请教了搞好这个"教授委员会"的办法。徐老和陈教授给我出了很多好主意，而且其中大部分在我的任期里都办到了。这个"教授委员会"是当时清华的第一个"教授委员会"。我记得成立不久，我被召到校长办公室谈话，一位常务副校长质问我："出了问题谁负责？是你负责，还是教授委员会负责？"我明确地回答他："我负责。"这才算得到认可。

我工作的风格是 Boiler Maker 式的普渡人，务实。我不喜欢无休止地开会，厌恶冗长的发言。因为时间对我们来说太宝贵了，浪费自己的时间有如"慢性自杀"，浪费别人的时间有如"谋财害命"。我宁可直接和每位教师单独谈话，也不愿搞那些"一般号召"的无效会议。在我任职期间，每一位青年教师都与我一起制定了个人的发展规划。为了扩大我们的结构试验室，我到香港通过香港的清华同学会募捐，筹集了资金建立了国内最大的三维拟动力结构试验室。为了适应社会的需要，我在清华成立了"房地产研究所"、"交通研究所"和"安全研究中心"。为了加强基础研究，我把系里的科研明确在"高性能混凝土"、"工程安全性与耐久性"、"工程防灾减灾"三个主攻方向上，让每一个方向都获得了国家科技部或国家自然科学基金重大项目的支持。为了保证清华土木系在全国同行中的领导地位，我坚决地采取措施，保证全系每年的经费有50％以上用于基础研究。

我不希望自己因为忙于行政工作而脱离教学和科研的第一线。1994 年我出任国家科技部在土木、水利领域唯一的"攀登计划"的首席科学家，组织全国 180 余名专家共同工作了五年，解决"重大工程安全性和耐久性的基础研究"问题。我一直坚持给研究生和本科生开设英文课。我的每个研究生每周都有一个小时可以和我面对面地讨论他们的研究工作，雷打不动。我不能放松自己的教学和科研工作，我必须考虑，不当系主任以后，在自己的专业上仍然能充满活力。

在清华，我不是一个十分听话的系主任。我一直认为，大学就是大学，不是官场。中国教育目前的最大问题就是：不是教育家在管教育，而是教育官员在管教育。我认为，学校的决定一定要充分听取教师，特别是教授的意见。清华有一段时间宣传，把 50％以上全国最优秀的高考生都召到清华来了，所以非常自傲，在媒体上宣传所谓"半国英才进清华"。我却建议，最好不要宣传这个，因为清华没有给国家培养出大于 50％的人才，这不是"赔本企业"吗？清华有一段宣传 2011 年清华在 100 周年校庆时要成为"世界一流大学"。我就在全校的系主任会上告诉校长，说："2011 年，我们今天在座的大多数人大概都可以活到，到那时您怎么交账呢？"

有一段时间，清华大学想把土木、水利系合并，叫"土木水利工程学院"；我感觉不妥，"水利"在清华的历史上、在专业上原本就是"土木"的一部分；如果从学科整合考虑，应该把土木、建筑、水利和环境四个系合并，称之"人居环境科学与工程学院"。最后，校务委员会决定土木、水利系合并，对国内叫"土木水利工程学院"，对国外叫"Civil Engineering School"，并决定由我出任第一届院长。我认为这不是学科的整合，而是行政的整合，是"把两个马铃薯放在一个筐里"。我回绝了。我没有批评任何领导，只是说："我能力不够，不能出任这个院长。"学校的校长和党委书记找我谈了两个晚上，我仍然坚持自己的意见。事后校长对我说："我真没有想到您这么坚持，如果早知道，校务委员会可以先不做这个任命决定。"他又说："您可是清华历史上第一个不执行校务委员会决定的系主任。"

1996 年，清华大学土木工程系建系 70 周年时，我写了一篇"面向 21 世纪的清华大学土木工程系"，里面回顾了我任系主任后的一些贡献和对清华土木系未来的展望。现在重新看过，仍然感慨万千。我自认为，我的看法是对的。一个系有没有活力就看教授们是否真正激活了；一个系能不能前进，就看领导者有没有远见卓识，全系有没有很好的组织。这就是：Well prediction 和 Well organization。历史会证明一切。

15.6 教师：一个崇尚奉献的职业 Teacher: a Sacred Profession

我很庆幸自己到现在还在忙碌地工作，这是一种享受，也是一种责任。在国内，我连续三届是全国的政协委员，甚至能在人民大会堂的讲台上为工程质量和教育质量的问题呼吁；在中国土木工程学会，我是外事工作委员会主任，常常代表学会出现在各种国际场合；同时我又是可靠度委员会的主任委员，直接参与和引导新规范的修订。在国际上，我做过英国结构工程师学会（IStructE）的副主席，是第一个任此职的中国大陆学者。在世界工程组织联合会（WFEO），我是中国委员会（WFEO-CHINA）的副主席，又是教育与培训工作委员会（WFEO－CET）和能力建设工作委员会（WFEO－CCB）唯一的中国代表，许多时候我必须代表中国在国际上发表意见。

我同时庆幸，自己到现在还在从事第一线的科研工作，目前正在完成国家"灾害环境下重大工程安全的基础研究"（973 计划）和国家自然科学基金的研究项目，一些非常吸引人的新想法正在实现中。我更庆幸，到现在我还能给本科生和研究生上课，不仅使我仍然保持一个清醒的头脑，同时让我一直和学生在一起。今年，政府、学校和学生给了我许多荣誉，我获得了全国宝钢优秀教师奖和上海交大"最受学生欢迎的教师奖"，成为"上海教学名师"。我喜欢初中时读过的一本小说《古丽娅的道路》中的一句话："让生命燃烧起熊熊的烈火，而不要光冒烟。"

人的一生，什么是最重要的？我又想起我中学时读过的一本小说《钢铁是怎样炼成的》里面的一句话："人最宝贵的是生命，生命对每个人来讲只有一次，一个人的生命应该这样度过：当他回首往事时，不会因虚度年华而悔恨，也不会因碌碌无为而羞耻，在临死的时候他能够说：我的整个生命和全部精力都已献给了世界上最壮丽的事业——为人类的解放事业而斗争。"这句话告诉我，人生最重要的是奉献，人生最享受的也是奉献，而教师就是一个崇尚奉献的职业。我的父母是教师，陈惠发老师也是教师，我庆幸自己的后半生也坚定地选择了教师这个职业。

2000 年作为英国结构工程师学会副主席出席 Annual Dinner.

2006 年与世界工程组织联（WFEO）主席 Kamel Ayadi 谈.

2005 年在波多黎各出席世界工程组织联合会（WFEO）工作会议. 2005 年刘西拉教授与他梯队的青年教师.

2006 年刘西拉教授在上课.　　　　2006 年刘西拉教授在试验室.

在这章结束前，我想起我 91 岁的父亲。他是中国药科大学的终身教授。2006 年 3 月，他病重在南京鼓楼医院 ICU 抢救。我为了第二天要给学生上课，20 日晚依依不舍地离开危在旦夕的父亲，赶回上海。我当时唯一的期望是，上完课就立即回到父亲身边，要陪伴他走完生命的最后一程。3 月 21 日一早，我与往常一样站在讲台上给学生上课，好像什么事也没发生。给本科生上完"土木工程概论"，又接着给研究生上"计算结构力学"。就在上午第三小节课快结束时，我的手机开始不断地振动，我预感到不幸。往常我在课间是不休息的，这次我破例休息了十分钟。休息时，一个人到楼道里看了弟弟发来的一连串短信，父亲的血压一直在下降，最后就病故在这课间休息的十分钟里。弟弟说他会代我亲吻亲爱的父亲！我的心里十分悲痛！我犹豫自己能不能把课上完…… 最后我还是回到讲台上，把这一切告诉了学生。课堂里有的学生发出惊叹，全班随之鸦雀无声。我对学生说了这么几句话："我的父亲是一名教师，非常非常遗憾在他最后的时刻，我不能在他的身边！我相信他会原谅我的，因为他知道，我是在上课。同学们，你们知道吗？讲台是一个多么神圣的地方！教师是一个多么神圣的职业！希望你们中间有志的青年也选择教师这个职业。现在继续上课……。"

我父亲最喜欢的是陶行知先生的一句话，也是我的座右铭，这句话就是：

"捧着一颗心来，不带半根草去。"

"Come with a whole heart,
Go away without any blade of grass."

Tao Xingzhi (1891-1946)
A great educationalist in China

16

NTU and I
臺大與我

By Y. B. Yang 楊永斌

When I was in the elementary school, my mom used to say to me: "If you cannot get the first grades, you can just quit." She said this not because she was strict, but because we were so poor and had no other choice. We had to study and win.

16.1 Born in the War Place 出生戰地

I have to admit that to use English to write an article for a topic like this is a challenge for me, even though I have written two engineering books and more than 130 journal papers in English. There are two reasons for this. First, writing an autobiography is more difficult than writing a scientific paper, as it is not merely a collection of all major events. Second, writing a paper about the author himself is more difficult than about inanimate objects such as structures and bridges, since egoism is going to play a role and sometimes over exaggeration is possible.

My date of birth is August 22, 1954, but I was actually born on May 25, 1954 of the Chinese lunar calendar, roughly two month earlier than the "official date". One reason for the late registration of birthday at that time is that all babies were born at home, as there were no hospitals or clinics for delivery. Therefore, no official record was kept for the birth of babies. The other reason is that most parents wanted to make sure their babies can really survive before submitting their names to the village administrative office.

Kinmen (金門), also known as Quemoy, is the place where I was born. It is an island located off the coast of Fukien (福建), the southeastern province of China (see Fig. 1), with a nearest distance of roughly two kilometers, but has been under the control of the Taiwan government since the Nationalist government（國民政府）retreated to Taiwan in 1949. During the cold war, this island had been a battlefront of Taiwan, as it was always the target of seizure by Mainland China. Besides some major confrontations that caused the death of tens of thousands of soldiers and civilians, the Mainland Chinese army would fire the cannonballs containing propaganda papers（宣傳單）at the island "every odd numbered day"（單打雙不打）, and the Nationalist army on the island would fire back every other day. This has been a notorious history for about two decades (1958-1978) between the two governments on both sides of the Taiwan Strait（臺灣海峽）.

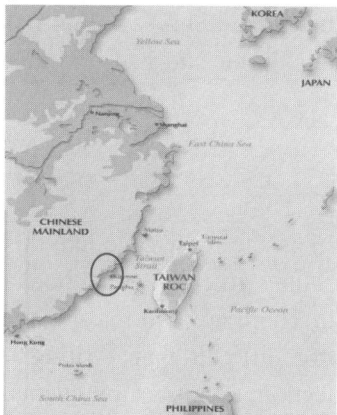

Location of Kinmen Island
(Modified from
http://maryknoll_taiwan.homestead.com/map-taiwan.jpg)

In my childhood, poverty was the norm of most villagers in Kinmen, simply because of endless wars and isolation from the outside world. There was no electricity or running water. The residents generally relied on sweet potatoes and local vegetables as the main ingredients of nutrition. All the rice, flour, fruit (e.g., oranges, bananas, etc.), and living materials had to be imported from Taiwan irregularly by the navy as part of their mission in supplying the military goods and arsenal using the landing crafts, a kind of tank and cargo carrier equipped with some defensive weapons.

I am the last child of my parents who have four sons and four daughters. My dad was a farmer, who could read some commonly used Chinese words. Our family owned only some small pieces of infertile lands for cultivation. The main crops at that time included sweat potatoes, peanuts, corn, and sorghum. Regardless of how hard my dad and brothers had been working, the crops grown each year were not sufficient to support the entire family. For this reason, not all my elder sisters and brothers were allowed to attend schools. I was lucky to be the youngest son and was given the privileged right of attending schools. When in the elementary school, I showed my tendency for mathematics.

I remember my mom used to say to me: "If you cannot get the first grades, you can just quit." She said this not because she was strict, but because we were so poor and had no other choice. We had to study and win. To my mother and many contemporary illiterate villagers, attending schools is a luxurious and unaffordable expenditure.

Both my dad and mom were generally healthy, before they passed away. My dad died naturally in January 2000 at the age of 93 and my mom died in a similar way in October 2006 at the age of 95. When I get older, I feel I inherited a large portion of my character from the mother side. Her words of "getting the first grades" actually shaped my attitude of learning in my early childhood. Because of this, I was serious at all stages of learning. If there is anything worthy to say about my mom in her way of bringing me up, I would say that a woman may not be able to change the life of her husband, but her words and attitude are likely to shape the future of her children.

16.2 From Junior to Senior High School 從初中到高中

Entering the Kin-Ning Junior High School (金寧國民中學) (1966-1969) was a kind of enchantment to me. The school is located some five kilometers away from home and I had to use bicycle as the tool of transportation. Ever since I was a kid, I preferred to stay quiet and ponder, rather than to play balls and games with schoolmates. My favorite courses include geometry, chemistry, physics and English. Whenever I had time, I would choose to stay in the classroom, thinking about some mathematical problems. Under the inspiration of my geometry teacher, I wished I could invent some theorems comparable to the Pythagorean Theorem. Thus, I often drew triangles, circles and the like for all kinds of connections, trying to search for any rules behind them. Of course, I failed in this regard eventually.

Behind the school buildings, there is an embankment sided with beefwood trees. I enjoyed reading English aloud there by myself, while strolling back and forth slowly in the breezy afternoon. The problem with me then was how to correctly pronounce each new word. Besides mimicking what the teacher said, the only way was to look for the pronunciation instruction available for each symbol of the K. K. Phonetic System (KK 音標) on the front pages of the Far Eastern English-Chinese Dictionary (遠東英漢辭典). At that time, radios were totally forbidden in Kinmen, not to mention the English-teaching programs now popular in the air, due to the restriction of broadcasting devices for military usage.

In 1969, I was admitted by the Kinmen High School (金門高中) as the first student through the entrance examination. I continued my habit of practicing English each morning in the rosebay garden on campus. For most courses, and especially for mathematics, I could always get a grade of 95 marks or higher for most exams. Thus, I enjoyed very much my "honorable" position in school, until one day the mathematics teacher Mr. Tsun-Shin Lee (李存鑫) called me to his office. He told me that he saw my life in school as something not to my utmost strength, since I could always get good grades with little effort. He thus persuaded me to move to Taiwan for more challenging schools. This was really a shock to me, as I had to leave home.

16.3 Starting to be Away from Home 揮別家鄉

After consultation with my family members, my eldest brother insisted that I should go for Taiwan, although this would significantly increase the financial burden of the family. Under the guidance of some senior countrymen, I took the landing craft in the summer of 1970 to cross the Taiwan Strait, arriving at Pier No. 13 of the Kaohsiung Harbor (高雄港). Then I went with other countrymen by the slowest but cheapest train to Taipei, which stopped at every station. This was my first acquaintance with trains, and therefore I was quite excited.

Fortunately, I passed the entrance exam of the Chien-Kuo High School (建國高中), obtaining a grade of 95 marks for mathematics. I was one of the only five students admitted to transfer to the second-year classes at Chien-Kuo in that year. This school was one of the most prestigious high schools in Taiwan. Before I went to Chien-Kuo, I didn't realize how competitive the students there were.

In the first semester, I found my good old days were over. I fell behind most of my classmates for most courses and began to have a feeling of frustration. There was nothing wrong with me, because I worked just in the same way as I did in the Kinmen High School. However, I observed most of my classmates went for some reinforcing classes (補習班) after they left school. Even though I had no money to pay for such extra classes, I thought I should do something by myself. I began to buy some exercise books that contained thousands of tricky problems from a nearby bookshop for courses such as mathematics, physics, chemistry, and English.

Everyday I solved as many problems as I could and checked the solutions I got with those available in the exercise books. By doing in this way for a couple of weeks, I found I digested much better the materials taught by teachers in classes and became smarter than ever in solving various problems for most courses. In July 1972, I joined the National Entrance Exam for universities (大專聯考) and earned a total grade that was good enough for me to choose the Department of Civil Engineering, National Taiwan University (台大土木系) as the destiny of study.

16.4 Undergraduate Study at NTU 進入台大

The National Taiwan University located at Taipei (台北市), Taiwan is a comprehensive university, which offers programs for all kinds of disciplines, including literature and arts, science, law, medicine, engineering, and agriculture. Being a freshman in such a university was really enjoyable. You could always attend lectures of various subjects, sometimes debatable, given by some admirable professors from different departments.

Unfortunately, my eldest brother, the key financial supporter of the family, got a brain tumor and passed away after an unsuccessful surgery during my second year of study. I began to search for part-time jobs, among which serving as a tutor to high school students was my favorite. I enjoyed seeing some originally "bad" students, as were referred to by their parents; turn out to be "good" students under my instruction. The income I earned in this regard enabled me not only to support myself, but to send part of the savings home. Actually, I started to be a contributing member of my family from this moment.

As part of the assigned national duty, I served in the Combined Service Forces (聯勤總部) for almost two years from 1976 to 1978. Besides the routine works, I spent most of the time left in reviewing the major course materials I learned during my four years of study at NTU, with special focus on mechanics of materials and structural theory. Similar to what I did at Chien-Kuo, I tried to enhance my understanding of the principles of each course by solving a large number of exercise problems.

Fortunately, I passed two important exams with very low passing rates. One is the entrance exam for graduate study at the NTU Department of Civil Engineering with a major in structural engineering. And the other is the Senior Civil Service Exam (公務人員高考) held by the Examination Yuan (考試院), a unique organization responsible for qualification and rating of government officers in Taiwan, by which I was awarded simultaneously the certificate for professional civil engineers (技師執照).

16.5 Graduate Study at NTU 台大研究生涯

I returned to NTU in September 1978 for study of Master's degree. Compared with the previous undergraduate study, I was generally relieved of the financial burden, since the fellowship I got from the university was sufficient to cover my living expenses. As such, I began to think of my future. My first wish was to go to the U.S.A. for advanced studies. To this end, I concentrated myself fully on the course works, making every effort to earn good grades for all the courses I took. I also tried to improve my listening capability in English by listening regularly to the tapes provided by a monthly English magazine. Eventually, I was able to get a TOEFL[1] grade that could meet the requirements of most top American universities for entry application.

For the dream to come true, we also need to be lucky, besides

---

[1] Test of English as a Foreign Language.

working hard. My luck arrived when Prof. William McGuire of Cornell University was invited to our department to give a speech. Under the encouragement of Prof. S. T. Mau (茅聲燾), I approached him after his lecture, expressing my desire to study at Cornell. This was the first time for me to communicate in English. I was extremely nervous and could only understand vaguely what Prof. McGuire said to me. If this was the first interview given by Prof. McGuire, I would say that my performance was rather unsuccessful. However, thanks to Prof. Mau's timely help, Prof. McGuire did understand my intention to go for advanced study at Cornell and particularly to work with him.

16.6 New Career at Cornell 康大歲月

To my surprise, I received a letter from Cornell University, Ithaca, New York informing me that I was admitted for doctoral study with a research assistantship. This was a turning point in my life. In the summer of 1980, I flew to Cornell to start my doctoral study at the School of Civil and Environmental Engineering, located at Hollister Hall. As my first priority, I went to Prof. McGuire's office to seek his permission for study under his supervision. Still I had some difficulty in managing English to express my thought. However, Prof. McGuire seemed to be aware of the linguistic problem with me and was patient in figuring out what I tried to mean in our conversation. It was amazing that he accepted me as one of his Ph.D. students.

The first work I completed at Cornell is the implementation of a general eigenvalue program for solving single or multiple, distinct or identical eigen roots for a system with symmetric matrices. It was written based on the approach proposed in a pretty new Ph.D. thesis from the University of Illinois at Urbana-Champaign, handed over to me by Prof. McGuire for reference. I later became acquainted with the author of the thesis Dr. In-Won Lee when he took a teaching job at KAIST, Korea through the *KKCNN Conference* to be mentioned later. This eigenvalue solver is quite versatile and has been used by most of my graduate students at NTU for solving the eigenvalues of various structural stability and dynamic problems.

As far as the doctoral research is concerned, the assignment from Prof. McGuire was to deal with the warping properties of steel sections and the nonlinear behavior of space frames composed of thin-walled sections. To understand the basic properties of thin-walled sections, I started reading some classical works such as Bleich (1952), Vlasov (1961), Timoshenko and Gere (1961), Chen and Atsuta (1977), Ziegler (1977), Gjelsvik (1981), etc., in addition to the steel book by McGuire (1970). But to derive a general nonlinear theory for three-dimensional beams, I needed to have a solid background on elasticity and continuum mechanics. Thus, I also read classical books such as Fung (1965), Malvern (1969), and Washizu (1982). Concerning the nonlinear analysis of structures, I also traced hundreds of technical papers, as well as finite element books, on subjects related to the virtual work formulation, buckling of thin-walled beams, incremental-iterative methods, and so on. This period of study at Cornell helped me lay the foundation for nonlinear mechanics theories and analyses.

In the 1970s to 80s, Cornell was quite famous for its research on interactive computer graphics, which was carried out mainly in the laboratory located at Rand Hall. The principal investigators for the projects on development of computer graphics techniques included Donald Greenberg of the Architecture Department and William McGuire and John F. Abel of Structural Engineering, among others. I was a student member of this research group. Upon the time of graduation, most students in this group not only came up with a thesis, but also some analysis programs, with pre- and/or post-processor for displaying their input data and analysis results through the cathode-ray-tube screens in the laboratory. Throughout the period of study at Cornell, I spent my time primarily in three places: Hollister Hall, Rand Hall, and Carpenter Hall, where Engineering Library is located.

Prof. McGuire is a great teacher. I used to consult him for issues far beyond the academic. Even after returning to NTU, I often write him about what I have done in the recent past and what I am going to do in the near future. He is willing to share with me any ideas he has. Sometimes he sent me articles cut down from the New York Times or other journals on something interesting of the time, e.g., comments given by some columnists on the Taiwan Strait policies or on the movies *Crouching Dragon and Hidden Tiger*（臥龍藏虎）directed by Li An （李安）, the Oscar Award receiver. I was flattered to hear his words like these: "A good teacher is one who is lucky enough to have some very brilliant students," a quotation from his former colleague Prof. Emeritus Seng-Lip Lee of the National University of Singapore, if I am correct. In fact, Prof. McGuire's name is quite popular among the graduate students of the Structural Engineering Division of the Department of Civil Engineering at NTU, because his book *Matrix Structural Analysis* (McGuire et al. 2000) has been adopted as the text for the course on Advanced Structural Analysis, ever since the first edition.

The other two members of the advisory committee for my doctoral study at Cornell were Prof. John F. Abel and Prof. Subrata Mukherjee of the Theoretical and Applied Mechanics Department. They both are good teachers and friends, offering me assistance far beyond courses and researches. In the past few years, we maintained frequent contacts with each other and often met in international conferences and academic meetings.

I got married with Ru-Wong Huang（黃如宛）in 1980 prior to my departure for the U.S.A. My acquaintance with my wife was quite simple. When finishing my undergraduate study at NTU in 1976, I was looking for a girl to dance with me in the farewell party. One of my classmates volunteered to bring one of his sisters to serve as my partner. That was the way I got acquainted with my wife. It was simple and non-romantic, but it is faithful and reliable. As a matter of fact, the dance I had with my wife in the farewell party thirty-one years ago is the only dance I have ever had.

By the government regulations for emigration at that time, which were abandoned some years later, my wife Ru-Wong could not fly directly with me to the U.S.A. She joined me the second year though, allowing me to concentrate fully on the Ph.D. study. We had our first daughter Judy born in Ithaca in April, 1982 and second daughter Carol born in February, 1984. When I got my Ph.D. degree in May, 1984, my wife was awarded the "Ph.T. degree" by some friends for her effort in Pushing-Husband-Through.

I hadn't thought of applying for any jobs in the U.S.A. after obtaining my degree from Cornell. In the summer of 1984, I flew with my wife and two little daughters back to Taipei to take a teaching position at the Department of Civil Engineering, NTU, my alma mater. On the way home, we stopped by Los Angeles to visit some friends, while enjoying a good time with our daughters in the Disneyland.

It should be added that after obtaining her Bachelor's and Master's degrees from the Department of Civil Engineering, NTU, my elder daughter Judy returned to her birthplace Cornell in the summer of 2006 to study for the doctoral degree, following basically the same track as her dad's some twenty-two years ago. This is amazing to many of our friends at Cornell and in Taiwan. Frankly speaking, I haven't persuaded Judy to go for civil engineering or even structural mechanics. But she has confidence in mechanics and engineering and she loves it. This may be partly due to my daily influence.

16.7 Research/Teaching Career at NTU 為人師表

In my first few years of service as an associate professor at NTU, I was eager to build up my capability in research. Therefore, I placed my efforts mostly on the teaching and research works, while maintaining a close relation with both undergraduate and graduate students. The first challenge I had was to see if we could have our papers published by prestigious journals, considering that in the mid 1980s, very few papers were published by authors from Taiwan in international journals. Fortunately, with the assistance of some brilliant, hard-working graduate students, we solved a number of problems related to the stability of space frames, curved beams, tapered beams, pretwisted bars, and yield surfaces of steel sections, some of which are of the benchmark nature, through a series of research projects sponsored by the National Science Council (國家科學委員會).

In solving any mechanics problem, we often took two different approaches. One is the analytical approach traditionally adopted for solving the boundary-value problems. This approach enables us to present the solutions in closed-form for some special problems. The other is the numerical approach based primarily on the finite element methods, which enables us to extend the theory to treating a wide range of problems encountered in practice. For problems of the benchmark type, we always compared the results obtained by the two approaches to ensure their consistency.

Some of the former students working with me at this stage often recall that I might call on them in their dormitory incidentally on weekends or at night time, handing over my comments to them on the derivations or results of computation they gave me the other day.

Thanks to the effort of the group members, in the first five years of my service at NTU, i.e., from 1984 to 1989, we published a total of 13 papers and discussions in *Journal of Structural Engineering* and *Journal of Engineering Mechanics* of the American Society of Civil Engineers, in addition to some others. This gave me the confidence that our works were well accepted by the international academic community.

After teaching at NTU for four years, I was promoted to the position of professor. In the same year, I was granted the Distinguished Research Award by the National Science Council for a two-year term. Later, I was granted consecutively for the same award for another four terms. In 1992, I was also elected as Ten Outstanding Young Persons (十大傑出青年) by the Junior Chamber International (國際青年商會), Taiwan. Another thing pleasing me in this year is the birth of my son Ted, who turns out to be ten years younger than his elder sister Judy. Starting from this time, our family has a total of five members (Fig. 2).

Because of the research results accumulated for the nonlinear behavior of space frames in the 1980s, I began to think about writing a book on this subject. After contacting some publishers, we signed the contract with Prentice Hall, Singapore, and had the book *Theory and Analysis of Nonlinear Framed Structures* published in 1994. My former student Dr. S. R. Kuo (郭世榮) is the co-author of this book.

In anticipation of the construction of high speed railways in Taiwan, I began to switch our major area of research in the 1990s to the vibration of bridges caused by the moving trains and vehicles, with emphasis on the vehicle-bridge interaction and passengers' riding comfort. Again, the research results accumulated along these lines were huge enough for us to write another book. Through a contract with World Scientific,

Singapore, I published the book *Vehicle-Bridge Interaction Dynamics – with Applications to High Speed Railways* in 2005. My former students Drs. J. D. Yau（姚忠達）and Y. S. Wu（吳演聲）are the co-authors of this book.

It is well known that structural stability and dynamics are not separate subjects, but share some fundamental links. However, there was a lack of a journal dealing specifically with the integrated subjects of structural stability and dynamics. My friends Prof. C. M. Wang of NUS and Prof. J. N. Reddy of Texas A&M University shared with me this point of view. So we decided to launch a new journal in 2001, entitled *International Journal of Structural Stability and Dynamics* with the support of World Scientific. After years of growth, this journal has been selected for coverage by SCI since the first issue of 2005.

Family Photo Taken in Our 2004 Summer Trip to Mainland China (From left to right: Ted, Judy, Ru-Wong, me, and Carol).

As of now (February 2007), a total of eleven students have completed their Ph.D. dissertations under my supervision, including S. R. Kuo（郭世榮, 1991), B. H. Lin (林炳宏, 1993), J. D. Yau（姚忠達, 1996), S. C. Yang（楊順欽, 1996), C. H. Chen（陳振華, 1997), J. T. Chang（張健財, 1997), Y. S. Wu（吳演聲, 2000), H. H. Hung（洪曉慧, 2000), C. C. Hsiao（蕭吉謹, 2002), C. W. Lin（林正偉, 2004), and L. C. Hsu（許琳青, 2006). Topics covered by these dissertations include stability of framed structures, nonlinear analysis based on rigid body considerations, nonlinear behavior of plates and shells, cable-stayed bridges under wind loads and moving loads, moving load problems, bridge frequency extraction, vehicle-bridge interactions and passengers' riding comfort, train-rail-bridge systems, ground and tunnel vibrations induced by moving trains, etc.

In 2006, I was awarded as a Distinguished Professor（終身特聘教授）at NTU. This was an honor created in 2006 for honoring those professors who have received at least three times of distinguished research award from the NSC or their equivalents. The other major honors and awards I have received so far include:

1. Outstanding Teaching Award （教學特優教師獎）, Ministry of Education（教育部）(1994),
2. Distinguished Engineering Professor Award （傑出工程教授）, Chinese Institute of Engineers（中國工程師學會）(1994),
3. Outstanding Scholar Award（傑出人才講座）, Foundation for the Advancement of Outstanding Scholarship, chaired by Nobel Laureate Dr. Y. T. Lee（李遠哲）(1998-03),
4. Fellow, American Society of Civil Engineers (2000),
5. Munro Prize (Best Paper Award), *Engineering Structures* (2003),
6. Fellow, Chinese Institute of Civil and Hydraulic Engineering（中國土木水利工程學會）(2003),
7. Fellow, Chinese Society of Theoretical and Applied Mechanics（中華民國力學學會）(2006),
8. Best Paper Award, Advances in Structural Engineering, an International Journal (2006), and
9. Distinguished Book Award（傑出學術專書獎）, NTU (2006).

16.8 The KKCNN Conference 五校聯合研討會

Starting from 1991, the NTU Civil Engineering Department was invited to join as the third partner of the annual seminar on civil engineering that had been conducted between Kyoto University (KU) and Korea Advanced Institute of Science and Technology (KAIST) since 1989. Such a membership was made possible through my coordination with Profs. Naruhito Shiraishi and Chang-Koon Choi, coordinators of KU and KAIST, respectively. This seminar was conceived primarily to improve the English presentation ability of the faculty members and graduate students of the three institutions for technical papers, as we are all from non-English speaking countries. It was hosted annually by each institution in rotation. Later, this seminar was further expanded to include two new members: National University of Singapore and Chulalonkorn University of Thailand. As a result, it was renamed as the *KKCNN Conference*, with each of the five letters indicating the first letter of the five member institutions.

In December, 2006, the *19th KKCNN Conference* held in Kyoto University attracted more than 160 participants from the five institutions, representing a great booming from the past conferences. This series of conferences has become an important forum for exchange of the research results for the faculty members and graduate students in the civil engineering departments of the five institutions. One direct benefit from the *KKCNN Conference* is that many of the faculty members of the five institutions have got acquainted with each other, forming a solid basis for cooperation between different institutions.

16.9 Earthquake Engineering Research Facilities 國家地震中心

In the summer of 1988, I visited the Institute of Industrial Science, the University of Tokyo for three months with other three colleagues from Taiwan. The purpose of this trip was to study the technique of pseudodynamic testing with Prof. K. Takanashi, a renowned pioneer for this technique. In March 1990, the National Center for Research on Earthquake Engineering (NCREE) （國家地震工程研究中心） was established under the joint effort of the NSC and NTU with Prof. Chau-Shioung Yeh （葉超雄） as the first Director. I was then appointed as Head of the Earthquake Simulation Laboratory of the NCREE. One major duty of this job was to design and construct the laboratory for housing the shaking table, pseudodynamic testing facilities, strong floor and reaction walls, through cooperation with the selected structural consulting firm, constructor and machine vendor. Before this could be done, we had to determine the architectural layout of the laboratory and the specifications for each major component of the facilities to be purchased and installed.

To this end, a delegation of seven persons, including the Director Prof. Yeh and myself, was organized to visit some of the most famous earthquake engineering research laboratories in Japan and the U.S.A. in June 1990, including the Institute of Industrial Science and Chiba Experimental Station of the University of Tokyo; Research Institute of Okumura Corporation, Laboratories of Public Works Research Institute, Tsukuba, Japan; Laboratories of the University of California, Berkeley; National Center for Earthquake Engineering Research, Buffalo, New York; and Center for Advanced Technology for Large Structural Systems of Lehigh University, Bethlehem, Pennsylvania.

Based on the data collected from this trip, some basic guidelines were drawn for sizing the shaking table and pseudodynamic testing facilities, as well as for designing the laboratory that houses all the equipment. First of all, all equipment and personnel should be allowed to move in the laboratory in a convenient way. For example, the entrances and cranes should be arranged such that large specimens can be transported to the desired place inside the test space without any difficulty. Secondly, easy access should be provided for the shaking table pit, pump room, and the surrounding area of the floating (reaction) mass. Further, the test specimens on the strong floor may be bolted either from above or beneath the strong floor. Finally, all the major oil pipes should be hidden but reachable from different points of the strong floor.

In 1992 I was appointed as Chairman of the Working Team for Construction of the NCREE Office Building and Laboratory (新建工程工務小組召集人). During the design and construction period of these buildings, we frequently called together all the related parties, i.e., the university accounting staff, the architect, Fei and Cheng Associates (宗邁建築師事務所), the structural consulting firm, Sinotech Engineering Consultants, Inc. (中興工程顧問公司), the constructor, Continental Engineering Corporation (大陸工程公司), and sometimes the shaking table manufacturer, MTS Systems Corporation, U.S.A., to resolve issues of integrated nature but not clearly stated in the design documents.

While the construction of the building and laboratory was proceeding well, we were concerned with the manufacture of the tri-axial, six degrees-of-freedom shaking table system by MTS. In terms of overall performance, this shaking table is among the top ones in the world for earthquake simulation, which has a plan dimension of 5 x 5 m, a payload of 50 tons, and is carried by a floating (reaction) mass of 4000 tons. The loads acting on the floating mass are transmitted through the air bags and viscous dampers to the fixed foundation, and then to the bedrock by concrete piles of tens of meters in length (Fig. 3). For the purpose of design check, I flew with Dr. L. L. Chung（鍾立來）in August 1995 to the MTS headquarter located in the vicinity of Minneapolis, Minnesota to discuss with the chief engineers about the logistics of design for the NCREE shaking table, and to confirm that each component of the system had been manufactured as scheduled.

Prior to installation of the shaking table, I asked one of my graduate students Mr. Chung-Chen Lee（李忠誠）to check the layout of the viscous dampers attached between the fixed foundation and floating mass for supporting the shaking table, as part of study of his thesis. We then came up with a more efficient design for the dampers in terms of energy dissipation. Such a proposal was finally adopted by MTS. The other thing to mention is that inspired by the occurrence of the Northridge Earthquake on January 17, 1994, I requested the MTS engineers to check if the actuators of the shaking table, as originally configured, are capable of simulating the three-dimensional ground motions of the earthquake and to have them adjusted for this purpose when necessary. Even for today, I believe the NCREE shaking table remains one of the fewest shaking tables in the world that is capable of simulating the three-dimensional motions of catastrophic earthquakes such as the Northridge Earthquake (1994), Kobe Earthquake (1995), and Chi-chi Earthquake (1999)（九二一大地震）, as few shaking tables can provide such a high flow capacity for the oil pumping system.

In 1995, I quitted my job as Head of the Earthquake Simulation Laboratory of NCREE because I was elected as Chairman of the Department of Civil Engineering, NTU, and I could not divide myself between two heavy-duty jobs.

16.10 Serving as Chairman and Dean at NTU 擔任系主任、院長

I was Chairman of the Department of Civil Engineering, NTU, from

August 1995 to July 1998 for a term of three years. In June 1996, I signed an agreement with Norwegian Geotechnical Institute, Oslo, for mutual exchange of scholars and cooperation of research projects. In July 1997, the first Cross-Strait Conference on Civil and Structural Engineering was held among the following four universities: NTU, Tongji University (Shanghai) (同濟大學), Tsinghua University (Beijing) (清華大學), and the University of Hong Kong.

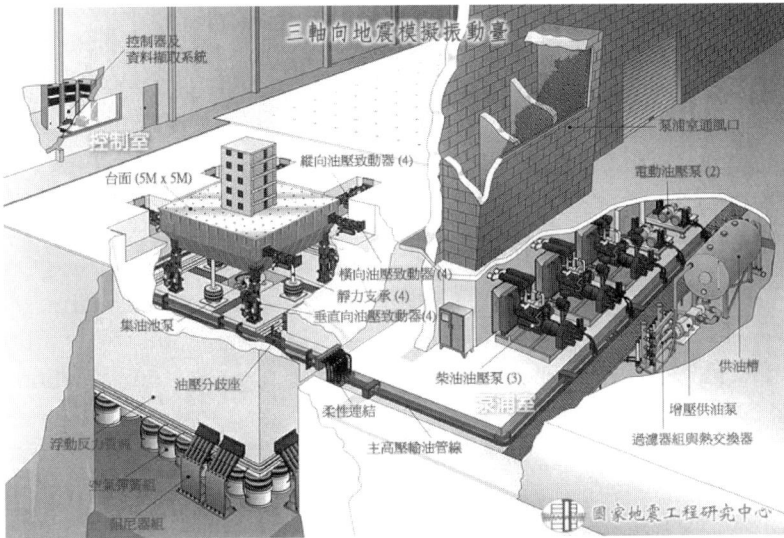

Shaking Table System of NCREE
(Obtained from http://www.ncree.org/ZH/ShakeTableArch.aspx)

Three other international conferences were held during this period:

1. The 5th KU-KAIST-NTU Joint Seminar/Workshop on Civil Engineering, November 26-29, 1995;

2. IASS International Colloquium on Computation of Shell and Spatial Structures, November 5-7, 1997; and

3. EASEC-6 Structural Engineering and Construction: Tradition, Present and Future, January 14-16, 1998.

In November 1996, I signed a memorandum with Dr. Kwet-Yew Yong, Head of the Department of Civil Engineering, the National

University of Singapore, for exchange of students between the two departments. By this agreement, the NUS will send each year five students or more to Taipei for industrial training for a period of three months under our arrangement, mainly in the summer season. In return, we will send an equivalent number of students to the consulting firms in Singapore under their arrangement. Each of the companies hiring the exchange students has to submit a final evaluation report for them to their parent department. Such a program has been quite inspiring to the participating students, as it provides an opportunity for them to be exposed to the profession. According to my observation, most students joining such a program had clearer planning of their future and therefore became more concentrated on their study.

In August 1999, I was elected Dean of the College of Engineering (工 學院) and started my service for a period of two terms, i.e., a total of six years. In 2002, I was also appointed as Chairman of the Civil Engineering Discipline (土木學門), NSC for a three-year term. The two jobs were basically overlapping, which made me rather busy all the time. Being the dean, one of my focuses was to promote instruction of courses by English within the college. I was glad that when I left the office, there were a total of 76 courses taught in English.

During my period of service, there were some organizational reconfigurations within the College of Engineering. First, under its request, the Department of Computer Science and Information Engineering (資訊工程學系) was approved in 2000 by the college meeting of representatives (院務會議) to switch its connection from our college to the newly formed College of Electrical Engineering and Computer Science (CEECS) (電機資訊學院). In 2001, the Department of Materials Science and Engineering (材料科學與工程學系) was established, actually upgraded from its original graduate institute. Partly because of the inauguration of this department in NTU, the materials science departments nationwide become a favorable choice for many high school students. Currently, the NTU Department of Materials Science and Engineering enjoys its ranking as the top three among all engineering departments in Taiwan.

In 2002, the Institute of Polymer Science and Engineering was launched（高分子科學與工程研究所）, while the Department of Shipbuilding and Ocean Engineering（造船與海洋工程學系）was renamed as Department of Engineering Science and Ocean Engineering（工程科學與海洋工程學系）. Also, the Nano-Electro-Mechanical Systems Research Center（奈米機電系統研究中心）was approved by MOE as a formal center at NTU, an upgrade from its original status as a regional center of the NSC. This center is co-supported by our college and CEECS. In addition to the above, several centers of the functional nature were established within the College of Engineering.

During my service as the dean, the College of Engineering was the key promoter for the cooperation with nine international institutions, including the University of Tokyo. Of particular interest is the establishment of a mechanism between NTU and Grenoble University of France for conferring dual Ph.D. degrees to students enrolling in the areas of mechanical engineering and materials engineering, if the students can meet the requirements of both sides. In 2004, we sent a delegation to visit the alumni leaders in Thailand and Malaysia to strengthen their relation with the alma mater, while attracting their children, friends and relatives to study at NTU.

The following are the international and cross-strait conferences held during my terms as the dean:

1. Recent Advances in Computational Mechanics Workshop, February 5, 2000;
2. 1st International Conference on Structural Stability and Dynamics, December 7-9, 2000;
3. Symposium on High Speed Railways and Bridge Dynamics, October 16-17, 2002;
4. IASS-APCS 2003: International Symposium on New Perspectives for Shell and Spatial Structures, October 22-25, 2003; and
5. 3rd Cross-Strait Conference on Structural and Geotechnical Engineering, October 23-24, 2003.

16.11 Presidential Election of NTU in 2005 台大校長選舉

I joined the race for the presidency of NTU in early 2005, simply because my term for the dean terminated in July 2005 and I could contribute my energy, ideas and experiences to the university. I accepted this challenge and regarded it as a meaningful training in my life.

Most people in the academic society of Taiwan may recall that among all the six candidates, I was the only one winning a majority of the university representatives（校務代表）in the first run of voting held on March 20. Actually, I got a nearly two-third majority, i.e., 201 votes out of the total of 319 votes, of the representatives present for voting. By the regulations of the Ministry of Education, each university should submit at least two qualified candidates to the ministry for their selection. Because of this, a second run of voting was conducted on April 24 for the remaining five candidates and a second qualified candidate was generated. The final decision by the MOE is that the second qualified candidate was selected as the NTU president. This is exactly all the story about.

As for myself, the story has been over and I am not going to remain in the same situation as I was in 2005. Everyone has to look ahead and move forward. As a matter of fact, I enjoyed a very pleasant year of sabbatical leave in 2006, spending more than three months at the Department of Civil Engineering, NUS, free of any administrative works, meetings, and social events. I concentrated my time on the writing of another book related to the ground and building vibrations caused by moving trains.

16.12 Accreditation of Engineering Programs 工程教育認證

On January 18-19, 2003, the National Engineering Education Conference was organized by the College of Engineering, NTU, under the support of both NSC and MOE in the Howard Resort Hotel on the northern coast of Taiwan. Most of the presidents and deans of engineering institutions and colleges, as well as senior professors, were invited to attend this conference. One hot topic of discussion of this conference was to establish a society dedicated to the advancement of engineering education, and I was appointed as the Chairman of the Preparatory Committee (籌備委員會召集人) for the establishment of such a society.

This happened to be the period when SARS (Severe Acute Respiratory Syndrome) was prevailing, causing a large number of death toll in Taiwan. As such, the inauguration of the Institute of Engineering Education Taiwan (IEET) (中華工程教育學會) was postponed until June 23, 2003. The IEET was formed as a non-governmental, not for profit, independent organization (http://www.ieet.org.tw/). Its major mission is to elevate the quality of engineering education in Taiwan through the accreditation of engineering programs. Professor C. H. Wei (魏哲和), then Chairman of NSC, was elected as the first President of IEET, and I was chosen as the Secretary General. All the functions of accreditation are carried out through the Accreditation Council (AC), which oversees the following three components: Engineering Accreditation Commission (EAC), Appeal and Review Committee (ARC), and Office of the Executive Director (OED) (Fig. 4). Currently, Prof. Wei is Chairman of AC and I am also appointed as Chief Executive Officer of OED. The EAC is the unit responsible for conducting the accreditation of engineering programs and for deciding whether each program has passed the accreditation. In the first and second years of operation, IEET was primarily funded by both MOE and NSC.

Actually, before the IEET was formed, I flew with Prof. S. H. Ou (歐善惠), then Vice President of National Cheng-Kung University (成功大學), and Prof. Andrew Wo (胡文聰) of NTU to Rotorua, New Zealand, on June 9-10, 2003 to attend the *6th Biennial Meeting of Washington Accord* under the support of NSC. At that time, the WA consisted of eight signatories, i.e., ABET (U.S.A.), CEAB (Canada), ECSA (South Africa), ECUK (U.K.), EA (Australia), EI (Ireland), HKIE (Hong Kong), and IPENZ (New Zealand). On behalf of IEET, we made it clear in the closing session that we were working currently on preparation of our accreditation system and we were determined to join this organization.

From 2003 to 2004, we had worked very hard in drafting and finalizing the *Accreditation Criteria*, now known as *AC2004*, and all the related procedures. The Criteria and Procedures Committee (CPC) chaired by myself prepared the first draft for all the documents. We revised iteratively all these documents, most of which were sent to EAC for second revision, and then to AC for final approval. By the time *AC2004* was approved on April 15, 2004 by the Accreditation Council, I had already chaired or participated in more than one hundred meetings. The time demanded in this regard was really huge.

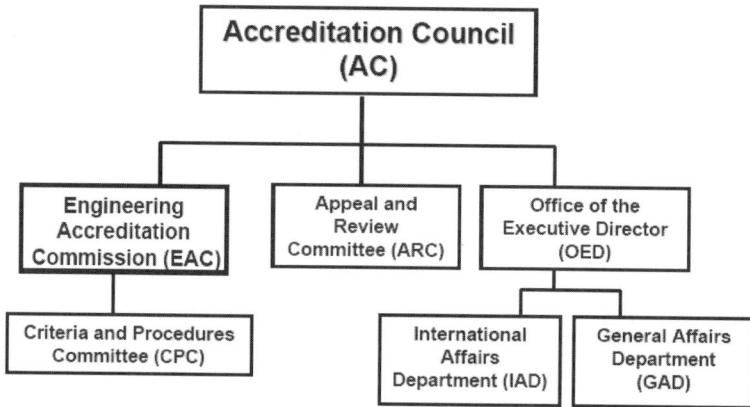

IEET Accreditation Council

The accreditation of the IEET is featured by the fact that it is substantially equivalent to those of the WA signatories, that it is outcomes-based, and that it is rooted in the mechanism of continuous improvement. One major task in conducting the accreditation from ground zero is to disseminate the principles and procedures to the programs seeking accreditation and to the evaluators for reviewing the programs. To this end, we have organized a number of workshops with different themes in each year. We also accept invitation from each institution for lecturing on accreditation and related procedures.

So far, the achievements we made are terrific. Thanks to the great team work of all the IEET staff. As of now (February 2007), a total of 88 engineering programs have been accredited, of which 12 programs were accredited in 2005, 35 programs in 2006, and 41 programs in 2007. The number of programs requesting accreditation from IEET continues to grow, due to the fact that many institutions have regarded IEET accreditation as their goal for pursuing excellence in teaching. It is expected that more and more programs will be accredited in 2008.

On June 15, 2005, IEET applied for the status of provisional signatory in the *7th Biennial Meeting of Washington Accord* held in Hong Kong and was approved by all eight signatories. In this meeting, JABEE of Japan was admitted as the 9th signatory. In 2006, IES of Singapore was approved as the 10th signatory. Due to the wide acceptance of accreditation in Taiwan and the continuing growth of the number of accredited programs, the IEET is ready to apply for the status of signatory in the *8th Biennial Meeting of Washington Accord* to be held in Washington, D.C., in June 2007.

We at IEET are also aware of the fact that the graduate education constitutes a significant part of higher education in Taiwan and that it is the leading edge of competition for national economy. A well-known fact is that one-fourth of the total number of degrees conferred to the engineering graduates each year belongs to Master's degree. To this end, a new criterion was added to the *AC 2004* to deal specifically with the accreditation of Master's programs, resulting in a total of nine criteria. Accordingly, the accreditation criteria are renamed as *AC2004⁺*. Starting from 2007, the IEET will accept application for accreditation of Master's programs. The other signatories of WA who are taking similar steps include JABEE of Japan and EI of Ireland.

16.13 Review of Major Civil Works 重大工程履勘

In the past decade, I have offered my service not only to the university, but to the engineering community and society. I was a member of the review team for the Zhonghe Line (中和線) of the Taipei Mass Rapid System in 1998, and for the Xindian Line (新店線) in 1999. Also, I was elected as President of the Chinese Society of Structural Engineers (中華民國結構工程學會) in 1998, and as President of the Chinese Institute of Civil and Hydraulic Engineering (中國土木水利工程學會) in 2005, each for a two-year term.

In 2005, I was invited by the Executive Yuan (行政院) to chair a review team for investigating the "necessity" of erecting a temporary steel bridge in one entrance of the Snow Mountain (or Hsuehshan) Tunnel (雪山隧道) for President Chen Shui-bian (陳水扁總統) of Taiwan to oversee the progress of construction inside the tunnel toward the stage of completion. The Snow Mountain Tunnel was one of the largest projects that have been carried out in Taiwan. It is a double-hole tunnel with a total length of 12.9 km, the second longest highway tunnel in Asia and the fourth longest in the world. Because of the varying geological conditions, it took a total of 15 years to finish this project. Prior to the above review, the "necessity" for erecting the temporary steel bridge for President Chen to pass was an issue criticized by the public. The conclusion from this review team was that the erection of the temporary steel bridge did enhance the progress of construction, and in that sense it was necessary, although the time of erection was interestingly coincident with the visit by President Chen.

The Snow Mountain Tunnel was regarded by Discovery Channel as *man made marvels*. Because of this eminent role, it receives from time to time concerns of the media for any problems it may have. My second connection with this tunnel was to serve as a member of the review team (履勘委員) for inspecting the readiness of the civil works, as well as the ventilation, control and signal systems, of the tunnel before it was open to the traffic. This had also been a hot subject of media for couples of months. This tunnel was finally open for traffic on June 16, 2006.

My third connection with the Snow Mountain Tunnel was due to the leaking of water at several cross sections starting from late September of 2006, which causes serious public concern about the safety of the tunnel and linings. Without any anticipation, I was called by Minister Duei Tsai (蔡堆) of the Ministry of Transportation and Communications (交通部) to form a team of experts to look into this problem, in my capacity as President of the Chinese Institute of Civil and Hydraulic Engineering. We formed a panel consisting of 26 experts of different disciplines, i.e., civil and structural engineering, geology, underground water, and tunneling. We also invited the tunneling expert Prof. Wulf Schubert from Graz University of Technology, Austria, and the underground water expert Prof. Makoto Nishigaki from Okayama University, Japan to join our effort.

The conclusion drawn unanimously by the panel was that the tunnel structure is rather safe under the present condition. However, we also suggested that proper methods should be adopted for discharging the leaking water such that no water pressure will be built up on the linings of the tunnel. Besides, some guidelines were proposed for treating the leaking problems in the short term and for installing the monitoring devices to ensure the safety of the tunnel in the long run.

The other gigantic project I got involved in 2006 is the final review of the high-speed railway for commercial operation, as invited by the Ministry of Transportation and Communications. The review team (履勘委員) consisted of 15 members divided into three groups as the civil works, buildings and rails; electro-mechanical systems; and operation and management. I was elected as chairman of the group on civil works, buildings, and rails. The high speed railway has a design speed of 250 km/hr and maximum speed of 350 km/hr, which can bring passengers from Taipei to Kaohsiung in 90 minutes. The entire railway lines of 345 km is carried mostly by bridges (73%) and tunnels (18%). The traditional embankments account for only 9% of the total length. This is the largest BOT (build, operate and transfer) project ever seen in the world, which was won in 1998 by the Taiwan High Speed Rail Corporation (THSRC) (台灣高速鐵路公司). The total cost of construction was 513.3 billion NT dollars (including the financial management cost), among which 105.7 billion NT dollars were offered by the government and the remainder by the company (including the financial management cost). Here, 32.5 NT dollars are roughly equal to 1 US dollar.

Unfortunately, the review of this project turned out to be a rather unhappy experience, as we felt strong political interferences behind the process. Though we had identified a number of outstanding issues for improvement prior to commercial operation, basically no evidences or data were provided by the THSRC to convince the review members that all these issues had been duly resolved. Without any support from the review members, the Ministry of Transportation and Communications announced by itself after an internal meeting that the high speed railway was ready for test running on December 24, 2006 and for commercial running on February 1, 2007.

16.14 Prof. W. F. Chen: A Model for Learning 學習的典範

As mentioned previously, I studied thoroughly the book by Prof. Chen on *Theory of Beam-Columns Vol. 2: Space Behavior and Design*, when I was a Ph.D. student at Cornell. This was my first private acquaintance with Prof. Chen. Compared with the previous beam theories, the approach presented in Prof. Chen's book is truly general, because equilibrium equations were written for the deformed configuration of the beam, taking into account the contribution of all kinds of actions and moments. As far as I know, this was the first time all actions of the beam, i.e., axial and shear forces, torsional and bending moments, were taken into account. I had actually re-derived most of the equations contained in the book, although sometimes ending with different results. There is no doubt that such an approach influenced strongly my later formulation of the buckling theory for three-dimensional beams.

As a Chinese, I always regard myself as a student of Prof. Chen, simply because I read many of his publications, and was influenced by all these publications. My second acquaintance with Prof. Chen was during the *ASCE Engineering Mechanics Conference* organized by him at Purdue University in May 1983. As a graduate student, I flew with Prof. McGuire to Purdue at West Lafayette to present our paper. I believe I shook hand with Prof. Chen, but he was so busy with the conference affairs that I could not find a time to talk to him personally.

Professor Chen was a legendary person for people of my age. When I attended the annual civil engineering (later renamed as KKCNN) conference in Kyoto in the 1990s, one of my friends from Kyoto or KAIST told me that there were three Chen's at Purdue. I was confused, but immediately I agreed with him, because we could see one Chen in steel structures, another Chen in materials constitutive laws, and the third Chen in soil mechanics. This friend also told me that Prof. Chen watched TV only when he was running on the treadmill. There was no wonder why he could be so efficient and so productive. He published more than 20 engineering books and more than 320 peer-reviewed technical papers! This is an achievement that could hardly be made by most scholars in the same profession.

I used to correspond with Prof. Chen by e-mails and was honored to be included in his list for distribution of messages and ideas. We had a significantly long period of overlapping in the position of dean, when he served from 1999 to 2006 at the University of Hawaii at Manoa and I served from 1999 to 2005 at NTU. During this period, I benefited a lot from Prof. Chen, as he always shared with me all the good ideas, often through the articles he wrote, about management of a good engineering college. Among others, I was highly inspired by his thoughts expressed in articles such as "How to build a first-rate engineering college?" and "On the reform of higher-education in China." But some very stimulating ideas from Prof. Chen can only be had in a casual manner, say, through the family gatherings.

16.15 Concern about Civil Engineering 土木工程的未來

In the past two decades, there has been a continuing decline in the morale of civil engineering practitioners, partly due to the shrinkage in the scale of major public constructions in Taiwan. This has seriously affected the intention of high school students to select civil engineering as their future profession. From my communications with friends overseas, I realize that this is not problem local to Taiwan. It also exists in many other countries, except Mainland China, where civil construction continues to boom at a fast pace.

Photo for Two Families in 2006 (From left to right, sitting: Lily and W. F.; second row, standing: Judy, Ru-Wong, me and Carol).

Actually, civil engineering is a very good profession. All the civil works you have built sustain a long time, compared with the life cycle of computers, and you will be proud to be a member of the design or construction team for that work. Since I joined the NTU in 1984, I have always tried to convince students, parents, and engineers that we need first-class students for civil engineering. The reason is quite obvious. All the problems to be encountered by humans in the near future are much more related to civil engineering than to other professions, for instance, the depletion of oil resources, deterioration in environment, reduced and polluted water resources, damaged ecological system, global climate changes, aging structures, and even aging population. However, all these problems are of integrated nature and require interdisciplinary efforts. This means that some changes have to be made in our civil engineering education. We should jump out of the original "box" of over-partitioning. We should place more emphasis on the concept of systems, as the future profession will rely much more on integration and cooperation with experts of different disciplines.

16.16 Closing Remarks 結語與致謝

Being an engineering professor, I always think from the point of "problem-solving." This has been the way for managing my life. Each day I keep a list of works to be done. After a while, some of the items in the list will be deleted, while others will be added. Also, some urgent matters will come in and the priority of each item will be changed accordingly.

I am 52 now, already passing the age of "knowing the destiny arranged by God" (五十而知天命) as noted by Confucius (孔子). At this point of reflection, I would say that I will continue my previous simple way of living, e.g., jogging twice a week for half an hour, playing badminton with my family members on weekends, reading some good books before sleeping, etc., besides my normal duty of teaching and researching. This is consistent with my belief that keeping a good habit for a long time is by itself a kind of strength.

I was quite hesitating when Prof. Chen invited me to write a chapter in his memoir under the title "NTU and I", because my journey at NTU is still continuing. However, after three weeks of intensive work on the chapter, I found it worthy to conduct a "mid-term self study report" such as this for a person having served at an institution for more than twenty-two years, because it gives you the moment of self examination. I would like to thank Prof. Chen for offering me such an invaluable opportunity and for honoring and encouraging me in this way.

References
參考文

Bleich, F. (1952), *The Buckling Strength of Metal Structures*, McGraw-Hill, New York, N.Y.

Chen, W. F., and Atsuta, T. (1977), *Theory of Beam-Columns, Vol. 2: Space Behavior and Design*, McGraw-Hill, New York, N.Y.

Fung, Y. C. (1965), *Foundations of Solid Mechanics*, Prentice Hall, Englewood Cliffs, N.J.

Gjelsvik, A. (1981), *The Theory of Thin Walled Bars*, John Wiley, New York, N.Y.

Malvern, L. E. (1969), *Introduction to the Mechanics of a Continuous Medium*, Prentice Hall, Englewood Cliffs, N.J.

McGuire, W. (1970), *Steel Structures*, Prentice Hall, Englewood Cliffs, N. J.

McGuire, W., Gallagher, R. H., and Ziemian, R. D. (2000), *Matrix Structural Analysis*, 2nd ed., John Wiley & Sons, New York, N.Y.

Timoshenko, S. P., and Gere, J. M. (1961), *Theory of Elastic Stability*, 2nd Ed., McGraw-Hill, New York, N.Y.

Vlasov, V. Z. (1961), *Thin-Walled Elastic Beams*, Israel Program for Scientific Translation, Jerusalem, Israel.

Washizu, K. (1982), *Variational Methods in Elasticity and Plasticity*, 3rd ed., Pergamon Press, Oxford, England.

Yang, Y. B., and Kuo, S. R. (1994), *Theory and Analysis of Nonlinear*

Framed Structures, Prentice Hall, Singapore.

Yang, Y. B., Yau, J. D., and Wu, Y. S. (2004), *Vehicle-Bridge Interaction Dynamics – with Applications to High Speed Railways*, World Scientific, Singapore.

Ziegler, H. (1977), *Principle of Structural Stability*, 2nd ed., Birkhauser, Stuttgart, Germany.

17

Kawasaki and I

By Toshio Atsuta

A Life with much Luck – New Challenges with Many Excellent Leaders

17.1 My family and the Atomic Bomb

I was born in Hiroshima Prefecture Japan on January 3, 1940 as the fourth child of Sataro, an engineering Cardinal of Japanese Navy, and Matsuko, a daughter of an only doctor in a local village.

My family lived in suburbs of Hiroshima when we got attack of the atomic bomb on August 6, 1945. At the time, I was five years old and stayed in a house with my mother. The house was broken by the shock of the explosion and I got hurt on my knee. The house was located a little far from the center of Hiroshima and so we were free from the contamination of radioactivity.

My elder brother Kishio was a student of Hiroshima University located at the center of Hirosima. Fortunately his group was sent to a factory of Mitsubishi Company for labor service and saved. My sister Yoshiko was a student of a woman high school in Hiroshima. She got a stomachache on the very day and was absent from school. She got saved but lost all her school friends.

On August 15, nine days after the atomic bomb, the war was finished and my father lost job. My family moved to Tokyo for a job. I attended primary school and middle school there in Tokyo but playing every day with no studying. Therefore I could not enter a decent high school and finally accepted by the least competitive Tokyo municipal high school. One year after, the school was decided to be a commercial high school, which means we have to study accountings and an abacus, the Japanese traditional hand calculating tool. I decided to quit the school to some other school, which was found easily near my newly moved house.

The new school, Ogikubo Municipal High School, was a woman high school formerly and the three quarter of students was women, who would not naturally study hard. The minority boy group had to study by themselves. I also stayed away from the school as much as possible and studied in a near public library.

Among the group of library students, there was an unusual boy, M. Miyauchi, who was very good in mathematics and a little bit violent. He always forced us to solve difficult math problems. If we could not solve the problems correctly, he hit us on the head. Thanks to him, I got interested in mathematics. He said he aimed to enter the University of Tokyo. Hearing that, I decided to go also to the university. Miyauchi went to USA before entering the university.

In1958, I took the entrance examination of the University of Tokyo, in which we had to take 8 subjects: Japanese, English, 2-mathematics, 2-sciences and 2-sociologies. I did good in math and science but bad in sociologies because I had never learned these subjects in the high school. I failed in the examination and entered a preparatory school for the university. I attended the school for two months then found to be suffering from pulmonary tuberculosis on both lungs.

I entered the hospital of the University of Tokyo. I needed total rest. I felt a little relief from hard work of studying. What I was allowed to do then were only eating nutritious food and lying in bed whole day. In secret, I studied in bed history and geography for entrance examination. A doctor of the hospital, one of my relatives, recommended me to go to medical school rather than engineering school because of my disease.

17.2 The University of Tokyo - From Medical to Engineering

In spring 1959, I took the entrance examination for medical course of the university and successfully passed this time. After two years study on biology and dissection of rabbits, I felt unsatisfied with the medical course which needed no mathematics. One of the classmates, M. Fujino, who also wanted to transfer to engineering school, said that Japanese shipbuilding would be the best in the world and he was going to enter the Department of Naval Architects in the Engineering School. I followed him. Later, he became a professor of the department.

In Department of Naval Architects, I took lecture of *"Plastic Design"* from Dr. Yuzuru Fujita, who was just returned from Lehigh University after finishing his PhD. It was the first lecture about plastic design in Japan and I was very impressed by the design concept. This experience became very helpful for my later life including to study at Lehigh University.

As for my graduate thesis, I took *"Buckling Strength of Embedded Cylindrical Shell"* under the supervision of Dr. Y. Yamamoto. Since the outward deflection of embedded shell is restricted, the buckling strength analysis was not straightforward but required careful calculation by energy method using Tiger mechanical calculator. In the experiment also, we had hard time in applying external pressure on the embedded shell.

As for the selection of company after graduation, Dr. Y. Yamamoto recommended me Kawasaki Heavy Industries, Ltd., because the company was strengthening the steel structure business. Three students wanted to get into Kawasaki. All were denied. We asked Prof. Hiroshi Kihara, an authority of shipbuilding in Japan, and he kindly called Dr. Toshio Yoshida, the director of Steel Structure Division of Kawasaki, and got OK answer. One was accepted by Kawasaki Aircraft Co., one was accepted by Shipbuilding Division of Kawasaki and I was successfully accepted by the Steel Structure Division of Kawasaki.

Kawasaki was basically a shipbuilding company when I entered in 1963 and aimed to go into the business of steel structures on land such as steel bridges, steel building frames, steel storage tanks, and so on. Most engineers there came from Shipbuilding Division.

Dr. T. Yoshida sent me to the University of Tokyo to study Civil Engineering as an associate researcher. I joined the Applied Mechanics Laboratory of Civil Engineering under Prof. T. Okumura. In the laboratory, there are many researchers from Japanese major companies including Hitachi Corp. and Nippon Steel. Among them, M. Hoshiya was from Hitachi Shipbuilding Company who later went to Stanford University and finished his PhD in one year. This fact impressed me very much. Dr. Hoshiya is now a professor of Musashi Technical University in Tokyo.

First, in the Department of Civil Engineering, I designed a floating gate for repairing dam gates of hydraulic power station. It was a good example for me to apply plastic design learned from Dr. Y. Fujita. Then I made a research on the ultimate strength of steel slab of a bridge, in which the effective width of reinforced steel plate after buckling was studied. Further an experimental study on lateral buckling of an arched pipe bridge was done which was to be constructed by Kawasaki.

17.3 Kawasaki Heavy Industries - From Shipbuilding to Civil Engineering

In 1965, I joined the Structure Developing Department of Kawasaki led by Dr. Sansei Miki who graduated from Aero-space engineering Dept of the University of Tokyo and attended Lehigh summer seminar that year.

At that time, construction of tall building became popular in Japan, and Kawasaki got orders of many steel structures of tall buildings such as Hotel New Ohtani, Kasumigaseki Building, etc. In these designs, analysis of beam-to-column connections was very important, and we held the research meetings with Prof. H. Umemura of Dept. of Architects of the University of Tokyo every month.

On the other hand, the construction of long suspension bridge was discussed in Japan. Kawasaki formed a study committee meeting with Prof. T. Okumura. I contributed in designing of tall steel tower of suspension bridges. Kawasaki received orders of many suspension bridges later including the world longest bridge, Akashi-Kaikyo Bridge constructed in 1996, which can be seen from my house in Akashi.

Besides buildings and bridges, we supported design of container cranes at berth, blast furnace at steel mill, many type of storage tanks at oil refinery, etc. A liquefied fuel storage tank consisted of double walled cylindrical shell structure. Usually, particle insulating media was filled in between the walls, which was in time piled and compressed and caused buckling of inner wall. In this analysis, my graduate thesis *"Buckling of Embedded Shell"* was helpful.

Japan Railway Company ordered many steel bridges for the Sinkansen Bullet train between Tokyo and Osaka. At the time, the speed of train was 200 km/hour and caused public hazard of noise. After this, concrete bridges rather than steel bridges were recommended. We made many studies to reduce noise from steel bridges and developed noise reduction material with a rubber company.

In 1967, Japan Railway Company planned construction of a double deck type of railway to increase transportation capacity. The problem was that the conventional train could not be interrupted during the construction period. The idea we had was that concrete columns were constructed earlier along both sides of the old railway, while steel beams were constructed in a separate workshop. At midnight, the steel beams were set on the concrete columns and fixed with long high strength bolts in a short time. For this purpose, we have studied strength of steel beam to concrete column connection and did also full scale tests by the 2000-

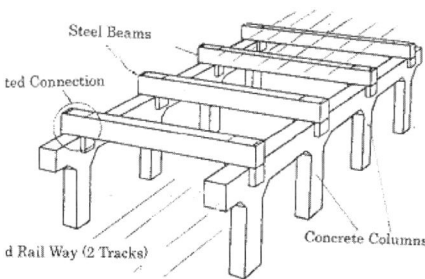

Double-Deck Railway.   Connections by Long Bolts.

ton testing machine in the University of Tokyo under the leadership of Prof. T. Okumura. The testing machine was made copying the Baldwin machine in Fritz Engineering Laboratory, Lehigh University.

17.4 Heading for U. S. A. - Lehigh University

In 1968, I heard from Prof. T. Okumura that Dr. H. Nishino returned from Lehigh University with his Ph.D., and the university was looking for his successor from Japan. I wanted to go to Lehigh because I was very much interested in plastic design learned from Prof. Y. Fujita. But at the time, I was busy in testing of the bolted beam-to column connections.

In May 3, 1969, I got married with Chika Imai, the third daughter of Shonosuke Imai, the chairman of a construction company in a county. We knew each other through the introduction by relatives of my father. At the time, I was 29 years old and she was a pretty young girl. Prof. T. Okumura played the role of go-between at the wedding ceremony.

When I finished the test of bolted connections, I decided to go to Lehigh and asked Kawasaki for a two-year leave to go to Lehigh University to get my MS degree in Civil Engineering. Actually, I was thinking to complete my MS in one year and my Ph.D. degree in another year following Dr. M. Hoshiya's experience at Stanford. I explained my plan to Prof. T. Okumura and received his agreement. I asked Dr. T. Yoshida, the Director of Steel Structure Division, about my plan of attending Lehigh University and got OK.

In July 10, 1969, I left Japan for New York. That was my first trip abroad. It was the very historical day that Captain Neal Armstrong put his foot prints on the surface of the moon for the first time as a human being. I dropped at NY Office of Kawasaki. Afterwards, I was completely alone in NY. I remember I had difficulty in buying a bus ticket to Bethlehem, Pennsylvania.

At the bus stop of Bethlehem, S. Morino, a Ph.D. candidate at Lehigh kindly came to see me. He was from Kyoto University. At Lehigh University, Dr. David VanHorn, the Department Head of Civil Engineering, was in a bad mood about my delay in joining Lehigh because of TOEFL test which I did not take earlier in Japan. He requested me to go to the University of Pennsylvania to take English language course for more than four weeks. I took train from Bethlehem to Philadelphia with Milan Vasec, who was also a new comer from Czechoslovakia. I shared a room of an apartment house with Milan and we became an intimate friend each other. Shortly, he had to go home because of the *"Spring of Prague"*. Milan is now a professor of Czech Institute of Technology.

17.5 Academic Life at Lehigh University —With Prof. W. F. Chen

In September 1969, I was accepted officially to the graduate school of Civil Engineering Department, Lehigh University. The meeting with

Prof. W. F. Chen there was the biggest event for me throughout my life.

Bethlehem was a quiet and beautiful city with much nature. Japanese were only few and being treated kindly by resident people. I rented an apartment house in Fountain Hill and invited my wife Chika to join me from Japan in October 1969. I could enjoy home-cooked Japanese meals every day, though Chika had difficulty in collecting Japanese foods.

As for my MS program, I took 5 courses 13 credit hours in Fall semester 1969, in Spring semester 1970, I took again 5 courses 13 credit hours, and in Summer semester 1970, I took 2 courses 6 credit hours. I got grade A in all 32 credit hours courses except the computer programming course. I had never used FORTRAN program in Japan so far.

As for my MS thesis, I developed the Column-Curvature-Curve (CCC) method for in-plane beam-column under the direction of Prof. W. F. Chen. The CCC-method is essentially a simplification of the Column-Deflection-Curve (CDC) method. The main difference is that analytically obtained curvature curve is used directly in the CCC-method instead of numerically integrated deflection curve as in the case in the CDC-method. My wife Chika helped me in typewriting the thesis as was the cases later on.

On October 11, 1970, I could attend the graduation ceremony in a black gown and hut. It was a happy and good experience for me and Chika. I got my MS diploma and S. Morino got his Ph.D. diploma from Dean R.D. Stout in a campus court. After this, Morino left Lehigh for Kyoto University and he is now the dean of the University of Mie in Japan.

The next target for me was Ph.D. degree. I started again to take required courses. I had already finished most of courses in Dept. of Civil Engineering, so I planed to take most courses from Dept. of Engineering Mechanics under the supervision of Prof. Chen.

In the Fall semester 1970, I took 4 courses 12 credit hours, three of them were from Engineering Mechanics Dept. including Fracture Mechanics by Prof. George Irwin. It was nice to hear a lecture from world famous authority like Dr. Irwin. In the Sprig semester 1971, I took 4 courses 12 credit hours all from Eng. Mech Department. I got grade of all A except the only course from Civil Eng. Department, i.e., Advanced Topics in Concrete Structure. I had never taken lectures on concrete before.

In the summer semester 1971, I took two courses, Special Problems by Prof. W. F. Chen, where I developed a computer program to design steel piles to keep slope stability of river dikes. It was an application of energy method in the theory of plasticity. It was said useful for actual application to Mississippi river dikes.

17.6 The Doctor of Philosophy - Struggle against Time

I passed the qualifying examination for my Ph.D. in February 1971. The period of my staying at Lehigh allowed by Kawasaki was two years from July 1969. This means that I had to go back to Japan in one year by July 1971. Obviously, it was not possible to finish the PhD dissertation in the rest of half year. I asked Kawasaki for extension but the answer was "Kawasaki business in steel structure is very busy. We need you. Come back as scheduled. You can continue your Ph.D. work in Japan".

I discussed this matter with Prof. Chen and he kindly wrote a letter requesting the extension of Atsuta's stay in Lehigh for more than one year to Mr. Y. Nakae, the Managing Director of Kawasaki, on 18 February, 1971.

While waiting the answer from Kawasaki, I was continuing the study for dissertation. The title of my dissertation was "*Analysis of Inelastic Beam-Columns*". The letter from Mr. Y. Nakae of Kawasaki came to Prof. W. F. Chen in April 1971, saying that Kawasaki accept the extension of Atsuta's stay in Lehigh until March 1972. This was really a big relief for me.

On September 16, 1971, my proposed program for the Ph.D. degree was approved by the Special Committee consisting of Prof. J. H. Daniels, Chairman, Prof. W. F. Chen, Thesis Advisor, Prof. G. C. Driscoll, Dept. of Civil Engineering, Prof. D. P. Updike, Dept. of Mechanics and Prof. D. A. Van Horn, Ex-Officio.

Meanwhile, Chika was pregnant and on January 4, 1972, a baby boy was born at St. Lukas Hospital in Fountain Hill. I named him Toll which means perfect in Japanese. My son Toll got an American citizenship and later, he got a job in Kawasaki Robotics Inc., in Detroit as an American sales engineer after graduating from a state university in Utah.

By the time, Prof. Chen and I had already published several papers to ASCE on inelastic beam-columns. These papers helped me make my dissertation greatly. There were no word-processors or PC at the time. All equations I derived had to be typewritten by hand. It was a hard job for my wife, Chika, who was pregnant at the time. I Xerox-copied the equation parts from my published papers and put them on the dissertation. This was discovered by Dr. VanHorn. He said that a dissertation must be original and Xerox-copy was not acceptable.

Time was limited until the end of March which Kawasaki allowed me to stay in Lehigh. I asked Prof. Chen to help again. He collected a group of secretaries and let them re-type all 317 pages of my dissertation using the same IBM typewriting machine.

The dissertation was completed by the beginning of March 1972 as promised with Kawasaki. It consisted of 9 chapters of separate subjects on inelastic beam-columns with biaxial bending. The field of study was wide. There were a few competing Ph.D. candidates who were studying similar subject for a long time before I came to Lehigh. They were not happy with my taking Ph.D. degree before them. Dissertation papers must be opened at the library of the Department two weeks before the defense, a public hearing meeting for Ph.D. My competitors for Ph.D.'s examined my dissertation paper closely and said to me, "We will give you a hard time at your defense."

In my dissertation defense, I showed a derivation of equilibrium equations of a general beam-column in three-dimensional behaviors. In the defense, the competitor students asked many tough questions including the validity of those equations. The committee of my Ph.D. required me to meet and answer those questions.

From that day, my busy life started again. Every night, I went to computer center to calculate the ultimate strength of an inelastic beam-column to show the validity of my equations in the dissertation. During the time, I had already bought air tickets preparing to go back to Japan with my wife and young baby. I had to rewrite the air tickets three times before I got the agreement of the competing candidates. On April 25,

1972, my dissertation was approved by the committee.

17.7 The Books "Theory of Beam-Columns"

Joining Lehigh from September 1969 till April 1972, I could finish my MS and Ph.D. degrees in two years and eight months. The graduation ceremony was going to be held on the founder's day, June 11, 1972 but I could not attend it as the matter of course. Just before I leave Lehigh, Prof. W. F. Chen proposed me to write textbooks together since we have done so many studies about beam-column analysis. We signed the contract with McGraw-Hill Book Company to write a two-volume book, *"Theory of Beam-Columns"*. Vol.1: *In-plane Behavior and Design, and* Vol.2: *Space Behavior and Design*.

The completion of the books took five years, while I was working for Kawasaki in various business divisions: Steel Structure Division, Offshore Equipment Department in Shipbuilding Division and Crushing Plant Division. The draft of the books was made every night, Chika type-wrote and mailed to Prof. W. F. Chen. Vol.1 was published in 1976 and Vol.2 in 1977.

Prof. Shoji Toma of Hokkaigakuen University Japan, who got his Ph.D. at Purdue University under the supervision of Prof. Chen, proposed me to translate those books together into Japanese because it would be good textbooks for Japanese civil engineers. The books were too big for us to translate though, then, we decided to publish a small book, *"Story of Buckling - in order to avoid accident"* from Kajima Publishing Company Japan.

17.8 Steel Structure Division

On May 6, 1972, I came back to Japan with my wife Chika and the four month baby Toll. In Kawasaki, I returned to the Steel Structure Division led by Dr. Miki, where I joined a design group of blast furnace at Mizushima Mill Factory of Kawasaki Steel Corporation which was a sister company of Kawasaki Heavy Industries. At that time Japanese steel mill business was very active and many new factories were constructed in Japan. It was a striking contrast to the steel mill business in USA such as Bethlehem Steel I saw at Lehigh. Shell plates of blast furnace were very thick and had many holes to hang insulating liner plates by bolts. I developed a design rule for the fatigue strength of perforated thick plates.

What we designed next was a big arched rib bridge named "Kobe Ohashi" which was the only access for the artificial island "Port Island" to the city of Kobe. The bridge was equipped with entire lifelines for the island: road, highway, railway, water line, and electric power line. The strength of bottom part of the arched member was the problem. I did stress concentration analysis and designed the part. When the big earthquake hit Kobe in 1995, the bridge got damaged but the bottom part I designed was safe and sound.

Arched Rib Bridge "Kobe Ohashi".

17.9 Offshore Equipment Department − Shipbuilding Division

In 1973, I moved to Offshore Equipment Dept. of Shipbuilding Division led by Dr. S. Miki who had also moved from the Steel Structure Division. We designed a self elevating platform for marine construction use. It had four 100m long legs jacking up the 6,000 ton platform. The basic design came from IHC Gusto, Holland under a license agreement. We had developed our own jack-up rigs in which I did fatigue analysis of legs. The work was published in the paper entitled *"The Fatigue Design of an Offshore Structure*" and presented at the Offshore Technology Conference (OTC) at Houston in May 1976. In this work, the course I took in Lehigh, Response of System to Random Loads by Prof. F.P. Beer was very useful.

Shipbuilding business at that time was in a boom in Japan and building docks were full. Kawasaki had no space to build offshore structures and decided to close our Offshore Equipment Department in March 1976. All members were separated. S. Toma and N. Sugimoto moved to Shipbuilding Division. They later went to Purdue University to study their Ph.D. degree under Prof. W. F. Chen. I wanted to go to Technical Institute of the company. Dr. T. Yoshida, the senior managing director, needed me to support Crushing Plant Division which was on the verge of bankruptcy. I joined the Division with Dr. H. Oba, the director and Mr. T. Kamei, the manager. Both became the President and the Chairman of Kawasaki later.

Before moving to Crushing Plant Division, I asked Dr. T. Yoshida for one moth leave of absence from the company. I wanted to go to Lehigh again to finish Vol.1 of the book, "*Theory of Beam-Columns*" with Prof. W. F. Chen because the contracted time limit with McGraw-Hill was coming. Dr. Yoshida kindly permitted me to go to Lehigh for one month as a business trip to finish the rest of my Ph.D. work. We finished Vol.1 of the two-volume book in this period. On the way back to Japan, I stopped at Houston and gave a speech "*Fatigue Design of an Offshore Structure*" at the OTC in May 1976.

17.10 Crushing Plant Division

Crushing Plant Division was mainly making machinery to crush rocks into pebbles for concrete material. I checked design of all crushing machines and achieved considerable reduction of the manufacturing cost. At the same time, I developed several new products for the Division. Portable Crushing Plants were developed for exporting crushing machines to Middle East countries. I visited Iran, Iraq, Saudi Arabia, Egypt, Dubai, UAE, Quatar, for market research and opened the market. During these trips, I continued to write the draft of Vol.2 of the book

"*Theory of Beam-Columns*". Studies were also made for a plant to make pollution-free pellets from incinerated garbage.

Casting facility was there in the factory for producing anti-wear liners to protect the crushing machine. I proposed to make more valuable cast products, and developed cast connections of pipe-to-pipe members for offshore structures. The cast connection named "PRENODE" can have smooth corner surface and reduce stress concentration thus the offshore tubular structure can have enough fatigue strength even in a rough sea condition. We presented PRENODE at the Offshore Technology Conference (OTC) in Houston in May 1978 and received the Meritorious Award for Engineering Innovation.

A Jack-up Type Offshore Rig.   Cast Pipe Connection "PRENODE".

The shortage of oil raised the need of coal and the use of new type of fuel, coal-oil-mixture (COM) was expected because it can be handled like the conventional oil. We developed a ball mill which crushes coal, mixes with oil and at the same time dehydrate the coal. This technology was patented and the mill was used by the Japanese Government. The Crushing Plant Division had revived in three years.

17.11 Welding Research Laboratory

In April 1979, I was requested to move to Welding Research Laboratory of Technical Institute led by Dr. S. Susei, who became later the Vice

President of Kawasaki. I started the development of corrosion resistant double walled pipe with Dr. T. Yoshida, who just retired from the Senior Managing Director, and Dr. S. Matsui who just came to Kawasaki from Associate Professor of Osaka University. The double walled pipe was a high strength carbon steel pipe lined with a thin corrosion resistant pipe inside. First, SIPM, the Royal Dutch Shell, tried a double walled pipe as tubing for a high pressure gas well. They experienced a sudden buckling of liner tube due to inside negative pressure which was called "implosion". For analysis of implosion, my graduate thesis *"Buckling Strength of Embedded Cylindrical Shell"* in 1962, worked very well.

We had developed the Thermo-hydraulic Fit Method to produce a tightly fitted double walled pipe named "TFP". A thin inner pipe is expanded plastically by water pressure onto inside of the heat-expanded outer pipe. After cooling of the outer pipe, the tight fit pipe, TFP, is produced and this was proved to be implosion free by series of tests.

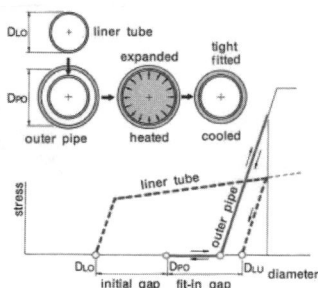

| Implosion. | TFP Production Method. | TFP Pipeline. |

Evaluation project by the Battle Memorial Institute USA approved our TFP as No.1 among corrosion resistant double walled pipes collected all over the world. We presented TFP at OTC 1980 in Houston and again received the Meritorious Award for Engineering Innovation.

Concerning the production process of our double walled pipe TFP, we applied 130 patents and sold the right to Nippon Steel Corporation, which constructed a production line for TFP in their Hikari Works later.

Another requirement was a ceramics lined wear resistant pipe. For this double walled pipe, we developed another process, the thermal shrink fit method. In this method, every time when we apply a ring heating and ring cooling successively on a double walled pipe, the carbon steel outer pipe shrinks two percent of the diameter. The method was also patented and actually applied for manufacturing of mud pump liners for oil production.

At Welding Research Laboratory, new joining technologies were studied such as diffusion welding, friction welding, electron beam welding (EBW), etc. EBW was very powerful and could weld very thick plates at a time, but it needed a vacuum chamber, which limited the size of work pieces. In order to fill the gap of EBW, we studied the use of laser for welding and cutting of materials which could be done in atmospheric condition with no chamber needed. Since then, I felt necessity of laser oscillator suitable for material processing.

17.12 Opto-Engineering Laboratory — My Dream

I proposed to Dr. A. Nakamura, Director of Technical Institute and Vice President of Kawasaki later, to establish Opto-Engineering Laboratory to study application of laser and to develop optical equipments including laser oscillators. It was approved in the spring 1987 and I was the manager of the laboratory. I was very happy because the study of optics was my lifelong dream. We started the study of developing a high power CO_2 laser and obtaining technical support from Ferranti, UK. We had developed 10kW axial flow type CO_2 laser equipment generating high power beam with good beam quality, which was sold to Nissan Automobile Company.

Besides lasers, I tried to make optical equipments to be new products of Kawasaki. First I selected was a video projector with RGB liquid crystal panels. We made several types for trials. For our newly designed projector, we needed the world best liquid crystal panels (LCPs), which we could introduce from the No.1 LCP maker, Sharp Corporation, under the joint research agreement. Sharp was also developing its own LCP projector. Therefore the agreement restricted Kawasaki projector to be sold only for transportation equipments, such as train, ship, aircraft, etc. which were Kawasaki's main business units.

Japan Railway Company adopted Kawasaki projectors "PAXVISION" for their Shinkansen bullet trains between Osaka and Hakata with 300 km/h speed. A problem was found that cabin of the train could not be darkened which was different from that of airplane. The projected picture was not clear and no passengers saw the screen. Another big problem was dirt coming into the projector along with cooling air, which put on the LC panels and its expanded image was shown on the screen. We had to go to Hakata train shed to clean the LC panels almost every night. To solve this problem, I invented a closed type projector with heat-sink inside and radiator outside; both are connected with heat pipe. It was patented again.

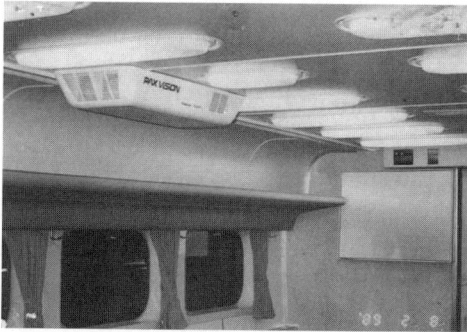

Kawasaki Video Projector "PAXVISION" amounted on JR Bullet Train.

We next studied the development of Chemical Oxygen Iodine Laser (COIL), which was under development only by Rockwell International Inc. (now Boeing) for Weapon Laboratory of US Air force. COIL can emit high power laser beam of 1.315 micrometer wave length which can go through the optical fiber with small losses. In 1990, we developed 1kW high power COIL for industrial purposes and supplied to the Advanced Laser Engineering Center (ALEC). This is the only COIL produced for industrial purpose in the world. For this achievement, I received the Innovative Research Award from the Ministry of Science and Technology of Japan in 1995.

In the early morning on January 17, 1995, a big earthquake hit Kobe area. About 6,500 persons were killed and many buildings collapsed or burned down. My house was OK with slight damage on brace members.

Kawasaki had a big damage especially at the shipbuilding dockyard and the employee had difficulties because of the lack of lifelines: electricity, gas, and water and transportation system.

Chemical Oxygen Iodine Laser "COIL".

"COIL" for Industry.

In order to investigate the damage of the city, many civil engineers came to Kobe city, including Prof. L. W. Lu from Lehigh University and Prof. S. Morino from University of Mie.

Earthquake hit Kobe.

Prof. Lu Inspecting Damage.

17.13 Kanto Technical Institute - New Products by Photon Technologies

On February 1, 1995, two weeks after the earthquake, Kawasaki opened Kanto Technical Institute (KTI) near Tokyo. It had been scheduled as the centennial anniversary memory of Kawasaki. About 100 researchers moved from Kobe. Some were wearing bandages on heads due to the earthquake. I was promoted to be an associate director and appointed as the head of KTI. The mission of KTI was to develop innovative products for Kawasaki utilizing advanced technologies.

KTI recruited young and active researchers in fields of applied physics, optics, nuclear science, new material, nano-technology, etc. First, we developed sterling magnets to control direction of electron beam for synchrotron radiation. In total, 570 units were made and supplied to "Spring-8", the world largest synchrotron radiation facility. We also made a new X-ray beam line facility in "Spring-8". An electron beam size monitoring system was also developed and shipped to Stanford Linear Accelerating Collider (SLAC). Next, we developed free electron laser (FEL) which can change wave length continuously. The equipments were supplied to Science University of Tokyo for research on nano-technology and bio-technology.

Table top size equipments were also developed. One was tunable solid state laser DFG useful for medical surgery which was supplied to the Institute of Physical and Chemical Research. The other was laser plasma X-ray microscope which could see bio samples as alive. Meanwhile, the chemical laser, COIL, had been improved and high power beam of 10kW was emitted. This machine was moved to Aircraft and Space Division of Kawasaki and tests are being continued.

17.14 Akashi Technical Institute — Technical Support to All Products

In April, 1999, I was promoted to be a Director of Kawasaki and appointed as the head of Akashi Technical Institute (ATI) which was to support all business divisions of Kawasaki keeping basic technologies for Kawasaki products: ships, aircrafts, rolling stocks, motor cycles, industrial robots, gas turbine, incinerators, steel structures, plants and so on.

About 200 researchers there were smart and worked hard supporting engineers from business divisions. They developed many new products including fluidized bed combustion furnace for waste. This air floating technology was unique and so we utilized it for development of belt conveyers carrying large amount of coal with high speed quietly. The products were supplied to power stations and steel mill companies and highly appreciated. The technology was also applied to traveling system for a natural lawn ground of sucker field out of the baseball stadium for sunshine. The air lifted sucker field more than 10,000 square meters must be very flat and the total weight was about 8000 tons.

As for rolling stock business, boggy of train cars used to be cast steel avoiding stress concentration. Kawasaki proposed welded boggy structure, but the American Society of Railway would not approve it. We consulted with Prof. John Fisher of Lehigh University who was the chairman of the committee for fatigue strength of railway cars. He evaluated car boggy structure and persuaded the committee members. Finally, we were approved and started shipping of subway cars to the USA.

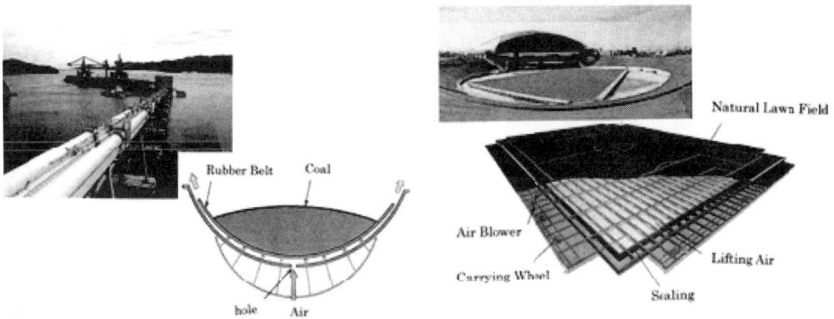

Air Floating Belt Conveyer.

Air Movable Soccer Ground.

Every time when the model of cars was changed, the Port Authority of N.Y. required a full scale collision test on the first production. If the test results differ from the original calculation, the design was required to be changed. This was a very severe requirement for us because change of design lead delay of delivery and to the payment of big amount of penalty. The strength research group of ATI developed a simulation program for estimating collapse failure of subway cars. Our simulation results predicted precisely the full scale test results, thus the Port Authority agreed Kawasaki not to do the full scale collision test any more. So far, Kawasaki has shipped more than 3500 cars including 2000 subway cars to the Port Authority of New York.

As the head of the Institute, I concentrated myself to develop new products in new business fields for Kawasaki. I advocated "Patent Oriented Business Strategy (POBS)". According to my request, about 100 researchers wrote patents by themselves and some of them became new products of Kawasaki such as fatigue sensor, photo catalyst, new type battery, and so on. The battery was the nickel hydrogen type and the electrodes were not solid but granular thus the capacity could be large and the shape of battery could be very free. I expect that a structural member can be a battery such as roof of automobile, frame of bicycle, even a floor of building. This battery named "Gigacell" is about to be sold in the market.

Full Scale Collision Test of Subway Car.   Simulated Analysis.

In April, 1999, I got additional post of the visiting professor of Himeji University of Technology and gave some lectures about business innovation. In the same year, I was registered as a consulting researcher of the University of Tokyo. In March, 2003, I retired from the director of Kawasaki Heavy Industries, Ltd. So far, I had applied 350 patents for Kawasaki and 108 of them were registered by the Japanese Patent Office.

17.15 The New Industry Research Organization — Regional Innovation

After the big earthquake attacked Kobe, the recovery of regional industry had been sluggish. The governor of Hyogo Prefecture, Mr. T. Kaihara, requested Dr. H. Ohba, the chairman of Kawasaki, that industry sector of big companies should contribute to the revitalization of regional industry of medium and small-sized business. Dr. Ohba agreed and the New

Industry Research Organization (NIRO) was founded in April, 1997.

NIRO was organized by regional industry donated 10 million dollars and the local governments donated 5 million dollars. Dr. Oba was the chairman and Dr. S. Matsui was the Executive Managing Director both from Kawasaki. We invited Dr. Hiroyuki Yoshikawa who was the President of the University of Tokyo, as the president of the NIRO Institute. Since then, I was serving as the Vice President.

NIRO had totally cooperative agreement with MIT. I visited Boston several times and made hearing of technology seeds from about 50 professors in total. Some of the seeds were transferred to NIRO and became the new products of the regional small companies, such as ring sensor for health monitoring, wheel chair with balls to move all direction.

The similar cooperate agreement was signed with Cambridge University, UK. We sent a researcher there to search good technology for innovation.

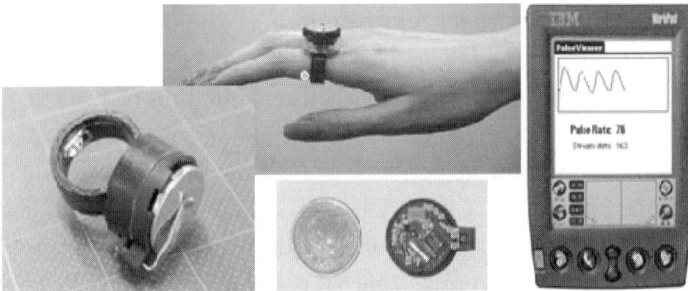

Wearable Sensor for Health Monitoring .

In 2004, I got an additional post of the executive consultant of the Hyogo Prefectural Institute of Technology (HITEC) and the visiting professor of Kobe University. This means that I am to play in all three sectors; Industry (NIRO), Academia (Kobe University) and the Government (HITEC).

17.16 Still Flying Life — Do the Best and Enjoy Life!

When I entered Kawasaki, I decided to collect my own ideas and put them on a notebook. The target was 1000 ideas by the time of retiring the company so that I can select some for my own business. When I retired Kawasaki actually in 2003, my ideas on the notebook were about 1300. Several of them are what I want to do business myself.

Business of Kawasaki is very good now. The total sales last year 2006 was about 13 billion dollars including motorcycles 3.5, aircrafts 2.3, trains 1.7, gas turbines 1.6, ships 1.1, plant and structures 1.7. They are busy with these original products started 100 years ago by the first president of the company, Kojiro Matsukata.

Kawasaki opened their own patents to NIRO for the revitalization of regional industry, in which included are products by photon technology developed in KTI and many patents devised as POBS at ATI. At NIRO, we are using these patents for forming national projects such as the low dose X-ray image scanner for airport security, the solid state tunable laser for gallstone surgery, etc. Some patents are being transferred to regional small companies such as portable solar co-generation system, noise reduction system for steel structure, sound proof wall for Highway Bridge with photo catalyst, etc.

Through the long experience in Kawasaki, I acknowledged the importance of patent. I also felt importance of power of individuals rather than power of organization. Good idea comes from wisdom of an individual person not from a group of people. The bigger the company is, the less idea comes out. Once a big company goes into a business, one can not stop it and goes into the endless competition for product enhancement or cost reduction, forgetting about the consumer's satisfaction. This is clear if you see the fierce competition in digital camera, cellular phone or personal computer today.

My hobbies are thinking invention, playing magic, playing go-game and flying with para-glider. These are all based on my desire to do something different. Flying alone high in the sky with para-glider changes your view on life. I am trying to be interested in anything around me and trying to improve it. It is important also to be moved emotionally when you meet something excellent.

My usual belief is "Do the Best and Enjoy Life".

My first solo flight from Mt. Iwaya 600 meters high (August 26, 2004).

Index

Index in Chinese (Chapter 15)